D0889926

ESSENTIALS OF MODERN ALGEBRA

Essentials of Modern Algebra

Cheryl Chute Miller
The State University of New York at Potsdam

Mercury Learning and Information
Dulles, Virginia
Boston, Massachusetts
New Delhi

Publisher: David Pallai
Mercury Learning and Information
22841 Quicksilver Drive
Dulles, VA 20166
info@merclearning.com
www.merclearning.com
1-800-758-3756

This book is printed on acid-free paper.

C.C. Miller. *Essentials of Modern Algebra.*
ISBN: 978-1-936420-18-6

Library of Congress Control Number: 2012943785

131415321

Printed in the United States of America

For my family that believed in me, cheered for me, and always encouraged me, especially my daughter, Rebecca, whose constant question, "Is it done?" kept me writing, and my husband, Jeffrey, without whose love, patience, and constant support I could not have completed this book.

Contents

Preface

Over many years of teaching an undergraduate course in Modern Algebra (also known as Abstract Algebra) I slowly determined what, for me, seemed to be the "essential" topics to cover and the order in which the material most easily flowed through a course, or sequence of courses. Most importantly I began to find more for my students to work on in class together, with help from one another and from me. Slowly over the years a variety of examples or theorems became common for me to use this way so I began to categorize them into what are called Projects in this book. Each Project is a set of related problems to solve, some are calculations for practice using concepts like operations on sets, permutations, cosets, or field extensions, and others require proofs. Each chapter (other than the preliminaries) has several that can be used when covering that specific material.

This book can be used for a course in Modern Algebra in a variety of ways, some suggestions are given for a one semester course or two semester sequence at the undergraduate level. Chapter 0 is intended to be a resource, not material to cover in detail. The material in Chapter 2 can be covered along with Chapter 1 if time is short. It is put in a separate chapter since I have found that students enjoy the break from proofs with calculations involving the groups of units $U(n)$, Cyclic Groups, and Permutation Groups for a short time. The text is written to use the information from previous chapters to assist with new material, so if a course was given only about Rings, the facts about groups in Chapter 1 that are referenced can be discussed in only additive notation with commutativity.

If students have only one course in Modern Algebra it is important to include material relating to groups *and* rings, Chapters 1 and 4. Including portions of Chapter 6 (Domains), makes it possible to cover much about Polynomial Rings (Chapters 7 and 8) as well. It would be difficult to also include information from Chapter 9 (Extension fields) without discussing quotient rings and ideals from Chapter 5, since several critical proofs refer to maximal ideals and quotient rings.

For a two semester sequence the order of topics in the text can be followed directly with some optional pieces being excluded, such as Ordered Integral Domains and The Sylow Theorems as time allows. Just as the material in Chapter 2 can be weaved into the basics about groups in Chapter 1, it is possible to include material about polynomials (Chapter 7) while introducing rings rather than later.

The notation used is fairly standard, but I use $ord(a)$ to refer to the order of a group element instead of $|a|$, to keep it distinct from the cardinality of a set. Eventually in Chapter 2 it is shown that $ord(a)$ and the cardinality of the cyclic subgroup generated by a are equal. Since the symbol $\mathbb{R}$ is reserved for the set of real numbers, I use A to refer to a general ring instead of R. Similarly K is used to denote a field throughout the text instead of F to avoid notion involving extensions that looks too similar to inappropriate language out of context. This is a personal preference that evolved after having remarks made by students in my classes over the years.

Many of my colleagues at SUNY Potsdam should be thanked for their assistance and support throughout this project, but I will only mention a few. Harold Ellingsen used drafts of the text in his class which helped me refine and correct many details, the text is significantly improved due to his assistance. Joel Foisy gave his support by helping keep my teaching schedule manageable during the writing of the text, while he and Jason Howald also read portions finding critical changes that needed to be made. Jason Howald, Derek Habermas, and Kerrith Chapman cannot be thanked enough for their help while I struggled learning to create this final product in LaTeX, a nontrivial exercise. Finally, Gerald Ratliff and Nancy Dodge-Reyome convinced me to contact publishers to set this entire project into motion, this would never have happened without their encouragement.

Chapter 0

Preliminaries

The material in this text assumes that you have already been exposed to specific mathematical topics. This chapter will serve as a reminder of some key ideas referred to throughout the text. In general, most theorems in this section do not include proofs, as they are meant as reminders of previous information.

0.1 Sets

All of the ideas in this text are based on the concept of a set with certain properties. You should already have experience with sets written both as lists in brackets, $\{1, 2, 3, \ldots\}$, or in *set-builder* notation such as $\{x : x \text{ is an integer and } x > 0\}$. In the first we expect to see and follow a pattern while the second clearly describes which elements are in the set. One main difference is that when a set is written with set-builder notation we must determine if any elements are in the set at all!

Example 0.1 The set defined by $\{x : x \text{ is an integer, } x^2 = 1, \text{ and } -3 < x < -2\}$ has no elements in it, since the only integers satisfying $x^2 = 1$ are 1 and -1, neither of which is between -3 and -2.

We will follow conventional notation, often naming a set with a capital letter and using $x \in A$ to mean that x is an element of the set named A, and $x \notin A$ to mean the opposite. The usual notation for the empty set $\emptyset$ is used, and for two sets A and B:

- $A = B$, if the two sets have exactly the same elements.
- $A \subseteq B$, (A is a subset of B) if every element of A is also an element of B. We often prove that $A = B$ by showing that both $A \subseteq B$ and $B \subseteq A$ are true.
- $A \subset B$ (A is a proper subset of B) when $A \subseteq B$ is true but $A \neq B$.

- $A \cup B = \{x : x \in A \text{ or } x \in B\}$.
- $A \cap B = \{x : x \in A \text{ and } x \in B\}$.
- $A - B = \{x : x \in A \text{ and } x \notin B\}$.
- If $A \cap B = \emptyset$ we say that A and B are disjoint.
- $A \times B = \{(a, b) : a \in A \text{ and } b \in B\}$.
- $|A|$ denotes the number of elements in A, i.e., the cardinality of the set A.
- If A and B are finite and disjoint then $|A \cup B| = |A| + |B|$.
- The Power Set of a set A, $\wp(A)$, is a new set whose elements are the subsets of A. Notice that both A and $\emptyset$ are always elements of $\wp(A)$.

Example 0.2 Consider the set $A = \{1, 2, 3\}$. The subsets of A are $\emptyset$, $\{1\}$, $\{2\}$, $\{3\}$, $\{1, 2\}$, $\{1, 3\}$, $\{2, 3\}$, and $\{1, 2, 3\}$.

$$\wp(A) = \{\emptyset, \{1\}, \{2\}, \{3\}, \{1, 2\}, \{1, 3\}, \{2, 3\}, \{1, 2, 3\}\}.$$

Notice that $\wp(A)$ has 8 elements. In general, if $|A| = n$, then $|\wp(A)| = 2^n$.

There are familiar sets we will refer to frequently and thus reserve notation for them; the set of integers $\mathbb{Z}$, natural numbers $\mathbb{N}$, real numbers $\mathbb{R}$, rational numbers $\mathbb{Q}$, and the empty set $\emptyset$. Note that $\mathbb{N}$ is the same as the set of positive integers. For more basic information about sets see [8].

0.2 Mathematical Induction

One of the most important tools in mathematics is called *Mathematical Induction*. This proof technique is common to almost every mathematics course and is frequently used when a statement refers to an arbitrary natural number. There are several ways to state Mathematical Induction, often referred to as PMI or the "Principal of Mathematical Induction." See [8] for more about Mathematical Induction.

Theorem 0.3

Principal of Mathematical Induction Suppose for each $n \in \mathbb{N}$, $P(n)$ is a statement involving n. If:

(i) $P(1)$ is true (the base case), and
(ii) For an arbitrary $k \in \mathbb{N}$, if $P(k)$ is true then $P(k + 1)$ is also true (the induction step),

then $P(n)$ is true for all $n \in \mathbb{N}$.

Example 0.4 Consider the statement "for any natural number n, $2+4+6+\cdots+2n = n(n+1)$." Thus for each $n \in \mathbb{N}$, $P(n)$ represents the statement "$2+4+6+\cdots+2n = n(n+1)$." To prove that $P(n)$ is true for every natural number n we need to verify (i) and (ii) of the definition.

(i) The statement $P(1)$ is "$2 = 1(1+1)$." Since we know $2 = 2$, then $P(1)$ holds.
(ii) Now consider an arbitrary $k \in \mathbb{N}$ and assume that $P(k)$ is true, or in other words assume that $2+4+\cdots+2k = k(k+1)$ (this is referred to as the induction hypothesis). For the induction step we must prove that $P(k+1)$ is also true. $P(k+1)$ is the statement "$2+4+\cdots+2(k+1) = (k+1)(k+2)$." Notice that $2+4+\cdots+2(k+1) = 2+4+\cdots+2k+2(k+1)$, so using $P(k)$ we have:

$$\begin{aligned} 2+4+\cdots+2(k+1) &= (2+4+\cdots+2k)+2(k+1) \\ &= k(k+1)+2(k+1) \\ &= (k+1)(k+2) \end{aligned}$$

Thus $P(k+1)$ is true and so by the PMI, $P(n)$ is true for every natural number n.

Theorem 0.3 can be modified if the statement $P(n)$ is only true when $n \geq m$ for a fixed natural number m.

Theorem 0.5

Suppose for each $n \in \mathbb{N}$, $P(n)$ is a statement involving n. If:

(i) $P(m)$ is true (the base case), and
(ii) For an arbitrary $k \in \mathbb{N}$ with $k \geq m$, if $P(k)$ is true then $P(k+1)$ is also true (the induction step),

then $P(n)$ is true for all $n \in \mathbb{N}$ with $n \geq m$.

Example 0.6 We will prove that for all $n \in \mathbb{N}$ with $n \geq 4$, $2^n \leq n!$ (recall $n! = 1 \cdot 2 \cdot 3 \cdots (n-1) \cdot n$).

This time we have the statement $P(n)$ as $2^n \leq n!$. $P(4)$ says that $2^4 \leq 4!$, so since $2^4 = 16$ and $4! = 24$, $P(4)$ is true.

Now assume that $P(k)$ is true for some $k \in \mathbb{N}$ with $k \geq 4$, i.e., assume $2^k \leq k!$. Using $P(k)$ and $2 > 0$ we have:

$$2^{k+1} = 2(2^k) \leq 2(k!)$$

As $k \geq 4$ we know that $k+1 \geq 2$, or $2 \leq k+1$. Hence

$$2(k!) \leq (k+1)(k!) = (k+1)!$$

Hence by $2^{k+1} \leq (k+1)!$ we have that $P(k+1)$ is true, so by PMI $P(n)$ is true for all $n \in \mathbb{N}$ with $n \geq 4$. Note that in this case $P(1), P(2)$, and $P(3)$ are all false.

Another modification is called ***Strong Induction***, which assumes that for all $t \leq k$, $P(t)$ is true as the induction hypothesis instead of just $P(k)$.

Theorem 0.7

Suppose for each $n \in \mathbb{N}$, $P(n)$ is a statement involving n. If:

(i) $P(1)$ is true (the base case), and
(ii) For an arbitrary $k \in \mathbb{N}$, if $P(t)$ is true for all $t \leq k$ then $P(k+1)$ is also true (the induction step),

then $P(n)$ is true for all $n \in \mathbb{N}$.

0.3 Partitions and Equivalence Relations

A relation $\sim$ on a nonempty set A is a way to match its elements based on some criterion. For example, if $A = \mathbb{Z}$ we can define $n \sim m$ if and only if $n = m + 3$. Then $1 \sim -2$, $3 \sim 0$, and $10 \sim 7$. We will use relations with very specific properties later in the text, called *equivalence relations*. For more details about relations on sets see ([8]).

Definition 0.8 Let A be a nonempty set and $\sim$ a relation on A. $\sim$ is an ***equivalence relation*** if and only if it satisfies the following properties:

(i) For every $a \in A$ we have $a \sim a$. ($\sim$ is *reflexive* on A)
(ii) For every $a, b \in A$ if $a \sim b$ then $b \sim a$. ($\sim$ is *symmetric* on A)
(iii) For every $a, b, c \in A$ if $a \sim b$ and $b \sim c$ then $a \sim c$. ($\sim$ is *transitive* on A)

The example of a relation on $\mathbb{Z}$ before this definition is not an equivalence relation. Notice that (i) fails as $3 \sim 3$ is clearly wrong. You are encouraged to see if either (ii) or (iii) hold for this relation.

Example 0.9 Define a relation on $\mathbb{Z}$ by $n \sim m$ if and only if $n - m$ is even. Notice that since 0 is even, $n \sim n$ for any integer and $\sim$ is reflexive. Also if $n - m$ is even, then $m - n$ is even, so $n \sim m$ tells us $m \sim n$, and $\sim$ is symmetric. Finally, if $n \sim m$ and $m \sim s$, then $n - s = n - m + m - s$, and since both $n - m$ and $m - s$ are even we have $n - s$ even and $n \sim s$, making $\sim$ transitive. Thus we have an equivalence relation.

The equivalence relation in the previous example breaks the set of integers into parts. For each integer x we can look at which integers are equivalent to x, defining sets of the form $A_0 = \{n : n \sim 0\}$, $A_1 = \{n : n \sim 1\}$, $A_2 = \{n : n \sim 2\}$, and so on. It should be clear that since $\sim$ is reflexive, for each n, $n \in A_n$. Also if $n \sim 1$ then $n-1$ is even, say $n-1 = 2k$. Thus $n + 1 = 2k + 2$ so $n - (-1)$ is also even and $n \sim -1$ as well. This gives us $A_1 \subseteq A_{-1}$. You should show $A_1 \supseteq A_{-1}$ to see that $A_{-1} = A_1$. This same argument holds for each i, so $A_{-i} = A_i$. Can you tell if $A_3 = A_{25}$?

Definition 0.10 Let A be a nonempty set and suppose we have a collection of subsets of A denoted by $A_1, A_2, \ldots, A_n$ (possibly an infinite collection with subscripts in an index set I). This collection of subsets of A forms a ***partition of A*** if and only if the following hold:

(i) For each $i \neq j$ either $A_i = A_j$ or $A_i \cap A_j = \emptyset$.
(ii) $\bigcup_{i=1}^{n} A_i = A$.

In Example 0.3 we created a collection of sets, $\{A_n : n \in \mathbb{Z}\}$. We noticed that each integer $n \in A_n$, so we clearly have $\bigcup_{i=1}^{\infty} A_i = \mathbb{Z}$ and (ii) true. Also suppose we have $i, j \in \mathbb{Z}$ with $i \neq j$ and $A_i \cap A_j \neq \emptyset$, then there must be some $a \in A_i \cap A_j$. Thus $a \sim i$ and $a \sim j$. Using symmetry we have $i \sim a$ and $a \sim j$, so by transitivity we know $i \sim j$. Now for any $x \in A_i$, $x \sim i$ and $i \sim j$ so $x \sim j$ using transitivity. Thus $x \in A_j$, showing $A_i \subseteq A_j$. Similarly, if $x \in A_i$, $x \sim j$, and $j \sim i$ so $x \sim i$ using transitivity. Thus $x \in A_i$. Now $A_i \supseteq A_j$, so $A_i = A_j$, and (i) holds in the definition. Thus the relation $\sim$ also defines a partition on A.

Definition 0.11 Suppose A is a nonempty set and $\sim$ is an equivalence relation on A. For each $a \in A$ define $[a] = \{x \in A : x \sim a\}$. The set $[a]$ is called the *equivalence class* of a.

The critical connection between partitions and equivalence relations is seen in the next theorem.

Theorem 0.12

Suppose A is a nonempty set and $\sim$ is an equivalence relation on A. The collection of equivalence classes, $\{[a] : a \in A\}$, forms a partition of A.

0.4 Functions

Functions map one nonempty set to another. We say f is a function from set A to set B (denoted $f : A \to B$), if it matches every $a \in A$ with a *uniquely defined* $f(a) \in B$. The uniqueness only means that there cannot be two different answers to $f(a)$ when a specific $a \in A$ is selected; it does not however prevent two different elements in A from having the same answer, i.e., for $a, c \in A$ it is possible to have $f(a) = f(c)$. A is called the *domain* of the function ($dom(f)$) and the set $f(A) = \{b \in B \colon f(a) = b \text{ for some } a \in A\}$ is called the *range* or *image* of f.

Example 0.13 Consider the sets $A = \{0, 1, 2, 3\}$ and $B = \{x, y, z, w, u, v\}$. Suppose we define:

$$f(0) = z,\ f(1) = x,\ f(2) = w, \text{ and } f(3) = x$$

f is a function since each element of A has a uniquely defined element designated by f as its image. There is only one answer for each of $f(0)$, $f(1)$, $f(2)$, and $f(3)$. Notice that $f(1) = f(3)$ which is not prohibited by the definition. Also we do not require that each element of B occurs as an answer.

Definition 0.14 Let A and B denote nonempty sets, and let f be a function $f : A \to B$.

(i) If for all $a, c \in A$ with $a \neq c$ we must have $f(a) \neq f(c)$, then we say f is ***one to one*** (or *injective*).
(ii) If for every $b \in B$ there exists $a \in A$ with $f(a) = b$, then we say f is ***onto*** (or *surjective*).
(iii) If f is both injective and surjective we say that f is ***bijective*** (or f is a *bijection*).

In many cases the strategy for proving that a function is one to one uses the contrapositive of the statement in (i). Thus we assume that $f(a) = f(c)$ for some $a, c \in A$ and prove that $a = c$. Looking back at Example 0.13 we notice that f is not one to one since $f(1) = f(3)$. Also f is not onto since there is no element of A that maps to y.

Theorem 0.15

Let A and B be nonempty sets.

(i) $|A| = |B|$ if and only if there exists a bijection $f : A \to B$.
(ii) If A and B are finite sets and $|A| = |B|$, then any injective function $f : A \to B$ is also surjective.
(iii) If A and B are finite sets and $|A| = |B|$, then any surjective function $f : A \to B$ is also injective.
(iv) If A and B are finite sets, $|A| = |B|$, and $A \subseteq B$, then $A = B$.

The next example relies on the fact that every integer is either even (of the form $2k$) or odd (of the form $2k+1$), but cannot be both at the same time.

Example 0.16 Consider the natural numbers $\mathbb{N}$ and integers $\mathbb{Z}$. Define $f : \mathbb{N} \to \mathbb{Z}$ by

$$f(x) = \begin{cases} n & \text{if } x = 2n, \\ -n & \text{if } x = 2n+1 \end{cases}$$

For each natural number x, x is either even or odd, and thus can be written as either $2n$ or $2n+1$ for some integer n. Thus $f(x)$ is defined for each $x \in \mathbb{N}$. As x cannot be both even and odd, nor can we have $x = 2n$ and $x = 2m$ unless $n = m$, the answer $f(x)$ is uniquely defined and f is a function.

Also if we had $x, z \in \mathbb{N}$ with $f(x) = f(z) = n$, then $x = 2n$ and $z = 2n$, and so $x = z$. Similarly if $x, z \in \mathbb{N}$ with $f(x) = f(z) = -n$, then $x = 2n+1$ and $z = 2n+1$, and so $x = z$ again. Thus f is one to one (injective).

Now suppose we had any integer $k \in \mathbb{Z}$. If $k = 0$, then as $1 = 2(0)+1$ we have $f(1) = 0$. If $k > 0$, then $2k \in \mathbb{N}$ so that $f(2k) = k$. And finally, if $k < 0$, then $-k > 0$, and so $2(-k)+1 \in \mathbb{N}$ giving us $f(2(-k)+1) = -(-k) = k$. Thus for any $k \in \mathbb{Z}$ we found $n \in \mathbb{N}$ with $f(n) = k$, and f is surjective.

Hence, we know that f is a bijection and so we can conclude that $|\mathbb{N}| = |\mathbb{Z}|$, which is not an obvious conclusion since $\mathbb{N} \subset \mathbb{Z}$!

For two functions $f : A \to B$ and $g : B \to C$ we can define $g \circ f : A \to C$ (g composed with f) so that for any $a \in A$, $(g \circ f)(a) = g(f(a))$. Notice that this definition requires $f(A) \subseteq B$ since we cannot calculate $g(f(a))$ unless $f(a)$ is in the domain of the function g. The opposite composition $f \circ g$ can only be defined if $g(B) \subseteq A$.

We will refer to the next theorem many times. One part of the proof is provided to illustrate proof techniques.

Theorem 0.17

Suppose A, B, and C are nonempty sets and f, g are functions with $f : A \to B$, $g : B \to C$.

(i) If f and g are both one to one (injective) then $g \circ f$ is also one to one (injective).
(ii) If f and g are both onto (surjective) then $g \circ f$ is also onto (surjective).
(iii) If f and g are both bijections then $g \circ f$ is a bijection.

Proof We will prove (i) of this theorem and leave (ii) and (iii) as exercises at the end of the chapter.

(i) Suppose we have A, B, and C nonempty sets and $f : A \to B$, $g : B \to C$ functions. Assume that f and g are both one to one (injective). To show $g \circ f$ is injective suppose there are elements $u, v \in A$ with $(g \circ f)(u) = (g \circ f)(v)$. Since f is a function there are $y, z \in B$ with $f(u) = y$ and $f(v) = z$. Thus $(g \circ f)(u) = g(y)$ and $(g \circ f)(v) = g(z)$. By our assumption we have $g(y) = g(z)$, but g is one to one, so we must have $y = z$. Now we know $f(u) = f(v)$, but f is also one to one which tells us that $u = v$ as needed. Hence $g \circ f$ is injective as well. $\square$

If f is a bijective function we can also define the inverse function $f^{-1} : B \to A$ so that for every $a \in A$, $f^{-1}(f(a)) = a$ and for every $b \in B$, $f\left(f^{-1}(b)\right) = b$. Thus $f^{-1} \circ f = id_A$ and $f \circ f^{-1} = id_B$. Where id stands for the identity function and the subscript is the set it maps both to and from.

For more about functions see [8].

0.5 Some Number Theory Results

You are only expected to be familiar with some basic concepts of Number Theory, in fact we have already used one in the previous section, namely that every positive integer is either even or odd but cannot be both. More about Number Theory can be found in ([7]).

Theorem 0.18

The Division Algorithm Let n and m be integers. If $m \neq 0$ then there exist unique integers q and r so that $n = mq + r$, where $0 \leq r < |m|$. Note that $|m|$ refers to the absolute value of m.

We will often apply the theorem above when $m > 0$ so we will be able to say $0 \leq r < m$. If we have $r = 0$, i.e., $n = mq$, we say m ***divides*** n or m is a ***factor*** of n. We may occasionally write $m \mid n$ to mean m divides n. Notice from this definition that every positive integer m divides 0 since we have $0 = m0 + 0$.

Definition 0.19

(i) The positive integer p is ***prime*** if $p > 1$ and the only positive integers that divide p are 1 and p.
(ii) The positive integer p is ***composite*** if there exist positive integers $x > 1$ and $y > 1$ with $p = xy$.
(iii) Two integers n, m are ***relatively prime*** if no prime evenly divides both m and n.

Notice that the definition of prime only mentions *positive* integers. For example, 2 is prime even though we can divide it evenly by -2 and -1 as well as $1, 2$. Thus whenever a number is assumed to be prime, it is also assumed to be positive.

Theorem 0.20

Every positive integer $n > 1$ can be factored into a product of powers of distinct prime factors, $n = p_1^{k_1} p_2^{k_2} \cdots p_s^{k_s}$. Each p_i is a different prime and $s, k_i \in \mathbb{N}$. This factorization is unique up to the order of the factors.

Theorem 0.21

Suppose p is a prime number and a, b are integers. If $p \mid ab$ then $p \mid a$ or $p \mid b$.

Theorem 0.22

If integers n, m are **relatively prime** then there exist integers x, y so that $nx + my = 1$.

Example 0.23 Consider the integer $n = 98$. We see that the prime number 2 evenly divides 98, as does the prime number 7. As in Theorem 0.20 we can factor $98 = 2 \cdot 7^2$. Also $9 = 3^2$ is relatively prime to 98, so by Theorem 0.22 there should be integers x, y with $98x + 9y = 1$. In fact, $98(-1) + 9(11) = 1$ using $x = -1$ and $y = 11$.

0.6 A Few Results From Linear Algebra

Matrices are only used as examples in this text, but the concept of a vector space becomes important in Chapter 9. See more about both in ([5]).

Definition 0.24 An $n \times n$ ***matrix*** A is an array (n rows and n columns) with n^2 many entries, from a specific set. We write $A = [a_{ij}]$ where the entry a_{ij} is in the ith row and jth column. The set of all $n \times n$ matrices whose entries are from the set B is denoted $M_n(B)$.

Matrix addition and matrix multiplication are defined on the set $M_n(B)$ as long as B is a set for which addition and multiplication are defined. If $A, C \in M_n(B)$ then $A + C = D$ where $d_{ij} = a_{ij} + b_{ij}$ for each i and j. However, multiplication is more complicated. If $A, C \in M_n(B)$ then $AC = S$ where for each i and j, $s_{ij} = \sum_{k=1}^{n} a_{ik} \cdot c_{kj}$.

Example 0.25 Consider the 3×3 matrices A and C with entries from $\mathbb{Z}$, i.e., $A, C \in M_3(\mathbb{Z})$ shown below:

$$A = \begin{bmatrix} 0 & -1 & 2 \\ 3 & 1 & 0 \\ 1 & 2 & 2 \end{bmatrix} \qquad C = \begin{bmatrix} 2 & 1 & 1 \\ 0 & -2 & 1 \\ 1 & 3 & 2 \end{bmatrix}$$

$$A + C = \begin{bmatrix} 0+2 & -1+1 & 2+1 \\ 3+0 & 1-2 & 0+1 \\ 1+1 & 2+3 & 2+2 \end{bmatrix} = \begin{bmatrix} 2 & 0 & 3 \\ 3 & -1 & 1 \\ 2 & 5 & 4 \end{bmatrix}$$

$$AC = \begin{bmatrix} 0(2) - 1(0) + 2(1) & 0(1) - 1(-2) + 2(3) & 0(1) - 1(1) + 2(2) \\ 3(2) + 1(0) + 0(1) & 3(1) + 1(-2) + 0(3) & 3(1) + 1(1) + 0(2) \\ 1(2) + 2(0) + 2(1) & 1(1) + 2(-2) + 2(3) & 1(1) + 2(1) + 2(2) \end{bmatrix} =$$

$$\begin{bmatrix} 2 & 8 & 3 \\ 6 & 1 & 4 \\ 4 & 3 & 7 \end{bmatrix}$$

Notice in the calculation of AC above, the numbers in the parentheses are the same for each column while the numbers outside of the parentheses are the same in each row! Thus we find the entry in the ij-position by multiplying the entries of row i of A and the entries in column j of C, by position. You should find the matrix CA for practice, and verify that it is not the same as AC.

Definition 0.26 Let V be a nonempty set. V is a ***vector space*** over $\mathbb{R}$ if there exist an addition $\oplus$, and scalar multiplication $\otimes$ with the following properties:

(i) For all $u, v \in V$, $u \oplus v \in V$.
(ii) For all $u \in V$ and $r \in \mathbb{R}$, $r \otimes u \in V$.
(iii) For all $u, v, w \in V$, $u \oplus v = v \oplus u$, and $(u \oplus v) \oplus w = u \oplus (v \oplus w)$.
(iv) There exists an element $\mathbf{0}$ of V with the property that for any $u \in V$, $u \oplus \mathbf{0} = u$.
(v) For each $u \in V$ there is an element $w \in V$ so that $u \oplus w = w \oplus u = \mathbf{0}$.
(vi) For all $u, v \in V$ and $r, s \in \mathbb{R}$, $(rs) \otimes u = r \otimes (s \otimes u)$, $(r + s) \otimes u = (r \otimes u) \oplus (s \otimes u)$, and $r \otimes (u \oplus v) = (r \otimes u) \oplus (r \otimes v)$.
(vii) For all $u \in V$, $1 \otimes u = u$.

Example 0.27 Consider the set of 2×2 matrices with real number entries, $M_2(\mathbb{R})$. We can use this set as V in the definition above with the usual addition of matrices as described after Definition 0.24. For scalar multiplication we define

$$r \begin{bmatrix} a & b \\ c & d \end{bmatrix} = \begin{bmatrix} ra & rb \\ rc & rd \end{bmatrix}$$

where the multiplication inside the matrix is just real number multiplication. It should be clear that (i), (ii), and (iii) are true. Also in $M_2(\mathbb{R})$ we find

$$\mathbf{0} = \begin{bmatrix} 0 & 0 \\ 0 & 0 \end{bmatrix}, \text{ and } \quad 1 \begin{bmatrix} a & b \\ c & d \end{bmatrix} = \begin{bmatrix} a & b \\ c & d \end{bmatrix}$$

which make (iv), and (vii) true. By negating each entry of a matrix it is easy to see how (v) holds as well. To see that (vi) is true:

$$(rs) \begin{bmatrix} a & b \\ c & d \end{bmatrix} = \begin{bmatrix} rsa & rsb \\ rsc & rsd \end{bmatrix} = r \begin{bmatrix} sa & sb \\ sc & sd \end{bmatrix}$$

$$r\left(\begin{bmatrix} a & b \\ c & d \end{bmatrix} + \begin{bmatrix} u & v \\ w & z \end{bmatrix} \right) = r \begin{bmatrix} a+u & b+v \\ c+w & d+z \end{bmatrix} = \begin{bmatrix} ra+ru & rb+rv \\ rc+rw & rd+rz \end{bmatrix}$$

$$r \begin{bmatrix} a & b \\ c & d \end{bmatrix} + r \begin{bmatrix} u & v \\ w & z \end{bmatrix} = \begin{bmatrix} ra & rb \\ rc & rd \end{bmatrix} + \begin{bmatrix} ru & rv \\ rw & rz \end{bmatrix} = \begin{bmatrix} ra+ru & rb+rv \\ rc+rw & rd+rz \end{bmatrix}$$

Thus, $M_2(\mathbb{R})$ is a vector space over $\mathbb{R}$.

Definition 0.28 Let V be a vector space over $\mathbb{R}$. A subset $W \subseteq V$ is a ***basis*** for V if and only if the following properties hold:

(i) For every $u \in V$, there exists $w_1, w_2, \ldots, w_k \in W$ and $r_1, r_2, \ldots, r_k \in \mathbb{R}$ so that $u = (r_1 \otimes w_1) \oplus (r_2 \otimes w_2) \oplus \cdots \oplus (r_k \otimes w_k)$ (W spans V).

(ii) For each n and $w_1, w_2, \ldots, w_n$ distinct elements of W, if $r_1, r_2, \ldots, r_n \in \mathbb{R}$ with $(r_1 \otimes w_1) \oplus (r_2 \otimes w_2) \oplus \cdots \oplus (r_n \otimes w_n) = \mathbf{0}$ then $0 = r_1 = r_2 = \cdots = r_n$ (W is independent).

$|W|$ is called the dimension of V.

In the vector space $M_2(\mathbb{R})$ (Example 0.27), you should be able to verify that the set shown below is a basis. Recall however that while a basis is not unique (other bases could also exist), every basis will have the same cardinality. Thus the dimension of $M_2(\mathbb{R})$ is 4.

$$W = \left\{ \begin{bmatrix} 1 & 0 \\ 0 & 0 \end{bmatrix}, \begin{bmatrix} 0 & 1 \\ 0 & 0 \end{bmatrix}, \begin{bmatrix} 0 & 0 \\ 1 & 0 \end{bmatrix}, \begin{bmatrix} 0 & 0 \\ 0 & 1 \end{bmatrix} \right\}$$

Exercises for Chapter 0

List in brackets the elements of each set described below. If the set is empty write $\emptyset$ instead.

1. $A = \{x : x \text{ is an integer and } -2 < x < 3\}$
2. $B = \{x : x \text{ is a real number and } x^2 - 5x = 0\}$
3. $C = \{x : x \text{ is an integer and } 2x - 6 = 2\}$
4. $D = \{x : x \text{ is a natural number and } 3x + 8 = 5\}$
5. $E = \{x : x \text{ is a rational number and } 4x^2 + 1 = 10\}$
6. For the sets A, B, and C in Exercises 1,2,3 find the sets $A \cap B, B \cup C$.
7. For the sets A, B, and C in Exercises 1,2,3 find the sets $C \cup (A \cap B), \wp(B)$.
8. For the sets A, B, and C in Exercises 1,2,3 find the sets $C \cap B, B \cup A$.
9. For the sets A, B, and C in Exercises 1,2,3 find the sets $A \cup (C \cap B), \wp(A)$.
10. For the sets B, C, and D in Exercises 2,3,4 find the sets $C \cap D, B \cup D$.
11. For the sets B, C, and D in Exercises 2,3,4 find the sets $C \cup (D \cap B), \wp(C)$.
12. For the sets C, D, and E in Exercises 3,4,5 find the sets $E \cap C, D \cup E$.
13. For the sets A, B, and E in Exercises 1,2,5 find the sets $E \cup (A \cap B), \wp(E)$.

For each statement about nonempty sets A, B, C, D either prove that the statement is true or find a counterexample with nonempty sets that make it fail.

14. If $A \subseteq (B \cup C)$ then $A \subseteq B$ or $A \subseteq C$.
15. If $A \subset (B \cup C)$ then $A \subset B$ or $A \subset C$.
16. If $A \subseteq (B \cap C)$ then $A \subseteq B$ and $A \subseteq C$.
17. If $A \subset (B \cap C)$ then $A \subset B$ and $A \subset C$.
18. If $A \cap B \subset C$ then $A \subset C$ or $B \subset C$.
19. $A \cup (B \cap C) = (A \cup B) \cap (A \cup C)$.
20. $(A - B) \cup (B - A) = (A \cup B) - (A \cap B)$.
21. $A \cap (B \cup C) = (A \cap B) \cup (A \cap C)$.

Prove each of the following using Mathematical Induction.

22. Prove: For all $n \in \mathbb{N}, 1 + 3 + \cdots + (2n+1) = (n+1)^2$.
23. Prove: For all $n \in \mathbb{N}$ with $n \geq 2, n^2 > n + 1$.
24. Prove: For all $n \in \mathbb{N}$ with $n \geq 4, n! > n^2$.
25. Prove: For all $n \in \mathbb{N}$, $(2n+1) + (2n+3) + \cdots + (2n + (2n-1)) = 3(n^2)$.
26. Prove: For all $n \in \mathbb{N}$, $1^3 + 2^3 + \cdots + n^3 = \frac{n^2(n+1)^2}{4}$.
27. Prove: For all $n \in \mathbb{N}$, 4 evenly divides $3^{2n-1} + 1$.
28. Prove: For all $n \in \mathbb{N}$, $1(1!) + 2(2!) + \cdots + n(n!) = (n+1)! - 1$.

For each relation $\sim$ given, either prove that $\sim$ is an equivalence relation or find a counterexample showing it fails. If $\sim$ is a equivalence relation determine the equivalence classes.

29. Define $\sim$ on $\mathbb{Z}$ by $a \sim b$ if and only if $a - b = 3n$ for some $n \in \mathbb{Z}$.
30. Define $\sim$ on $\mathbb{Z}$ by $a \sim b$ if and only if $a = 2b$.
31. Define $\sim$ on $\mathbb{Z}$ by $a \sim b$ if and only if $a + b > 0$.
32. Define $\sim$ on $\mathbb{R}$ by $x \sim y$ if and only if $x^2 = y^2$.
33. Define $\sim$ on $\mathbb{R}$ by $x \sim y$ if and only if $x^2 + y^2 = 1$.
34. Define $\sim$ on $\mathbb{Z} \times \mathbb{N}$ by $(m, n) \sim (u, v)$ if and only if $mv = nu$.
35. Define $\sim$ on $\mathbb{Z} \times \mathbb{N}$ by $(m, n) \sim (u, v)$ if and only if $mu = nv$.
36. Define $\sim$ on the set $\mathbb{Z} \times \mathbb{N}$ by $(m, n) \sim (u, v)$ if and only if $m = u$.
37. Define $\sim$ on $\mathbb{R}$ by $x \sim y$ if and only if $x - y \in \mathbb{Z}$.
38. Define $\sim$ on $\mathbb{N}$ by $m \sim n$ if and only if m evenly divides n.

Determine if any function defined below is injective, surjective, or bijective. Prove your answers. Remember that elements of $\mathbb{Q}$ are of the form $\frac{a}{b}$ where $a, b \in \mathbb{Z}$ and $b \neq 0$.

39. $g : \mathbb{Z} \to \mathbb{Z}$, where $g(x) = |x|$. ($|x|$ means the absolute value of x.)
40. $r : \mathbb{Z} \to \mathbb{Z}$, where $r(x) = 7x$.
41. $t : \mathbb{Q} \to \mathbb{Q}$, where $t(x) = 5x - 3$.
42. $h : \mathbb{Z} \to \mathbb{Z}$, where $h(x) = 2x - 5$.

43. $k : \mathbb{N} \to \mathbb{N}$, where $k(x) = x^2 + x$.
44. $f : \mathbb{Q} \to \mathbb{Z}$, where $f\left(\frac{a}{b}\right) = a - b$. Note that $\frac{a}{b}$ is written in lowest terms in order to be sure f is a function!
45. $s : \mathbb{Q} \to \mathbb{Z}$, where $s\left(\frac{a}{b}\right) = ab$. Note that $\frac{a}{b}$ is written in lowest terms in order to be sure s is a function!

Prove each statement below:

46. Suppose X is a nonempty set with subsets A and B, and f is a function with $f : X \to X$. If f is injective then $f(A) \cap f(B) = f(A \cap B)$.
47. If f and g are functions with $f : A \to B$, $g : B \to C$ and $g \circ f$ is surjective then g is surjective.
48. If f and g are functions with $f : A \to B$, $g : B \to C$ and g, f are both surjective then $g \circ f$ is surjective.
49. If f and g are functions with $f : A \to B$, $g : B \to C$ and g, f are both bijections then $g \circ f$ is a bijection.
50. Find a bijection from $\mathbb{Z}$ to the set of even integers $\mathbb{E}$ (and prove it is a bijection).

Using the matrices A, B, C, D, E defined below, compute the indicated matrix.

$$A = \begin{bmatrix} 3 & -1 & 2 \\ 0 & 4 & -2 \\ 1 & 1 & 0 \end{bmatrix} \quad B = \begin{bmatrix} \frac{1}{2} & -1 & 0 \\ 2 & \frac{-2}{3} & 1 \\ 2 & -3 & -1 \end{bmatrix} \quad C = \begin{bmatrix} 1 & \frac{1}{3} & -1 \\ -1 & 0 & 4 \\ \frac{3}{4} & 0 & 1 \end{bmatrix}$$

$$D = \begin{bmatrix} 0 & \frac{2}{5} & 2 \\ 1 & 1 & \frac{1}{2} \\ \frac{3}{4} & -1 & 0 \end{bmatrix} \quad E = \begin{bmatrix} 2 & 2 & 2 \\ 3 & 1 & 1 \\ \frac{1}{2} & 0 & -1 \end{bmatrix}$$

51. $A + C$
52. $B + 2C$
53. AB and BA
54. BC and CB
55. $(AB) + (AC)$
56. $C - D$
57. $3E - 2C$
58. CD and DC
59. DE and ED
60. $(AD) + (BC)$
61. $E(2A + D)$

Chapter 1

Groups

We all look for structure in our world, from the simplest gestures to the complexities of the universe. In each field of mathematics such as Logic, Algebra, Geometry, Analysis, or Topology there is a particular type of structure under discussion. Each subject includes definitions and theorems for objects, sub-objects, similarity of objects, and functions between objects. In Modern Algebra the objects in question are groups, rings, and fields; i.e., nonempty sets with special operations.

1.1 Operations

Operations are familiar to everyone, even if the word operation is not. We use operations every day such as addition, multiplication, and exponentiation. For mathematical precision we first define an operation on a set.

Definition 1.1 Let A denote a nonempty set. An ***operation on*** A is a function $f : A \times A \to A$.

Note that this definition tells us:

(i) For every pair $(a, b) \in A \times A$, $f(a, b)$ is uniquely defined.
(ii) For every pair $(a, b) \in A \times A$, $f(a, b)$ is an element of A.

This definition is for a binary operation since it uses two elements from A to determine the answer. There are similar definitions for n-ary operations, but we will not consider them here.

Example 1.2 The most familiar examples of operations are addition, subtraction, and multiplication on the set of integers, $\mathbb{Z}$. We can write addition as a function $+ : \mathbb{Z} \times \mathbb{Z} \to \mathbb{Z}$ with $+(a, b) = c$, as described in Definition 1.1, however, this notation is not intuitive so we use the standard notation $a + b = c$ in its place.

The function $+$ forms an operation on $\mathbb{Z}$ since we can always add two integers, and when added, only one answer can occur for each pair, i.e., $a+b$ is uniquely defined for any integers a and b; and the sum of two integers always yields an integer.

Subtraction and multiplication are operations on $\mathbb{Z}$ as well, but division fails to be an operation since for the integers $a = 1$ and $b = 0$ there is no integer answer to $a \div b$. If instead we used the rational or real numbers, divisions still does not make an operation using the same two elements a and b. Is there a set on which usual division is an operation?

It is critical that we always mention the set and operation together. Having a rule that defines an operation on one set does not guarantee that it will still be an operation if the set is changed.

Example 1.3 Although subtraction is an operation on $\mathbb{Z}$, it is NOT an operation on $\mathbb{N}$ since 3 and 7 are elements of $\mathbb{N}$, but $3 - 7$ is not. It violates condition (ii) of the definition.

To denote a set and an operation together, such as the set of integers $\mathbb{Z}$ with addition, we often write $(\mathbb{Z}, +)$.

Example 1.4 Consider each pair below and determine if the function (as usually defined) is an operation on the set. Note that $\mathbb{R}^+$ is the set of positive real numbers in (c) and $\cdot$ denotes multiplication of matrices in (b).

(a) ($\mathbb{Z}$, exponentiation)
(b) $(M_2(\mathbb{R}), \cdot)$
(c) ($\mathbb{R}^+$, exponentiation)

An operation does not have to be defined on a set of numbers, it is simply a rule defined on a set. The set could be the students in a specific room at a specific time, or the set of hairs on your head at a given moment. It can be dangerous to rely on the intuitive operation of addition on the integers to drive your understanding or beliefs about operations, since there will be properties we discuss that do not hold in the integers with addition. The following example gives a set whose elements are specific names, and an operation between them, to illustrate this idea.

Example 1.5 Consider the set $A = \{John, Sue, Pam, Henry\}$ with the function $f : A \times A \to A$ defined by:

$$f(Pam, Henry) = Sue \qquad f(Henry, Pam) = Sue$$
$$f(x, John) = John \text{ for all } x \in A \qquad f(John, x) = John \text{ for all } x \in A$$
$$f(x, Sue) = x \text{ for all } x \in A \qquad f(Sue, x) = x \text{ for all } x \in A$$
$$f(x, x) = x \text{ for all } x \in A$$

It is often convenient to display the results of an operation in a ***Cayley*** table, a table showing the elements of the set as column and row headings and the result of the operation in the interior of the table. The Cayley table for our operation is shown below. Notice that $f(Henry, Sue)$ is found in the last row (labeled $Henry$ on the left) and second column (labeled Sue at the top), telling us $f(Henry, Sue) = Henry$.

f	$John$	Sue	Pam	$Henry$
$John$	$John$	$John$	$John$	$John$
Sue	$John$	Sue	Pam	$Henry$
Pam	$John$	Pam	Pam	Sue
$Henry$	$John$	$Henry$	Sue	$Henry$

When working with numbers, and operations like addition or multiplication, properties can be so familiar we never expect them to fail. Since our operations need not be defined on sets of numbers, we must carefully define the properties we are interested in, i.e., desirable structure and patterns.

1.2 Properties of an Operation

Suppose we have a nonempty set A, and an operation $*$ defined on A. Instead of writing $*(a, b)$ we will use the more standard notation of $a * b$, later shortening it to simply ab with the understanding that the operation is not necessarily usual multiplication.

The next definition describes many familiar properties of addition on the set of integers. However, the question of whether the same properties hold for another operation or on a different set always requires either a proof or counterexample. The terms defined here are among the most commonly misspelled words in Modern Algebra, so write them carefully!

Definition 1.6 Let A be a nonempty set and $*$ an operation on A.

(i) $*$ is ***commutative*** on A if $a * b = b * a$ for every $a, b \in A$.
(ii) $*$ is ***associative*** on A if $(a * b) * c = a * (b * c)$ for every $a, b, c \in A$.
(iii) A contains an ***identity*** for $*$ if there exists some $e \in A$ for which $a * e = a$ and $e * a = a$ for all $a \in A$.
(iv) An element $a \in A$ has an ***inverse*** in A under $*$ if there is some $b \in A$ for which $a * b = e$ and $b * a = e$ (where e is the identity found in (iii)).

To show that an operation fails to have a property there must be at least one instance where the property fails, as we found in Example 1.3. The technique of using counterexamples is as important to learn as the proper methods of proof. The notion of testing a conjecture, to decide if there is evidence that it is true, can help minimize the frustration of trying to prove a false conjecture. A counterexample is the most common way that a conjecture is shown to fail in mathematical research.

Addition and multiplication on the integers are associative and commutative operations, as most students learn in grade school mathematics. However, subtraction is not associative on $\mathbb{Z}$. Consider the counterexample with $a = 1, b = 2$, and $c = 3$:

$$(a * b) * c = (1 - 2) - 3 = -1 - 3 = -4$$

$$a * (b * c) = 1 - (2 - 3) = 1 - (-1) = 2$$

Thus for these values we have $(a*b)*c \neq a*(b*c)$ and so we have shown that subtraction is not associative on $\mathbb{Z}$.

Example 1.7 Even though usual addition is associative on $\mathbb{Z}^+$, the positive integers, there is no identity for this operation. In order to have an identity there would need to be $e \in \mathbb{Z}^+$ for which $e + x = x$ for all $x \in \mathbb{Z}^+$. However, $e + 2$ cannot equal 2 under usual addition unless $e = 0$, and 0 is not a member of $\mathbb{Z}^+$.

When it is shown that there is no identity for an operation on a set A, then the question of inverses cannot even be discussed. Thus if property (iii) fails for an operation, then (iv) is not applicable. Even if an identity exists there is no guarantee there will be inverses for any elements of the set. Usual multiplication in the integers, $\mathbb{Z}$, is associative and has 1 as its identity. In order for 2 to have an inverse we would need an integer x with $2x = 1$, but no such integer exists. (Only two integers have inverses under multiplication. Which ones?)

All of the operations in this section have been 'usual' in the sense that you have seen them for many years and hence know most of the basic properties they have. Consider now the operation defined in Example 1.5 and the properties of Definition 1.6.

Example 1.8 Associativity is often the hardest property to verify for an unusual operation. If no specific formula is given, or we are not dealing with numbers, then the only way to prove associativity, is to check *every* combination of elements a, b, c. This can be time consuming as the size of the set increases. Using the set A in Example 1.5, we would have 43 possible equations to test since each of $a, b,$ or c has 4 choices. Some shortcuts can help in this situation. We will use $a * b$ to denote the operation instead of $f(a, b)$ for the rest of this example to make the notation easier to read.

If $John$ is in any position the answer must be $John$. To see why recall that $John * x = John$ and $x * John = John$ for any $x \in A$. Thus for any $x, y \in A$ we have:

$$
\begin{array}{ll}
(x * John) * y = John * y = John & x * (John * y) = x * John = John \\
(John * x) * y = John * y = John & x * (y * John) = x * John = John \\
John * (x * y) = John & (x * y) * John = John
\end{array}
$$

Also if all three are the same, then knowing $x * x = x$ for any $x \in A$ tells us $(x * x) * x = x * x = x$ and $x * (x * x) = x * x = x$.

When one of the three is Sue the answer seems to just ignore Sue and so:

$$
\begin{array}{ll}
(Sue * x) * y = x * y & Sue * (x * y) = x * y \\
(x * Sue) * y = x * y & x * (Sue * y) = x * y \\
(x * y) * Sue = x * y & x * (y * Sue) = x * y
\end{array}
$$

The final cases to check involve only Pam and $Henry$, which leaves six of them. We must determine if:

$$
\begin{array}{lcl}
(Pam * Pam) * Henry & = & Pam * (Pam * Henry) \\
(Pam * Henry) * Pam & = & Pam * (Henry * Pam) \\
(Pam * Henry) * Henry & = & Pam * (Henry * Henry) \\
(Henry * Pam) * Henry & = & Henry * (Pam * Henry) \\
(Henry * Henry) * Pam & = & Henry * (Henry * Pam) \\
(Henry * Pam) * Pam & = & Henry * (Pam * Pam)
\end{array}
$$

You will be asked to check these in an exercise at the end of the chapter. An advantage to writing out every possible combination is that a counterexample is automatically found if the operation is not associative. Is $*$ commutative? There is an identity in A (what is it?). Do all of the elements have inverses in A under $*$? Look for these questions as exercises at the end of the chapter as well.

If a formula is given for an operation on a set of numbers, then algebraic rules can be used to check for associativity. However, if the operation is not associative then a counterexample must still be produced.

Example 1.9 Consider the set of rational numbers $\mathbb{Q}$ with the rule $x * y = \frac{x+y}{2}$ where $+$ is usual addition and division by 2 is also as usual in $\mathbb{Q}$. You should verify that $*$ is an operation on $\mathbb{Q}$, and $*$ is commutative. To determine if $*$ is associative on $\mathbb{Q}$ we use algebraic manipulation:

$$(x * y) * z = \left(\frac{x+y}{2}\right) * z = \frac{\left(\frac{x+y}{2}\right) + z}{2} = \frac{x+y+2z}{4}$$

$$x * (y * z) = x * \left(\frac{y+z}{2}\right) = \frac{x + \left(\frac{y+z}{2}\right)}{2} = \frac{2x+y+z}{4}$$

This seems to imply that the two are not equal, but it depends on the values of x, y, z, so we give one specific counterexample. Let $x = 1$, $y = 2$, and $z = 3$, which gives $(x * y) * z = \frac{9}{4}$ and $x * (y * z) = \frac{7}{4}$. Hence $*$ is not associative on $\mathbb{Q}$. Is there an identity for $*$? If so, do any of the elements have inverses?

1.3 Groups

Once we have an operation on a nonempty set, we then choose which properties we require to define our mathematical structure.

Definition 1.10 Suppose G is a nonempty set and $*$ is an operation on G. $(G, *)$ is called a ***group*** (or G is a group under $*$) if it satisfies the following three properties:

(i) $*$ is associative on G.
(ii) G contains an identity for $*$.
(iii) Every element of G has an inverse under $*$, which is also in G.

If, in addition to these properties, $*$ is commutative then we say that $(G, *)$ is an ***abelian group***. Using the notation of sets from Chapter 0, we write $|G|$ to denote the cardinality of G, i.e., number of elements in G.

Will addition on the sets $\mathbb{Z}$, $\mathbb{Q}$, $\mathbb{R}$, and $M_2(\mathbb{R})$ create abelian groups? What about the operation of multiplication?

A special collection of groups will frequently be used in examples and are often referred to as clock groups. These groups are often defined in terms of equivalence classes of integers (we will see that connection in Chapter 3), or as moving around a clock, but we will take a more arithmetical approach. The operations on these sets involve modular arithmetic.

Definition 1.11 Let n denote an integer with $n > 1$. For any integer a, $a(\text{mod}n)$ is defined to be the *nonnegative remainder* less than n, found when a is divided by n.

For example, $7(\text{mod}2) = 1$ since $7 = (2)(3)+1$, and $-14(\text{mod}3) = 1$ since $-14 = (3)(-5) + 1$. We would not use $-14 = 3(-4) - 2$ since the remainder, by definition, must be *nonnegative.*

Let $\mathbb{Z}_n = \{0, 1, 2, \ldots, n-1\}$ (called the ***integers modulo n***). Define the operation ***addition modulo n***, denoted $+_n$, on $\mathbb{Z}_n$ by $a +_n b = (a + b)(\text{mod}n)$. To prove $\mathbb{Z}_n$ is a group, we must first assure ourselves that $+_n$ is an operation on our set. For any $a, b \in \mathbb{Z}_n$, $a + b$ is uniquely defined from usual addition in $\mathbb{Z}$, thus $(a + b)(\text{mod}n)$ is again uniquely defined since there will always be a unique nonnegative remainder when dividing by n (Theorem 0.18). Thus $a +_n b$ is also uniquely defined. Since we are dividing by n and requiring a nonnegative answer less than n, $a +_n b$ must be in $\mathbb{Z}_n$. Hence $+_n$ is an operation on $\mathbb{Z}_n$. The Cayley table for $(\mathbb{Z}_4, +_4)$ is shown below:

$+_4$	0	1	2	3
0	0	1	2	3
1	1	2	3	0
2	2	3	0	1
3	3	0	1	2

For each n, 0 is the identity, 0 is its own inverse, and the inverse of a nonzero element a is $n - a$ since $n(\text{mod}n) = 0$. Associativity, however, is more involved.

Assume we have $a, b, c \in \mathbb{Z}_n$. We need to show that $a +_n (b +_n c) = (a +_n b) +_n c$. With usual addition on $\mathbb{Z}$ and Theorem 0.18, we can write $a + b = xn + y$ with $0 \leq y < n$ for some $x, y \in \mathbb{Z}$. Thus we have $a +_n b = y$. Similarly $y + c = sn + t$ with $0 \leq t < n$ for some $s, t \in \mathbb{Z}$, and $y +_n c = t$. Therefore $(a +_n b) +_n c = t$.

We can also find $u, v \in \mathbb{Z}$ with $b + c = un + v$ and $0 \leq v < n$, as well as $q, r \in \mathbb{Z}$ with $a + v = qn + r$ and $0 \leq r < n$. Thus $a +_n (b +_n c) = r$. To see why $t = r$, consider the following equations and the fact that usual addition is both commutative and associative on $\mathbb{Z}$:

$$(a+b)+c = (xn+y)+c = xn+(y+c) = xn+(sn+t) = (x+s)n+t$$
$$a+(b+c) = a+(un+v) = un+(a+v) = un+(qn+r) = (u+q)n+r$$

Thus $((a+b)+c)(\text{mod } n) = t$, and $(a+(b+c))(\text{mod } n) = r$. But in $\mathbb{Z}$ $(a+b)+c = a+(b+c)$, and $(a+b+c)(\text{mod}n)$ is uniquely defined, so we have $t = r$. Hence $a +_n (b +_n c) = (a+b+c)(\text{mod}n) = (a +_n b) +_n c$, and $+_n$ is associative.

We now know $(\mathbb{Z}_n, +_n)$ is a group for each $n > 1$. In fact, the definition of $+_n$ shows it is an abelian group as well since $a+b = b+a$ under usual addition. These groups will be very useful examples throughout the text, so we include the result just proven as a theorem.

Theorem 1.12

For every $n > 1$, $\mathbb{Z}_n = \{0, 1, \ldots, n-1\}$ is an abelian group under the operation of $+_n$.

A similar operation of ***multiplication modulo n*** can be defined where $a \cdot_n b$ is the nonnegative remainder less than n after the product ab is divided by n, but will it form a group for each n? You should try different values of n as examples! This operation is discussed more in section 2.1.

Another way to create a group whose elements are not just numbers, is to use functions as elements. In the next example notice that the addition defined between two functions again creates a *function.*

Example 1.13 Define $\Im(\mathbb{R})$ to be the set of all functions of the form $f : \mathbb{R} \to \mathbb{R}$. For the operation on $\Im(\mathbb{R})$ we use addition of functions, thus for $f, g \in \Im(\mathbb{R})$ the function $f+g$ is defined by $(f+g)(x) = f(x)+g(x)$ for all x in $\mathbb{R}$. You will be asked to verify that this forms a group in an exercise at the end of the chapter.

We are now ready to prove a theorem about arbitrary groups, which is also one of the most important.

Theorem 1.14

Suppose G is a group under the operation $*$.

(i) There is exactly one element in G that has the property of an identity.
(ii) For each element $a \in G$ there is exactly one element of G that satisfies the property of the inverse of a.

Proof Assume G is a group under the operation $*$.

(i) Since G is a group we know there exists an element of G that satisfies the property of an identity, call it e. Suppose that another element of G, u, also satisfies the property of an identity. We need to show that $e = u$. Consider the element $e * u$ which must be in G. Since e is an identity $e * u = u$. Similarly, by u an identity $e * u = e$, so $e = e * u = u$, or $e = u$ as we wanted to show. Thus exactly one element of G is an identity, and we call it the identity of G and denote it by e_G.

(ii) Let $a \in G$. As G is a group there exists some $b \in G$ with $a * b = e_G = b * a$. Suppose there is another element $c \in G$ with $a * c = e_G = c * a$. We need to show that $b = c$. Consider the element $b * (a * c)$ in G. As $a * c = e_G$ then $b * (a * c) = b * e_G = b$. Also $b * a = e_G$, so $(b * a) * c = e_G * c = c$. But by associativity, $b * (a * c) = (b * a) * c$, thus $b = c$ as needed. Therefore there is exactly one inverse in G for a, which we denote as a^{-1}. □

Another type of group used throughout the text is created from previously discussed groups. Suppose we have two groups $(G, \bullet)$ and $(H, \oplus)$. The symbols $\bullet$ and $\oplus$ are used for the operations here to emphasize that they may not be the same. Recall that for any two sets G and H, $G \times H$ is the set of ordered pairs with the first of the pair from G and the second of the pair from H, i.e., $G \times H = \{(u, v) : u \in G \text{ and } v \in H\}$.

Definition 1.15 The set $G \times H$ is a group under the operation $*$ defined by $(a, b) * (c, d) = (a \bullet c, b \oplus d)$. We call the new group $G \times H$ the ***direct product*** of G and H.

Notice that the operation defined above on $G \times H$ simply uses the correct operation for each group to calculate the answer in each coordinate of the pair. $*$ is clearly an operation since both $\bullet$ and $\oplus$ are operations on their respective sets. Associativity follows easily, as well as the existence of an identity and inverses, since they depend only on the original operations. Thus (e_G, e_H) is the identity and $(u, v)^{-1} = (u^{-1}, v^{-1})$ in the group $G \times H$.

Example 1.16 If we use the group $(\mathbb{Z}_2, +_2)$ as both G and H in the previous definition we can create the group $\mathbb{Z}_2 \times \mathbb{Z}_2 = \{(0,0),(1,0),(0,1),(1,1)\}$. The Cayley table for the operation is:

$*$	$(0,0)$	$(1,0)$	$(0,1)$	$(1,1)$
$(0,0)$	$(0,0)$	$(1,0)$	$(0,1)$	$(1,1)$
$(1,0)$	$(1,0)$	$(0,0)$	$(1,1)$	$(0,1)$
$(0,1)$	$(0,1)$	$(1,1)$	$(0,0)$	$(1,0)$
$(1,1)$	$(1,1)$	$(0,1)$	$(1,0)$	$(0,0)$

The identity is (0,0) and each element has an inverse as can be seen in the table; for example, $(1,0)^{-1} = (1,0)$. You should find the other inverses for practice.

We can also define direct products with more than two coordinates if needed, again using the appropriate operation in each coordinate. See Project 1.3, for example.

1.4 Basic Algebra in Groups

We will now write ab to mean the result of whatever operation our group uses on the elements a and b. We drop the use of $*$ only to make the equations in theorems and examples easier to understand, but unless it is specifically stated you cannot assume this means any sort of "usual" multiplication. The phrase "G is a group" assumes there is an operation on G, but does not specify the operation, so we will still write ab to denote the result. The next two theorems help with algebraic calculations in a group.

Theorem 1.17

Suppose G is a group.

(i) For every $a \in G$, $(a^{-1})^{-1} = a$.
(ii) For every $a, b \in G$, $(ab)^{-1} = b^{-1}a^{-1}$.

Proof Suppose G is a group.

(i) Assume that $a \in G$. As G is a group there is an identity $e_G \in G$, and $a^{-1} \in G$ as well. Since $aa^{-1} = e_G$ and $a^{-1}a = e_G$, then a satisfies the definition as the inverse for the element a^{-1}. Thus $(a^{-1})^{-1} = a$ as needed.

(ii) Assume that $a, b \in G$, then $a^{-1}, b^{-1} \in G$ as well. $(ab)(b^{-1}a^{-1}) = a(bb^{-1})a^{-1}$ by associativity. Using $bb^{-1} = e_G$, $(ab)(b^{-1}a^{-1}) = a(e_G)a^{-1} = aa^{-1} = e_G$. Similarly you should show that $(b^{-1}a^{-1})(ab) = e_G$ for practice. Thus $b^{-1}a^{-1}$ is the inverse of ab. Hence $(ab)^{-1} = b^{-1}a^{-1}$. □

Example 1.18 In our group $(\mathbb{Z},+)$ the identity is 0, and for each element $a \in \mathbb{Z}$ we have $a^{-1} = -a$. Thus we can say that $3^{-1} = -3, (-4)^{-1} = 4$, and even $0^{-1} = 0$. It is critical to know what operation is being used here since with multiplication on $\mathbb{Z}$ the element 0 does not have an inverse and 0^{-1} is undefined. Thus if applying a theorem to a group which uses addition as its operation we must be very careful. The previous theorem applied to $(\mathbb{Z},+)$ would say that $(3+4)^{-1} = 4^{-1} + 3^{-1} = -4 + -3$, i.e., $7^{-1} = -7$.

Theorem 1.17 also helps us when we need to solve equations or prove equalities about arbitrary elements of a group, as in the next example.

Example 1.19 Let G be a group and $a, b, c \in G$. We will show that if $c(ab) = a$ then $b(a^{-1}c) = a^{-1}$.

Assume G is a group, $a, b, c \in G$, and $c(ab) = a$. Taking the inverse of both sides we have $(c(ab))^{-1} = a^{-1}$. Now using Theorem 1.17 we can say $(ab)^{-1}c^{-1} = a^{-1}$. Using this theorem again to calculate $(ab)^{-1} = b^{-1}a^{-1}$ we can conclude that $(b^{-1}a^{-1})c^{-1} = a^{-1}$. Multiplying by c on the right of each side gives $(b^{-1}a^{-1})c^{-1}c = a^{-1}c$, and then by associativity we have $(b^{-1}a^{-1})(c^{-1}c) = a^{-1}c$. By definition $c^{-1}c = e_G$ and $(b^{-1}a^{-1})e_G = (b^{-1}a^{-1})$, so we now have $(b^{-1}a^{-1}) = a^{-1}c$. Multiplying by b on the left of each side gives $b(b^{-1}a^{-1}) = b(a^{-1}c)$ so using associativity we have $(bb^{-1})a^{-1} = b(a^{-1}c)$. Again, we use $bb^{-1} = e_G$ and $e_Ga^{-1} = a^{-1}$ to conclude $a^{-1} = b(a^{-1}c)$ as needed.

It is important in algebraic calculations to use the rules that are known to apply, not to try to use methods from previous experience. In particular the commutative property cannot be used unless it is somehow known that the group is abelian. In the proof above it would have been wrong to multiply by c^{-1} on the left side in one half of the equation and on the right side in the other half, such as $c^{-1}cab = ac^{-1}$.

A theorem that is familiar from our experience with usual operations on the integers is called ***The Cancellation Law***, and its proof is an exercise at the end of the chapter.

Theorem 1.20

The Cancellation Law Suppose G is a group and $a, b, c \in G$.

(i) If $ab = ac$ then $b = c$.
(ii) If $ba = ca$ then $b = c$.

Notation for exponentiation becomes necessary when repeated algebraic calculations are involved. In groups, we can define exponentiation only using <u>integer</u> exponents, as shown next.

Definition 1.21 Let G be a group, $a \in G$, and $n \in \mathbb{Z}$.

$$a^n = \begin{cases} \underbrace{aa \cdots a}_{n\ many} & if\ n > 0 \\ e_G & if\ n = 0 \\ \underbrace{a^{-1}a^{-1} \cdots a^{-1}}_{-n\ many} & if\ n < 0 \end{cases}$$

This notation is for any operation and group, so if we use $(\mathbb{Z}, +)$ then the symbol 3^4 is interpreted as $3 + 3 + 3 + 3 = 12$, and 3^{-2} denotes $-3 + -3 = -6$. Do not be fooled into assuming it is usual exponentiation, which uses the operation of multiplication.

Example 1.22 Consider the group $\mathbb{Z} \times \mathbb{Q}^+$, where $\mathbb{Z}$ uses addition and $\mathbb{Q}^+$ uses multiplication:

$$\left(2,\ \tfrac{1}{3}\right)^4 = \left(2 + 2 + 2 + 2,\ \tfrac{1}{3} \cdot \tfrac{1}{3} \cdot \tfrac{1}{3} \cdot \tfrac{1}{3}\right) = \left(8,\ \tfrac{1}{81}\right)$$
$$\left(2,\ \tfrac{1}{3}\right)^{-4} = (-2 - 2 - 2 - 2,\ 3 \cdot 3 \cdot 3 \cdot 3) = (-8,\ 81)$$
$$\left(2,\ \tfrac{1}{3}\right)^0 = (0, 1)$$

We will frequently need to use some of the standard rules that apply to exponents, stated in the next theorem.

Theorem 1.23

Suppose G is a group, $a \in G$, and $m, n \in \mathbb{Z}$.

(i) $a^n a^m = a^{n+m}$
(ii) $(a^n)^m = a^{nm}$
(iii) $a^{-n} = (a^{-1})^n = (a^n)^{-1}$

Our definition of exponents means that we must prove each part of this theorem in cases, since each of m or n might be positive, negative, or 0. Technically, the positive cases are proven by two applications of Mathematical Induction (Theorem 0.3), then the positive cases can be used to prove the negative ones. We give a more intuitive "proof" for one specific situation and leave others for exercises at the end of the chapter.

Proof Suppose G is a group, $a \in G$, and $m, n \in \mathbb{Z}$.

We will consider part (i) in the case when $m > 0$ and $n < 0$. Thus using Definition 1.21 we have:

$$a^n = \underbrace{a^{-1}a^{-1} \cdots a^{-1}}_{-n \; many} \text{ and } a^m = \underbrace{a^{-1}a^{-1} \cdots a^{-1}}_{m \; many}$$

To multiply a^n and a^m we make use of associativity, and put the "middle" terms together.

$$a^n a^m = (a^{-1} \cdots a^{-1})(a \cdots a) = a^{-1} \cdots a^{-1}(a^{-1}a)a \cdots a$$

Since $a^{-1}a = e_G$ we shorten the list by one a and one a^{-1}. As long as we have both a and a^{-1} next to each other to become e_G we continue this process. Thus the process ends in the smaller of $-n$ or m steps. If $-n > m$ then we are left with:

$$a^n a^m = \underbrace{a^{-1}a^{-1} \cdots a^{-1}}_{-n-m \; many} \text{ or}$$

$$a^n a^m = a^{-(-n-m)} = a^{n+m}.$$

Similarly if $m > -n$ then:

$$a^n a^m = \underbrace{aa \cdots a}_{m-(-n) \; many} \quad \text{or}$$

$$a^n a^m = a^{m-(-n)} = a^{n+m}.$$

Finally, if $-n = m$, then no a or a^{-1} is left so $a^n a^m = e_G = a^0 = a^{n+m}$. Thus we have shown that $a^n a^m = a^{n+m}$. □

With exponentiation we can define the order of a group element. This definition will be frequently referred to in many of the chapters of the text.

Definition 1.24 Suppose we have a group G, and an element $a \in G$. If there exists a positive integer n with $a^n = e_G$, then the smallest positive n for which $a^n = e_G$ is called the ***order of a***, denoted $ord(a)$. If no positive integer n has $a^n = e_G$ then we say that a has $infinite\ order$.

Example 1.25 Let our group be $(\mathbb{Z}_7, +_7)$, and $a = 3$. Suppose there is an integer n so that $3^n = 0$.

$$(3 + 3 + \cdots + 3)(\text{mod} 7) = 0 \qquad (n \text{ many 3s})$$

Thus we would have $3n(\text{mod} 7) = 0$ so that $3n$ must be evenly divisible by 7 to leave no remainder. If we use $n = 49$ then $3^{49} = 0$, also $3^{-14} = 0$. But we must look for the *smallest positive* integer which makes $3n$ evenly divisible by 7, namely $n = 7$. Hence $ord(3) = 7$. In fact, every nonzero element of $\mathbb{Z}_7$ has order 7 (you should verify this).

Consider instead the group $(\mathbb{Z}_8, +_8)$.

$$2^4 = 2 +_8 2 +_8 2 +_8 2 = (2 + 2 + 2 + 2)(\text{mod} 8) = 0$$

This however is not enough to guarantee us $ord(2) = 4$ unless we know that 4 is the smallest positive integer with $2^n = 0$. Thus we must check to see if any of $2^1, 2^2$, or 2^3 are 0. As seen below none of these are equal to 0. Thus $ord(2) = 4$.

$$2^1 = 2 \qquad 2^2 = 2 +_8 2 = 4 \qquad 2^3 = 2 +_8 2 +_8 2 = 6$$

What are the orders of the other elements in $(\mathbb{Z}_8, +_8)$?

Finally, consider the group $(\mathbb{Q}, +)$, and let $a = \frac{2}{3}$. To determine the order of a, we suppose there is a positive integer n with $\left(\frac{2}{3}\right)^n = 0$, since 0 is the identity of this group. Using addition it means that we are adding $\frac{2}{3}$, n many times and get 0, or $(\frac{2}{3})n = 0$. However, in $\mathbb{Q}$, there is no integer other than 0 that can be multiplied by $\frac{2}{3}$ to arrive at 0. Hence $\frac{2}{3}$ has infinite order.

Theorem 1.26

Suppose G is a group and $a \in G$ with $ord(a) = n$. For any integer t, $a^t = e_G$ if and only if n evenly divides t, i.e., $t = nq$ for some integer q.

Proof Suppose that G is a group and $a \in G$ with $ord(a) = n$.

$(\rightarrow)$ Let t be an integer with $a^t = e_G$. We must show that n evenly divides t. Using Theorem 0.18 there exist unique integers q and r so that $t = nq + r$ and $0 \leq r < n$. Now $a^t = a^{nq+r} = a^{nq}a^r = (a^n)^q a^r$. But since $ord(a) = n$ we know $a^n = e_G$ and thus $a^t = (e_G)^q a^r = a^r$. Since $ord(a) = n$ and $0 \leq r < n$, we must have $r = 0$ because n is the least positive integer for which $a^n = e_G$. Thus $t = nq$ for some integer q, i.e., n evenly divides t.

$(\leftarrow)$ An exercise at the end of the chapter. □

How is this result useful? Suppose G is a group and $a \in G$. If we need to calculate a^{73} in G, knowing $ord(a)$ is a big help. If $ord(a) = 11$, we know that $73 = 66 + 7$ so $a^{73} = a^{66}a^7 = (a^{11})^6 a^7 = (e_G)^6 a^7 = a^7$. Hence we only needed to compute a^7, a much simpler task. What is a^{73} if $ord(a) = 8$ instead?

1.5 Subgroups

Now that groups and their operations are familiar objects, we look at subsets of the groups and decide if they are also groups, i.e., subgroups.

Definition 1.27 Let G be a group. A nonempty subset $H \subseteq G$ is called a ***subgroup*** of G if and only if H is also a group using the same operation that is defined on G.

Example 1.28 Clearly, $\mathbb{Z}$ is a subgroup of $(\mathbb{Q},+)$, since with the same addition we already know that $(\mathbb{Z},+)$ is a group. In the same way $\mathbb{Z}$ and $\mathbb{Q}$ are subgroups of $(\mathbb{R},+)$. It is easy to be led astray using this thought process however; for example, $\mathbb{Z}_3$ is not a subgroup of $\mathbb{Z}_5$, even though $\{0, 1, 2\} \subseteq \{0, 1, 2, 3, 4\}$. $\mathbb{Z}_3$ is a group, but **not** under addition modulo 5 since $2 +_5 2 = 4$, which is not in $\mathbb{Z}_3$.

Example 1.29 Consider $G = M_2(\mathbb{R})$ under matrix addition, and

$$H = \left\{ \begin{bmatrix} a & b \\ a & b \end{bmatrix} \ : \ a, b \in \mathbb{R} \right\}.$$

Is H a subgroup of G? We will look at every step that is needed to determine this, which will help us understand Theorem 1.30 more easily. Recall for H to be a group under the operation of G we must have:

- $H \neq \emptyset$
- The operation of G is also an operation on H.
- The operation of G is associative on H.
- H contains an identity under the operation of G.
- Every element of H must have an inverse that is also in H.

<u>H is nonempty</u>: Since we can choose $a = 0$ and $b = 0$ as real numbers, we know $\begin{bmatrix} 0 & 0 \\ 0 & 0 \end{bmatrix} \in H$. Thus H is a nonempty subset of G.

Matrix Addition is an operation on H: We know that addition of matrices is well defined as it was an operation on G, but to be an operation on H we must be sure that the answers after using this operation are back in H, i.e., if $x, y \in H$ then $x + y \in H$. Suppose $x, y \in H$, then by definition of H we have real numbers a, b, c, d as shown.

$$x = \begin{bmatrix} a & b \\ a & b \end{bmatrix} \text{ and } y = \begin{bmatrix} c & d \\ c & d \end{bmatrix}$$

Using matrix addition we compute $x + y$ next.

$$x + y = \begin{bmatrix} a & b \\ a & b \end{bmatrix} + \begin{bmatrix} c & d \\ c & d \end{bmatrix} = \begin{bmatrix} a+c & b+d \\ a+c & b+d \end{bmatrix}$$

Now $a + b$ and $c + d$ are real numbers and so $x + y \in H$. When for each $x, y \in H$ we have $x + y \in H$, we say that ***H is closed under addition***.

Matrix Addition is associative: Since we have not changed the operation here, and G is a group, we know that addition must continue to be associative. Any elements used from H are still in G, so the associativity of the operation cannot change.

There is an identity in H under addition: We already showed that the identity of G, namely $\begin{bmatrix} 0 & 0 \\ 0 & 0 \end{bmatrix}$, is in H and thus it must be the identity in H as well since any $x \in H$ must still have $x + \begin{bmatrix} 0 & 0 \\ 0 & 0 \end{bmatrix} = x$.

Every element of H has an inverse in H: Because G is a group there is an inverse for any element of H, however, the inverse might not be in H. Thus we need to show that if $x \in H$ then $x^{-1} \in H$. Suppose we have $x = \begin{bmatrix} a & b \\ a & b \end{bmatrix}$ in H. Since our operation is addition then $x^{-1} = \begin{bmatrix} -a & -b \\ -a & -b \end{bmatrix}$, which is clearly in H as $-a$ and $-b$ are real numbers. When for each $x \in H$ we have $x^{-1} \in H$, we say that ***H is closed under inverses***.

Therefore H is a subgroup of G.

Notice that not all parts of the definition of a group needed to be verified in the example above, only that H is nonempty, H is closed under the operation of G, the identity of G is in H, and H is closed under inverses. This leads us to the next theorem.

Theorem 1.30

Let G be a group and $H \subseteq G$. H is a subgroup of G if and only if the following hold:

(i) H is nonempty.
(ii) H is closed under the operation of G.
(iii) H is closed under inverses.

Proof Let G denote a group and $H \subseteq G$.

($\rightarrow$) Assume H is a subgroup of G. Thus by definition of a subgroup we must have H a nonempty set. Also by H a subgroup, the operation of G must be an operation on H, thus H is closed under the operation of G as needed. Now every element of H must have an inverse in H for H to be a group, so H is closed under inverses as well.

($\leftarrow$) Suppose H satisfies the three properties given. We must show that H is a subgroup of G, so is a group under the operation of G. Since H has elements by (i) we thus know H is a nonempty subset of G. Also H uses the operation on G, so ab must be uniquely defined for all elements $a, b \in H$. Also (ii) guarantees that the answers are back in H, so the operation on G is also an operation on H.

As the operation is associative on the group G and $H \subseteq G$ then it must be associative on H, and by (iii) every element of H has an inverse in H. The only step that really needs proof is that H contains an identity!

Since H is nonempty, there exists some $x \in H$. By (iii) we have $x^{-1} \in H$ as well. Now by (ii) $xx^{-1} \in H$, or $e_G \in H$. Since for every $x \in H$ we have $x * e_G = x = e_G x$ then e_G is the identity in H. Therefore H is a group under the operation of G, or H is a subgroup of G. □

Example 1.31 Consider the group $(\mathbb{Z}_{16}, +_{16})$. Is the set $H = \{0, 4, 8, 12\}$ a subgroup of $(\mathbb{Z}_{16}, +_{16})$? To use Theorem 1.30 we need to prove each of the parts (i), (ii), (iii) for our set H.

Since we are given that $0 \in H$ then clearly $H \neq \emptyset$. Now to verify (ii) and (iii) it helps to look at a Cayley table, which only uses the elements $0, 4, 8, 12$ and the operation $+_{16}$:

$+_{16}$	0	4	8	12
0	0	4	8	12
4	4	8	12	0
8	8	12	0	4
12	12	0	4	8

Notice that all of the answers in the table are back in H, so H is closed under $+_{16}$. Also $0^{-1} = 0$, $4^{-1} = 12$, $8^{-1} = 8$, and $12^{-1} = 4$ so the inverse of every element of H is back in H, i.e., H is closed under inverses. Thus by Theorem 1.30 H is a subgroup of $\mathbb{Z}_{16}$.

Is $H = \{0, 3, 6, 9, 12, 15\}$ also a subgroup of $\mathbb{Z}_{16}$?

Example 1.32 Recall from Example 1.16 the group $\mathbb{Z}_2 \times \mathbb{Z}_2 = \{(0,0), (1,0), (0,1), (1,1)\}$. Define $H_1 = \{(0,0), (0,1)\}$ and $H_2 = \{(0,0), (1,1)\}$. They are clearly nonempty subsets of $\mathbb{Z}_2 \times \mathbb{Z}_2$, but are they subgroups? Consider the two Cayley tables below that will help us see why these subsets are subgroups:

$*$	$(0,0)$	$(1,1)$
$(0,0)$	$(0,0)$	$(1,1)$
$(1,1)$	$(1,1)$	$(0,0)$

$*$	$(0,0)$	$(0,1)$
$(0,0)$	$(0,0)$	$(0,1)$
$(0,1)$	$(0,1)$	$(0,0)$

Each table shows that the subset is closed under both the operations on $\mathbb{Z}_2 \times \mathbb{Z}_2$ and inverses since $(0,0)^{-1} = (0,0)$, $(0,1)^{-1} = (0,1)$, and $(1,1)^{-1} = (1,1)$. Thus we have two subgroups of $\mathbb{Z}_2 \times \mathbb{Z}_2$.

Once we have subgroups we can consider the union or intersection of them as new subsets of our group and ask if either one will also form a subgroup. Consider the set $H_1 \cup H_2 = \{(0,0), (0,1), (1,1)\}$. We have $(0,1) * (1,1) = (1,0)$, but (1,0) is not back in our set. Thus $H_1 \cup H_2$ is not closed under the operation $*$. However, $H_1 \cap H_2 = \{(0,0)\}$ is a subgroup of $\mathbb{Z}_2 \times \mathbb{Z}_2$.

This leads to a theorem whose proof is left as an exercise at the end of the chapter.

Theorem 1.33

Let G be a group. If H_1 and H_2 are both subgroups of G then $H_1 \cap H_2$ is also a subgroup of G.

1.6 Homomorphisms

The final section in this chapter considers functions that map between groups. Since groups have a well-defined structure, we will be interested in functions that preserve this structure. Functions of this sort can also be used to discover if two groups are really "the same", i.e., they have the same number of elements and the operations act identically. In such a case the elements and operation of one group could simply be renamed and the other group is revealed.

Definition 1.34 Let $(G,*)$ and $(K,\otimes)$ be two groups. A function $f : G \to K$ is called a (group) ***homomorphism*** if $f(a*b) = f(a) \otimes f(b)$ for every $a, b \in G$. If the homomorphism f is also a bijection from G to K we say that f is an (group) ***isomorphism*** or that G and K are *isomorphic*.

Example 1.35 Consider the groups $(\mathbb{Z}, +)$ and $(\mathbb{Z}_4, +_4)$. Define $f : \mathbb{Z} \to \mathbb{Z}_4$ by $f(a) = a(\text{mod}4)$, thus each integer maps to its (nonnegative) remainder after being divided by 4. For example, we have $f(-25) = 3, f(10) = 2$ and $f(149) = 1$. To decide if the function is a homomorphism we need to determine if $f(m+n) = f(m) +_4 f(n)$ for all integers m, n.

Assume we have $m, n \in \mathbb{Z}$. Using Theorem 0.18 there are $k, s, p, t \in \mathbb{Z}$ with $m = 4k + s$, $n = 4p + t$, and $0 \leq s, t < 4$. This tells us $m(\text{mod}4) = s$ and $n(\text{mod}4) = t$, so $f(m) +_4 f(n) = (s+t)(\text{mod}4)$. Also $f(m+n) = (m+n)(\text{mod}4)$ and $m + n = 4(k+p) + (s+t)$, thus $(m+n)(\text{mod}4) = (s+t)(\text{mod}4)$. Therefore $f(m+n) = (s+t)(\text{mod}4)$, and we have $f(m+n) = f(m) +_4 f(n)$. Thus f is a group homomorphism. Is this function an isomorphism? Why or why not?

Example 1.36 Consider the groups $\mathbb{R}^+$ (positive real numbers) under multiplication and $(\mathbb{R}, +)$. Define $f : \mathbb{R} \to \mathbb{R}^+$ by $f(x) = e^x$. For any two real numbers x, y, we see that $f(x+y) = e^{x+y} = e^x e^y = f(x)f(y)$. Thus f is a homomorphism.

This function f is also one to one, since if we have $f(x) = f(y)$, then $e^x = e^y$, but taking the natural logarithm of both sides tells us $x = y$. Finally, f is onto, since given any positive real number x, $ln(x)$ is also a real number. Since $f(ln(x)) = e^{ln(x)} = x$ we have found the real number that maps to x. Thus f is onto, and hence f is an isomorphism.

A few facts that will be very important when we use homomorphisms are seen in the next theorem. Proofs of the last two statements are exercises at the end of the chapter.

Theorem 1.37

Let G and K be groups and suppose $f : G \to K$ is a homomorphism. Then the following must hold:

(i) $f(e_G) = e_K$.
(ii) $f(a^{-1}) = (f(a))^{-1}$ for each $a \in G$.
(iii) For every $n \in \mathbb{Z}$ and $a \in G$, $(f(a))^n = f(a^n)$.
(iv) For every $a \in G$, if a has finite order then $ord(f(a))$ evenly divides $ord(a)$.

Proof Assume G and K are groups and $f : G \to K$ is a homomorphism.

(i) Since e_G is the identity of G, we know $e_G e_G = e_G$. Thus using the function f, we have $f(e_G e_G) = f(e_G)$. Since f is a homomorphism $f(e_G)f(e_G) = f(e_G)$. In the group K, $f(e_G)$ has an inverse, so we can multiply by $f(e_G)^{-1}$.

$$f(e_G)^{-1}f(e_G)f(e_G) = f(e_G)^{-1}f(e_G)$$

In the group K, $f(e_G)^{-1}f(e_G) = e_K$, thus using associativity $e_K f(e_G) = e_K$ or $f(e_G) = e_K$.

(ii) Let $a \in G$. Thus a^{-1} is also in G and $aa^{-1} = e_G$. Using the function f we see that $f(aa^{-1}) = f(e_G)$. From the fact that f is a homomorphism we have $f(a)f(a^{-1}) = f(e_G)$. From (i) above we can conclude $f(a)f(a^{-1}) = e_K$. Similarly, we can show that $f(a^{-1})f(a) = e_K$. Thus $f(a^{-1})$ is the inverse of $f(a)$ in K, i.e., $f(a^{-1}) = (f(a))^{-1}$.

(iii) An exercise at the end of the chapter.

(iv) An exercise at the end of the chapter.

□

Notice that the previous theorem does not say "if and only if" thus verifying (i), (ii), (iii), and (iv) of the theorem does **not** prove that a function is a homomorphism. (See Project 1.9.) Theorem 1.37 can be very helpful when showing that a function is NOT a homomorphism as seen in the next example.

Example 1.38 Consider the group $(\mathbb{Z}, +)$ and the constant function $f : \mathbb{Z} \to \mathbb{Z}$ defined by $f(x) = 2$ for all $x \in \mathbb{Z}$. Theorem 1.37 easily shows us that f is not a homomorphism since $f(0) = 2$. Is any constant function from $\mathbb{Z}$ to $\mathbb{Z}$ a group homomorphism when the operation is addition?

Also notice that in (iv) of the theorem we only have $ord(f(a))$ evenly dividing $ord(a)$. Why can't we say they have the same order? Consider the function $f : \mathbb{Z}_8 \to \mathbb{Z}_4$ defined by $f(x) = x(\text{mod} 4)$. You should verify that f is a homomorphism; it is similar to the function in Example 1.35. If we choose $a = 1$ in $\mathbb{Z}_8$ we have $ord(a) = 8$, but $f(a) = 1$ in $\mathbb{Z}_4$ with $ord(f(a)) = 4$.

When the groups in question are small the Cayley tables of $(G, *)$ and $(K, \otimes)$ can be used to verify that a function $f : G \to K$ is a group homomorphism. Recall that in the Cayley table of group G, the result $a * b$ is the element found in row a and column b. If we replace a by $f(a)$, b by $f(b)$, and $a * b$ by $f(a * b)$ everywhere in the table, we would have $f(a*b)$ in the position for the result of $f(a) \otimes f(b)$ in the new table. If the results are correct in the table of K for all $a, b \in G$, we know that $f(a * b) = f(a) \otimes f(b)$ as needed.

Example 1.39 Consider the group $G = \mathbb{Z}_2 \times \mathbb{Z}_2$ defined in Example 1.16. The Cayley table was given as:

$*$	$(0,0)$	$(1,0)$	$(0,1)$	$(1,1)$
$(0,0)$	$(0,0)$	$(1,0)$	$(0,1)$	$(1,1)$
$(1,0)$	$(1,0)$	$(0,0)$	$(1,1)$	$(0,1)$
$(0,1)$	$(0,1)$	$(1,1)$	$(0,0)$	$(1,0)$
$(1,1)$	$(1,1)$	$(0,1)$	$(1,0)$	$(0,0)$

Now consider the group $(K, \otimes)$ to be $(\mathbb{Z}_8, +_8)$, and define $f : G \to K$ by:

$$\begin{array}{ll} f(0,0) = 0 & f(1,0) = 2 \\ f(0,1) = 4 & f(1,1) = 6 \end{array}$$

In the Cayley table for G above we replace (0,0) by 0, (1,0) by 2, (1,0) by 4, and (1,1) by 6 to get:

$+_8$	0	2	4	6
0	0	2	4	6
2	2	0	6	4
4	4	6	0	2
6	6	4	2	0

But this table does not follow the rules for K since $2 +_8 2 = 4$ and our table gives a result of 0 in the row labeled 2 and the column labeled 2. Since $f(1,0) +_8 f(1,0) \neq f((1,0) * (1,0))$ then f is **not** a homomorphism. Can you find a homomorphism between these groups?

When looking to see if groups are isomorphic, there are theorems that can help immediately. The proofs are left as exercises at the end of the chapter.

Theorem 1.40

Suppose G and K are groups and $f : G \to K$ is an ***isomorphism***.

(i) G is abelian if and only if K is abelian.
(ii) For every $a \in G$, if a has finite order then $ord(a) = ord(f(a))$.

There are many other properties, of course, but these can often be used to quickly find a reason two groups are not isomorphic.

Example 1.41 Recall the group $G = \mathbb{Z}_2 \times \mathbb{Z}_2$ whose Cayley table is seen below:

$*$	$(0,0)$	$(1,0)$	$(0,1)$	$(1,1)$
$(0,0)$	$(0,0)$	$(1,0)$	$(0,1)$	$(1,1)$
$(1,0)$	$(1,0)$	$(0,0)$	$(1,1)$	$(0,1)$
$(0,1)$	$(0,1)$	$(1,1)$	$(0,0)$	$(1,0)$
$(1,1)$	$(1,1)$	$(0,1)$	$(1,0)$	$(0,0)$

It is easy to see that G is not isomorphic to $(\mathbb{Z}_4, +_4)$ using property (ii) of Theorem 1.40. In $\mathbb{Z}_4$ the element 1 has order 4 but in G every element has $x^2 = e_G$. So no element in G has order 4, and so no function can satisfy (ii).

Homomorphisms will be important later, especially in Chapter 3 where the *Fundamental Theorem of Homomorphisms* will be discussed. Some of the exercises that follow will give you practice verifying functions are homomorphisms or determining why they are not.

Exercises for Chapter 1

Sections 1.1-1.2 Operations and Their Properties

Each of the following rules, denoted by $*$, is defined by a combination of the usual addition, multiplication, and division of real numbers. Be sure to notice what set the rule is being defined on!

1. Determine if the rule $x * y = \sqrt{xy}$ is an operation on the set of integers.
2. Determine if the rule $x * y = \frac{x}{y+1}$ is an operation on the set of real numbers.
3. Determine if the rule $x * y = xy - 1$ is an operation on the set of positive integers.
4. Determine if the rule $x * y = xy$ is an operation on the set of positive integers.
5. Determine if the rule $x * y = x + 3y$ is an operation on the set of positive integers.
6. Determine if the rule $x * y = 4x + 2y$ is an operation on the set of integers.
7. Determine if the rule $x * y = xy$ is an operation on the set of negative integers.
8. Determine if the rule $x*y = x+y$ is an operation on the set of nonzero rational numbers.
9. Determine if the rule $x * y = \frac{x+2}{y}$ is an operation on the set of nonzero real numbers.
10. Determine if the rule $x * y = x + y + xy$ is an operation on the set $\mathbb{R} - \{-1\}$.

    Determine if each operation $*$ in exercises 11 - 14 is associative, commutative, has an identity, and if each element has an inverse. Give a proof or counterexample for each property.

11. For $x, y \in \mathbb{R} - \{-1\}$, $x * y = x + y + xy$.

12. For $a, b \in \mathbb{Z}, a * b = a^2 - b^2$.
13. For $a, b \in \mathbb{Z}, a * b = a + 2b - 1$.
14. For $x, y \in \mathbb{Q}, x * y = \dfrac{y}{x^2 + 3}$.
15. For the operation defined in Example 1.5 on the set $\{John, Sue, Henry, Pam\}$, determine if $*$ is associative by determining if the final six equalities of Example 1.8 hold.
16. Determine if the operation defined in Example 1.5 is commutative, has an identity, and if each element has an inverse.

Section 1.3 Groups

17. Write out the Cayley table for the group $(\mathbb{Z}_6, +_6)$ and identify the inverse of each element.
18. Find the 10 elements of the group $\mathbb{Z}_5 \times \mathbb{Z}_2$ and write out the Cayley table. Recall that its operation uses $+_5$ in the first coordinate and $+_2$ in the second. Identify the inverse of each element.
19. What does Theorem 1.14 tell us about the entries in the Cayley table of a group?
20. Prove that a set with exactly one element, $A = \{a\}$, will always form a group since there is only one way to define an operation on the set. Be sure to show that the properties of a group all hold.
21. We can even create groups with games! Consider four cups placed in a square pattern on a table. If we have a penny in one of the cups there are four ways we can move it to another cup: *Horizontally, Vertically, Diagonally*, or *Stay where it is.* We will label them H, V, D, S. To define an operation, consider two movements in a row, i.e., $x * y$ means we first move the penny as x tells us to, then after that move the penny as y instructs. For example, $H * V = D$ since if we first move it horizontally and then vertically altogether we have moved it diagonally. Create the Cayley table for this group, identify the identity, and the inverse of each element.
22. Prove that the set $A = \left\{ \begin{bmatrix} 1 & 0 \\ 0 & -1 \end{bmatrix}, \begin{bmatrix} 1 & 0 \\ 0 & 1 \end{bmatrix}, \begin{bmatrix} -1 & 0 \\ 0 & 1 \end{bmatrix}, \begin{bmatrix} -1 & 0 \\ 0 & -1 \end{bmatrix} \right\}$ is a group under matrix multiplication. Is it abelian?
23. Verify that $\Im(\mathbb{R})$ is a group using the addition of functions as defined in Example 1.13.

Section 1.4 Basic Algebra in Groups

24. Prove the Cancellation Law (Theorem 1.20). Do not use any theorems that occur after it in the text.
25. Assume G is a group and $a \in G$. Consider a fixed integer k and use PMI (Theorem 0.3) to prove that for all $n \in \mathbb{N}, a^n a^k = a^{n+k}$. You will need to consider cases here since k can be positive, negative, or 0.
26. Prove Theorem 1.23 (ii) for the case of $n > 0$ and $m < 0$.

27. Assume G is a group and $a \in G$. Prove by PMI (Theorem 0.3) that for $n \in \mathbb{N}, (a^n)^{-1} = a^{-n}$.
28. Suppose G is a group with $a, b \in G$. Prove: If $a^3 = b$, then $b = aba^{-1}$.
29. Suppose G is a group and $a, b \in G$. Prove by PMI for all $n \in \mathbb{N}, (aba^{-1})^n = ab^n a^{-1}$. Do not assume G is abelian.
30. Suppose G is a group and $a, b \in G$. Prove: If $n \in \mathbb{N}$ and $ord(b) = n$, then $ord(aba^{-1}) = n$. Use the result of the previous exercise to help.
31. Find the order of each element in the group $(\mathbb{Z}_{12}, +_{12})$.
32. Find the order of each element in the group $(\mathbb{Z}_7, +_7)$.
33. Find the order of each element in the group $A = \left\{ \begin{bmatrix} 1 & 0 \\ 0 & -1 \end{bmatrix}, \begin{bmatrix} 1 & 0 \\ 0 & 1 \end{bmatrix}, \begin{bmatrix} -1 & 0 \\ 0 & 1 \end{bmatrix}, \begin{bmatrix} -1 & 0 \\ 0 & -1 \end{bmatrix} \right\}$ under matrix multiplication.
34. Find the order of each element in the group $G = \{Horizontally, Vertically, Diagonally, Stay\}$ from exercise 21 above (use the operation defined in that exercise).
35. Complete the proof of Theorem 1.26.
36. Suppose G is a group and $a \in G$. Prove: if $ord(a) = 6$ then $ord(a^5) = 6$. Do any other elements a^2, a^3, a^4 have have order 6? Explain.
37. Suppose G is a group and $a \in G$. Assume $a^{50} = e_G$ but $a^{75} \neq e_G$ and $a^{10} \neq e_G$. Find the order of a, and prove that your answer is correct.
38. Suppose G is a group and $a \in G$ with $a \neq e_G$. **Prove**: If $a^p = e_G$ for some prime number p, then $ord(a) = p$.
39. Suppose G is a group and $a \in G$. Prove: If $ord(a)$ is an even integer then for any odd $k \in \mathbb{Z}$, $a^k \neq e_G$. Would the statement still be true if the words even and odd changed places, i.e., if $ord(a)$ is odd then for any even $k \in \mathbb{Z}$, $a^k \neq e_G$? Give a brief justification of your answer.

Section 1.5 Subgroups

40. Prove or disprove that $H = \{0, 3, 6, 9\}$ is a subgroup of $(\mathbb{Z}_{12}, +_{12})$.
41. Prove or disprove that $H = \{0, 3, 6, 9\}$ is a subgroup of $(\mathbb{Z}_{10}, +_{10})$.
42. Prove that $H = \{0, 2, 4, 8, 16\}$ is a subgroup of $(\mathbb{Z}_{18}, +_{18})$.
43. Prove or disprove that $H = \{(0,0), (0,2), (1,0), (1,2)\}$ is a subgroup of $\mathbb{Z}_2 \times \mathbb{Z}_4$ under the operation using $+_2$ in the first coordinate and $+_4$ in the second coordinate.
44. Prove or disprove that $H = \{0, 2, 4, 6\}$ is a subgroup of $(\mathbb{Z}_7, +_7)$.
45. Determine if the set $\{\frac{n}{3} : n \in \mathbb{Z}\}$ is a subgroup of $(\mathbb{Q}, +)$. Either prove that it is or give a specific example of how it fails.
46. Determine if the set $\{\frac{1}{n} : n \in \mathbb{Z}, n \neq 0\}$ is a subgroup of $(\mathbb{Q}^*, \cdot)$ (nonzero rational numbers). Either prove that it is or give a specific example of how it fails.
47. Determine if $K = \{3n : n \in \mathbb{Z}\}$ is a subgroup of $(\mathbb{Z}, +)$. Either prove that it is or give a specific example of how it fails.

48. Determine if $K = \{3+n : n \in \mathbb{Z}\}$ is a subgroup of $(\mathbb{Z}, +)$. Either prove that it is or give a specific example of how it fails.
49. Consider the set $H = \left\{ \begin{bmatrix} a & b \\ -b & -a \end{bmatrix} : a, b \in \mathbb{R} \right\}$. Determine if H is a subgroup of $M_2(\mathbb{R})$ under usual addition of matrices.
50. Determine if the set $K = \{x^2 : x \in \mathbb{R}\}$ is a subgroup of $(\mathbb{R}, +)$. (Note that x^2 means usual multiplication.)
51. Determine if the set $K = \{x^2 : x \in \mathbb{R}^+\}$ is a subgroup of $(\mathbb{R}^+, \cdot)$. (Note that x^2 means usual multiplication, and $\mathbb{R}^+$ is the set of positive real numbers.)
52. Let G be a group, and define the set $H = \{a \in G : a^2 = e_G\}$. Prove: If G is abelian then H is a subgroup of G. Where does your proof break down if we do not know G is abelian?
53. Prove Theorem 1.33.

Section 1.6 Homomorphisms

54. Determine if the function $f : \mathbb{Z}_8 \to \mathbb{Z}_4$ defined by $f(x) = (2x)(\text{mod}4)$ is a homomorphism between groups $(\mathbb{Z}_8, +_8)$ and $(\mathbb{Z}_4, +_4)$. In this definition $2x$ means usual integer multiplication.
55. Prove that for any $n \in \mathbb{N}$ with $n > 1$, the function $f : \mathbb{Z} \to \mathbb{Z}_n$ defined by $f(x) = x(\text{mod}n)$ is a homomorphism between groups $(\mathbb{Z}, +)$ and $(\mathbb{Z}_n, +_n)$.
56. Prove that the function $f : \mathbb{Z} \to \mathbb{Z}$ defined by $f(x) = 4x$ is a homomorphism of $(\mathbb{Z}, +)$. (Note that $4x$ means usual integer multiplication.)
57. Determine if the function $g : \mathbb{R} \to \mathbb{R}$ defined by $g(x) = x^2 + 1$ is a homomorphism of $(\mathbb{R}, +)$.
58. Determine if the function $h : \mathbb{Z} \to \mathbb{Z}$ defined by $h(x) = x^3$ is a homomorphism of $(\mathbb{Z}, +)$.
59. Determine if the function $g : \mathbb{R}^+ \to \mathbb{R}^+$ defined by $g(x) = \frac{1}{x}$ is a homomorphism of $(\mathbb{R}^+, \cdot)$. Recall that $\mathbb{R}^+$ is the set of positive real numbers.
60. Determine if the function $f : M_2(\mathbb{R}) \to \mathbb{R}$ defined by $f\left(\begin{bmatrix} a & b \\ c & d \end{bmatrix}\right) = a + b$ is a homomorphism. Recall that both $M_2(\mathbb{R})$ and $\mathbb{R}$ use an operation of addition.
61. Use PMI to prove (iii) of Theorem 1.37.
62. Prove (iv) of Theorem 1.37.
63. Prove (i) of Theorem 1.40.
64. Prove (ii) of Theorem 1.40.
65. Suppose G and K are groups and $f : G \to K$ is a homomorphism. Define the set $f(G) = \{y \in K\text{: there is some } a \in G \text{ with } f(a) = y\}$. Prove that $f(G)$ is a subgroup of K.
66. Suppose that $f : G \to K$ and $g : K \to H$ are homomorphisms of the groups G, K, and H. Prove that the function $g \circ f$ is a homomorphism from G to H.

Projects for Chapter 1

Project 1.1

Consider a set A with exactly two elements, $A = \{a, b\}$. Suppose we have an operation on A called $*$, with the property that $a * a = b$.

1. There are exactly eight ways to complete the Cayley table for $(A, *)$; show each below:

$*$	a	b
a		
b		

$*$	a	b
a		
b		

$*$	a	b
a		
b		

$*$	a	b
a		
b		

$*$	a	b
a		
b		

$*$	a	b
a		
b		

$*$	a	b
a		
b		

$*$	a	b
a		
b		

2. Which Cayley tables shown above contain an identity for the operation and which do not? Be sure to explain why or why not for each.
3. To determine if the operation defined by one of our Cayley tables makes a group, we need to test whether the operation is associative. Using only the elements a or b, there are eight expressions of the form $x * (y * z)$ and $(x * y) * z$. Compute them for any table that has an identity!
4. For each table that has an identity and is associative, does it give us an operation on A forming a group? Explain why or why not.
5. If we assume that $a * a = a$ instead of $a * a = b$, would there be at least one table defining an operation on A forming a group?

Project 1.2

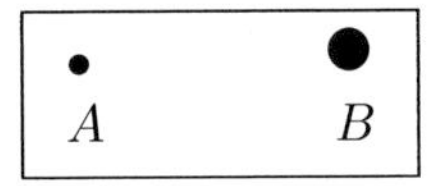

Consider two coins, one shown as a larger circle and the other a smaller circle, and imagine that each is black on the top and white on the bottom. They are on a board at positions labeled A and B as shown above. To keep the coins on the board we can only exchange coin positions or flip coins over. This gives rise to eight movements:

I : Don't change anything	S : Switch coin positions
F_A : Flip the coin at A	FS_A : Flip coin at A then Switch coin positions
F_B : Flip the coin at B	FS_B : Flip coin at B then Switch coin positions
F_{AB} : Flip both coins	FS_{AB} : Flip both coins then Switch coin positions

Define the set $G = \{I, F_A, F_B, F_{AB}, S, FS_A, FS_B, FS_{AB}\}$ and the operation $*$ on G where $x * y$ is the movement found by first performing the movement x followed by the movement y.

For example: If we first flip the coin at B and switch their positions, then the smaller coin is black in position B, and the larger coin is white in position A as in the left picture below. Then while still in that position, flip both coins as shown in the right picture to see the result.

A B

This is the same final position as first flipping over the coin at A then switching positions! So we have $FS_B * F_{AB} = FS_A$.

1. Complete the table for the operation $*$ on the set G. You may assume that $*$ is associative since it involves doing three moves in the order they appear, which will not depend on where the parentheses happen to lie.

$*$	I	F_A	F_B	F_{AB}	S	FS_A	FS_B	FS_{AB}
I	I			F_{AB}				
F_A	F_A	I	F_{AB}	F_B	FS_A	S	FS_{AB}	FS_B
F_B				F_A				
F_{AB}				I				
S				FS_{AB}				
FS_A				FS_B				
FS_B				FS_A				
FS_{AB}				S				

2. Determine the identity of G.
3. Find the inverse for each element of G, showing G is a group under $*$.
4. Determine if G is an abelian group. Either explain how we can tell from the table or find at least one example of how commutativity fails.

Project 1.3

(This was originally published in PRIMUS, see [6] for more.)

Let G be the set $\mathbb{Z} \times \mathbb{Z}_3 \times \mathbb{Z}_6 = \{(u, v, w) : u \in \mathbb{Z}, v \in \mathbb{Z}_3, w \in \mathbb{Z}_6\}$. Define the operation $*$ by:

$$(u, v, w) * (r, s, t) = (u + r + c_1 + c_2,\ \ v +_3 s,\ \ w +_6 t) \text{ where}$$

$$c_1 = \begin{cases} 0 & v + s < 3 \\ 1 & v + s \geq 3 \end{cases} \quad \text{and} \quad c_2 = \begin{cases} 0 & w + t < 6 \\ 1 & w + t \geq 6 \end{cases}.$$

The numbers c_1 and c_2 can be thought of as *carry the one* if the sum gets too large, just as we *carry the one* to the next digit when we get to 10 in usual addition. For example:

$$(-2,\ 1,\ 5) * (0,\ 1,\ 3) \ = (-2 + 0 + 0 + 1,\ \ 1 +_3 1,\ \ 5 +_6 3) = (-1,\ \ 2,\ \ 2)$$

Notice that $c_1 = 0$ since $1 + 1 = 2$ is less than 3, and $c_2 = 1$ since $5 + 3 = 8$ is larger than 6.

1. Calculate:
   $(-1, 1, 4) * (-2, 2, 3) =$
   $(2, 0, 2) * (-7, 2, 4) =$
   $(-4, 2, 5) * (2, 2, 5) =$
   $(-3, 1, 2) * (8, 1, 3) =$
2. Find the identity of this group.
3. Find the inverses of the elements $(2, 1, 2)$ and $(-3, 2, 5)$.
4. Solve the equation $(3, 1, 5) * x = (-3, 2, 4)$.
5. Is this group abelian? Why or why not?
6. Find the order of the element $(-1, 1, 4)$.

Project 1.4

(A variation of this was originally published in PRIMUS, see [6] for more.)

Define $G = \mathbb{Z} \times \mathbb{Z}_4$ and the operation $\circ^2$ by:

$$(a, b) \circ^2 (u, v) = (a + u + c, b +_4 v) \text{ where } c = \begin{cases} 0 & b + v < 4 \\ 2 & b + v \geq 4 \end{cases}$$

The number c can be thought of as *carry a two* if the sum gets too large, just as we carry the one to the next digit when we get to 10 in usual addition.

This group is denoted $\mathbb{Z} \circ^2 \mathbb{Z}_4$, to remind us that we carry a 2.

1. Calculate:
   $(-6,3) \circ^2 (2,3) =$
   $(-1,2) \circ^2 (-2,1) =$
   $(3,2)^2 = (3,2) \circ^2 (3,2) =$
   $(-1,3)^3 =$
2. Find the identity of this group.
3. Find the inverses of the elements $(-6,2)$ and $(5,1)$.
4. Solve the equation $(1,3) \circ^2 x = (-3,2)$.
5. Is this group abelian? Why or why not?

Project 1.5

Let G denote a group with identity e_G, and assume that $a, b \in G$.

Prove: If $ab^{-1}a^{-1} = b$ and $ab = b^{-1}$ then $a = e_G$.

1. Circle all of the logical flaws in the following "proof," and explain why any circled step is incorrect. Remember to look at the logic of each step to see if it follows from the previous steps. A mistake does not make each step following it incorrect.

   "Proof": Assume that G is a group with identity e_G and $a, b \in G$. Suppose that $ab^{-1}a^{-1} = b$, and $ab = b^{-1}$. Multiplying the first equation by a we get $a(ab^{-1}a^{-1}) = ba$. Using associativity, and the fact that $aa^{-1} = e_G$, we now have $ab^{-1} = ba$. Thus $b = b^{-1}$ by cancellation. Since we assumed $ab = b^{-1}$ then $ab = b$. Now $ab = e_G b$, so by cancellation $a = e_G$, and the proof is complete.

2. Give a correct proof of the statement above.
3. Again suppose that G denotes a group with identity e_G, and assume that $a, b \in G$. Prove: If $(ab)^3 = a^2$ then $(ba)^2 = a(b^{-1})$.

Project 1.6

Let G be a group with identity e_G, and $a \in G$.

1. Circle all of the logical flaws in the following "proof," and explain why any circled step is incorrect. Remember to look at the logic of each step to see if it follows from the previous steps. A mistake does not make each step following it incorrect.

   Prove: If $ord(a) = 10$ then $ord(a^3) = 10$.

   "Proof": Assume G is a group with identity e_G, $a \in G$, and that $ord(a) = 10$. Thus we know $a^{10} = e_G$ so we have $(a^{10})^3 = e_G$. By the power rules we know $(a^3)^{10} = a^{13}$. But $(a^3)^{10} = (a^{10})^3 = e_G$, and so $a^{13} = e_G$. Thus the order of a^3 must divide 30, and the divisors of 30 are 1, 3, 5, 10, and 30. We know that $a^3 \neq e_G$ and $a^5 \neq e_G$ since $ord(a) = 10$. Thus the first positive integer with $(a^3)^k = e_G$, is $k = 10$. Therefore $ord(a^3) = 10$.

2. Write a correct proof for the statement above.

Project 1.7

Let G denote a group with identity e_G and $a \in G$.

1. Assume that $ord(a) = 12$. What is the smallest positive integer k with $a^{8k} = e_G$? What is the order of a^8?
2. Again, assume that $ord(a) = 12$. Find $ord(a^5)$, $ord(a^7)$, and $ord(a^9)$. Can you guess which values of k between 1 and 12 will have $ord(a^k) = 12$?
3. Now suppose instead that $ord(a) = 6$ and $b \in G$ has $b^3 = a$. What are the possibilities for $ord(b)$? Use $ord(a) = 6$ to rule out all but one of the possibliities.
4. It is often thought that the following statement is true: *If* $ord(a) = n$ *and* $b^k = a$ *then* $ord(b) = nk$. But this is actually false in some groups! Let G be the group $(\mathbb{Z}_{12}, +_{12})$, $a = 2$ and $b = 5$. Show that $ord(2) = 6$, find k with $2 = 5^k$, and show that $ord(5) \neq 6k$.
5. Suppose $c, d \in G$, $ord(c) = n$ and $ord(d) = k$. Prove: If $cd = dc$, then $ord(cd)$ is a divisor of nk.

Project 1.8

Fill in the blanks for the following proof.

1. Let G be a group with identity e_G, and FIX the element $a_0 \in G$ for the rest of this problem. Define $H = \{x \in G : xa_0 = a_0x\}$.

   Prove: H is closed under inverses.

   Let $x \in H$. We need to show that _______ $\in H$, or in other words that _______________ = _______________. Consider the element $x(x^{-1}a_0)$ from G. Using _______________, $x(x^{-1}a_0) =$ (_______)_______. Since _______________ we have $x(x^{-1}a_0) = a_0$. Now look at $x(a_0x^{-1})$ in G. By _______________ we have $x(a_0x^{-1}) =$ (_______)_______. Since $x \in H$ we can say that ___________ = ___________, so then $x(a_0x^{-1}) =$ ___________. Using _______________ again we get $x(a_0x^{-1}) =$ _______(_______) so by the definition of inverses we have _______ = a_0. Since both elements $x(x^{-1}a_0)$ and $x(a_0x^{-1})$ equal a_0 we have that ___________ = ___________, so using the cancellation law, ___________ = ___________ as we needed to prove.

2. Is the set H defined above also a subgroup of G? Justify your answer!
3. Prove: If G is abelian then $H = \{x \in G : x = x^{-1}\}$ is a subgroup of G.

Assume that __________________, and define $H = \{x \in G : x = x^{-1}\}$. We need to show that ______________________. The identity of G, e_G, has the property that $e_G^2 = e_G$ so _______________. Thus _______________ and $H \neq \emptyset$.

Suppose that $a, b \in H$. We need to show that _______________ $\in H$. Now by Theorem _______, we know that $(ab)^{-1} = b^{-1}a^{-1}$. But G is abelian so we know _______________. Since a and b are from H we know ____________________ so then $(ab)^{-1} = ab$. Therefore _______________.

Suppose now that $a \in H$, and we will show that $a^{-1} \in H$. Since _______________, $a^{-1} \in H$ as well. Therefore H is a subgroup of G.

Project 1.9

Consider the groups $G = \mathbb{Q}^*$ (the nonzero rational numbers) under the operation of multiplication, and $K = \mathbb{Z}$ under the operation of addition. Note that every element of G can be written as $\frac{a}{b}$ where $a, b \in \mathbb{Z} - \{0\}$. Thus as integers we can find their greatest common divisor, $gcd(a, b)$, which is an integer and evenly divides both a and b.

Define the function $f : G \to K$ by $f\left(\frac{a}{b}\right) = \frac{a-b}{gcd(a,b)}$.

1. Show that the function above is well defined, by showing that for $a, b, n \in \mathbb{Z} - \{0\}$, we have $f\left(\frac{na}{nb}\right) = f\left(\frac{a}{b}\right)$.
2. One property of a homomorphism that turns out to be very useful is that $f(e_G) = e_K$. This is part of Theorem 1.37. Show that the function defined above does in fact have this property (even though we do not know if it is a homomorphism yet).
3. Another critical property from Theorem 1.37 is that for any $a \in G$ we have $f(a^{-1}) = (f(a))^{-1}$. Again, show that the function above does in fact have this property (even though we do not know if it is a homomorphism yet). Remember that we have $a \in \mathbb{Q}^*$ so $a = \frac{u}{v}$ for $u, v \in \mathbb{Z} - \{0\}$.
4. Show that f is NOT a homomorphism, with a counterexample.

   It is important to realize that the two properties discussed above do not guarantee that a function is a homomorphism, but can help show a function is not a homomorphism.
5. Using the same two groups, consider the function $g : G \to K$ defined by $g(\frac{a}{b}) = \frac{ab}{(gcd(a,b))^2}$. Show that both of the properties in 2 and 3 actually fail. Thus we automatically know the function is not a homomorphism.

Chapter 2

Special Groups

We have already seen one collection of groups that are particularly useful, the $\mathbb{Z}_n$ under $+_n$, but there are many others as well. In this chapter, we will consider groups of units, cyclic groups, permutation groups, and the symmetric and alternating groups.

2.1 Groups of Units

After Theorem 1.12, showing that each $\mathbb{Z}_n$ using the operation $+_n$ $(n > 1)$ is a group, the question was posed of whether a similar operation of multiplication *modulo* n on $\mathbb{Z}_n$ will also form a group. Recall $a \cdot_n b$ is the nonnegative remainder less than n after the product ab is divided by n. In fact, the answer is always **no** when using the set $\{0, 1, 2, \ldots, n-1\}$ and $\cdot_n$. The Cancellation Law (Theorem 1.20) fails since $1 \cdot_n 0 = 0 \cdot_n 0$ but $1 \neq 0$.

Definition 2.1 For $n > 1$, an element $x \in \mathbb{Z}_n$ is a ***unit*** under $\cdot_n$ if there exists $y \in \mathbb{Z}_n$ for which $x \cdot_n y = 1$.

Example 2.2 In $\mathbb{Z}_8$ $1 \cdot_8 1 = 1$, $3 \cdot_8 3 = 1$, $5 \cdot_8 5 = 1$, and $7 \cdot_8 7 = 1$. However, there is no solution to $2 \cdot_8 x = 1$, $4 \cdot_8 x = 1$, or $6 \cdot_8 x = 1$. Thus 1, 3, 5, and 7 are the units in $\mathbb{Z}_8$.

When we create the Cayley table for the units in $\mathbb{Z}_8$ under $\cdot_8$ we find a group. The identity is 1 and $1^{-1} = 1$, $3^{-1} = 3$, $5^{-1} = 5$, and $7^{-1} = 7$. (We will prove associativity soon.)

$\cdot_8$	1	3	5	7
1	1	3	5	7
3	3	1	7	5
5	5	7	1	3
7	7	5	3	1

Be sure you can find all units in $\mathbb{Z}_9$ for practice.

Definition 2.3 For each natural number $n > 1$, ***U(n)*** is the set of all units in $\mathbb{Z}_n$ under $\cdot_n$.

Theorem 2.4

For each natural number $n > 1$, $U(n)$ is a group under $\cdot_n$.

Proof Let n denote a natural number, $n > 1$. To show that $U(n)$ is a group under $\cdot_n$ we use Definition 1.10. First, it should be clear that 1 will always be a unit and thus $U(n) \neq \emptyset$. It is easier to prove that $\cdot_n$ is associative and commutative on $\mathbb{Z}_n$ first, to help us prove that $U(n)$ is a group. Theorem 0.18 is used frequently in the proof of associativity but not referenced each time.

Consider three elements $x, y, z \in \mathbb{Z}_n$. We need to prove that $x \cdot_n (y \cdot_n z) = (x \cdot_n y) \cdot_n z$. Using Theorem 0.18 we can write $yz = nk + r$ for some $k, r \in \mathbb{Z}$ and $0 \leq r < n$. Similarly, we can write $xr = nq + s$ for some $q, s \in \mathbb{Z}$ and $0 \leq s < n$. Thus $x \cdot_n (y \cdot_n z) = x \cdot_n r = s$.

Again, using Theorem 0.18 we can write $xy = nt + m$ for some $t, m \in \mathbb{Z}$ and $0 \leq m < n$. Similarly, we can write $mz = nw + u$ for some $w, u \in \mathbb{Z}$ and $0 \leq u < n$. Thus $(x \cdot_n y) \cdot_n z = m \cdot_n z = u$. If we can show that $s = u$ we will know that $x \cdot_n (y \cdot_n z) = (x \cdot_n y) \cdot_n z$. Consider the equations below using addition and multiplication in $\mathbb{Z}$:

$$x(yz) = x(nk + r) = xnk + xr = xnk + nq + s = n(xk + q) + s$$
$$(xy)z = (nt + m)z = ntz + mz = ntz + nw + u = n(tz + w) + u.$$

Thus $[x(yz)](\text{mod} n) = s$ and $[(xy)z](\text{mod} n) = u$. In $\mathbb{Z}$, $x(yz) = (xy)z$, so $s = u$ as we needed to show. Therefore, $x \cdot_n (y \cdot_n z) = (x \cdot_n y) \cdot_n z$, proving associativity.

Using similar steps it is easy to show that $\cdot_n$ is commutative since $xy = yx$ in the integers. You should write out the steps for practice.

Now we must show that $\cdot_n$ is an operation on the set $U(n)$. The key is to know that the product of two units is again a unit, as $\cdot_n$ is clearly well defined. Let $x, y \in U(n)$. Thus we have $a, b \in \mathbb{Z}_n$ with $x \cdot_n a = 1$ and $y \cdot_n b = 1$. Using associativity and commutativity of multiplication in $\mathbb{Z}_n$, we see below that $x \cdot_n y$ is also a unit and so $x \cdot_n y \in U(n)$.

$$(x \cdot_n y) \cdot_n (a \cdot_n b) = (x \cdot_n a) \cdot_n (y \cdot_n b) = (1 \cdot_n 1) = 1$$

Therefore $\cdot_n$ is an operation on the set $U(n)$.

It should be clear that since $1 \cdot_n x = x$ for every $x \in \mathbb{Z}_n$, 1 is the identity for the set $U(n)$. Finally we must show that the inverse of a unit is also a unit, a very simple task. If $x \in U(n)$ then there is $y \in \mathbb{Z}_n$ with $x \cdot_n y = 1$. But using commutativity of $\cdot_n$ we have $y \cdot_n x = 1$, or y is a unit, so $y = x^{-1} \in U(n)$.

Therefore $U(n)$ is a group under $\cdot_n$. In fact we also have that $U(n)$ is an abelian group. □

The most difficult part of working with $U(n)$ is finding it elements, for example how do we know which elements in $\mathbb{Z}_{38}$ are units?

Theorem 2.5

For each $n > 1$, the group $U(n)$ consists of the nonzero elements of $\mathbb{Z}_n$, which are *relatively prime* to n. (Definition 0.19)

Proof Let $n > 1$. Define $S = \{k \in \mathbb{Z}_n\colon k \neq 0,\ k \text{ relatively prime to } n\}$. We will show that $S = U(n)$, by subsets. Let $x \in U(n)$. We must show that $x \in S$. Since $x \in U(n)$ then there is $y \in \mathbb{Z}_n$ with $x \cdot_n y = 1$. If $x = 0$, no such y can exist since $0 \cdot_n y = 0$ for every $y \in \mathbb{Z}_n$. Thus $x \neq 0$.

Suppose now that x has common prime factor p with n, say $n = pb$ and $x = pc$ where b, c are positive integers less than n. Then the equation below under usual multiplication in $\mathbb{Z}$ shows us $b \cdot_n x = 0$.

$$bx = b(pc) = (bp)c = (pb)c = nc$$

But we have $x \cdot_n y = 1$, so multiplying by b gives us $b \cdot_n (x \cdot_n y) = b$. However, $(b \cdot_n x) \cdot_n y = 0 \cdot_n y = 0$. This contradicts the fact that $b > 0$. Thus x cannot have a common prime factor with n, so $x \in S$ and $U(n) \subseteq S$.

Suppose that $k \in S$, then $k > 0$, and k is relatively prime to n. By Theorem 0.22 there are integers a, b with $an + bk = 1$. Thus $bk = -an + 1$, so $(bk)(\bmod n) = 1$. Let $b(\bmod n) = z$ then $(bk)(\bmod n) = b(\bmod n) \cdot_n k$, so $z \cdot_n k = 1$ and $k \in U(n)$. Hence $S \subseteq U(n)$ and we have $S = U(n)$.

Thus $U(n)$ is the set of nonzero elements of $\mathbb{Z}_n$ which are relatively prime to n. □

Example 2.6 We can now determine the elements of $U(38)$ by knowing that $38 = 2(19)$. Thus $U(38)$ contains the elements of $\mathbb{Z}_{38}$ that do not have 2 or 19 as a factor. $U(38) = \{1, 3, 5, 7, 9, 11, 13, 15, 17, 21, 23, 25, 27, 29, 31, 33, 35, 37\}$.

Unlike the $\mathbb{Z}_n$, it is not trivial to discover $|U(n)|$. The formula to find $|U(n)|$ involves a function called *Euler's Function*, denoted by ϕ, which requires you to know all of the distinct primes that evenly divide n. We will take without proof the formula below, where Π refers to a product just as Σ denotes a sum:

$$\phi(n) = n \cdot \prod_{\substack{p \text{ a prime that} \\ \text{divides } n}} \left(1 - \tfrac{1}{p}\right)$$

In Example 2.6 above, $\phi(38) = 38\left(1 - \frac{1}{2}\right)\left(1 - \frac{1}{19}\right) = 38\left(\frac{1}{2}\right)\left(\frac{18}{19}\right) = 18$, and we found 18 elements of $U(38)$. This function is important in Number Theory but will not be investigated further here; see [7] for more information.

2.2 Cyclic Groups

Suppose G is a group (writing the operation as if it were multiplication).

Definition 2.7 For each $a \in G$ define the set $\langle a \rangle = \{a^n : n \in Z\}$.

Theorem 2.8

Suppose that G is a group and $a \in G$. Then the set $\langle a \rangle = \{a^n : n \in \mathbb{Z}\}$ is a subgroup of G, called the ***cyclic subgroup generated by a***.

Proof The set $\langle a \rangle$ is a subset of G, since G is closed under multiplication, thus we only need to verify the properties of Theorem 1.30 to complete the proof.

(i) Since 0 is an integer, a^0 is in the set $\langle a \rangle$, i.e., the identity of G is in $\langle a \rangle$ and $\langle a \rangle \neq \emptyset$.

(ii) Let $x, y \in \langle a \rangle$. Thus there exist integers m, n with $x = a^m, y = a^n$. Now $xy = a^m a^n = a^{m+n}$ by Theorem 1.23, and since $m + n$ is also an integer we have $xy \in \langle a \rangle$. Thus $\langle a \rangle$ is closed under the operation of G.

(iii) Let $x \in \langle a \rangle$. There exists an integer m with $x = a^m$. By Theorem 1.23, $x^{-1} = (a^m)^{-1} = a^{-m}$. Since $-m$ is also an integer, $x^{-1} \in \langle a \rangle$, so $\langle a \rangle$ is closed under inverses. Therefore, by Theorem 1.30, $\langle a \rangle$ is a subgroup of G. □

Example 2.9 In the group $(\mathbb{Z}_6, +_6)$, we can see that $\langle 2 \rangle = \{0, 2, 4\}$ since calculating powers gives us:

$$2^0 = 0 \text{ (the identity)},\ 2^1 = 2,\ 2^2 = 2 +_6 2 = 4,\ 2^3 = 2 +_6 2 +_6 2 = 0$$

You should verify that $\langle 3 \rangle = \{0, 3\}$ and $\langle 5 \rangle = \{0, 1, 2, 3, 4, 5\}$ for practice.

Definition 2.10 A group G is a ***cyclic group*** (or G is cyclic) if and only if there exists $a \in G$ with $G = \langle a \rangle$.

Example 2.11 The group $(\mathbb{Z}, +)$ is cyclic since $\mathbb{Z} = \langle 1 \rangle$. Recall that for $n \in \mathbb{N}$, the element 1^n means we are adding 1 to itself n times, so $1^n = n$, and $(-1)^n$ is found by adding -1 to itself n times so $1^{-n} = -n$. Also, 1^0 is defined to be the identity of our group, which is 0. Thus every integer can be written as $(1)^n$ or $(-1)^n$. But $\mathbb{Z} \neq \langle 2 \rangle$ since there is no integer n with $2^n = 3$. You should verify that $(\mathbb{Z}_n, +_n)$ is also a cyclic group for every $n > 1$.

Example 2.12 $(\mathbb{Q}, +)$ is not a cyclic group. For any $\frac{p}{q} \in \mathbb{Q}$, $\left(\frac{p}{q}\right)^n$ is the result of adding $\frac{p}{q}$ to itself n times, i.e., $\left(\frac{pn}{q}\right)$. There will be many fractions that are not in $\left\langle \frac{p}{q} \right\rangle$ since any prime number k that does not divide q has $\frac{1}{k} \notin \left\langle \frac{p}{q} \right\rangle$.

Cyclic groups have many properties that will be referenced later in the text. The following theorem gives two of them; the first part is left as an exercise at the end of the chapter.

Theorem 2.13

Let G be a cyclic group. Then:

(i) G is abelian.
(ii) Every subgroup of G is cyclic.

Proof Let G be a cyclic group. Thus $G = \langle a \rangle$ for some $a \in G$.

(i) is left as an exercise at the end of the chapter.

(ii) Suppose $H \subseteq G$ and H is a subgroup of G. If $H = \{e_G\}$ then $H = \langle e_G \rangle$ since $(e_G)^n = e_G$ for all $n \in \mathbb{Z}$, and H is already cyclic.

Thus we will assume that H contains at least one nonidentity element $c \neq e_G$. By G cyclic there is $m \in \mathbb{Z}$ with $c = a^m$ and $m \neq 0$. If $m < 0$ then $-m > 0$, and by Theorem 1.23 $c^{-1} = a^{-m}$. As H is a subgroup of G and $c \in H$ we know that $c^{-1} \in H$ as well. This tells us there exists at least one positive integer n with $a^n \in H$. Since every set of positive integers must have a least element, there must be a least positive integer k with $a^k \in H$. Let $x = a^k$.

Claim: $H = \langle x \rangle$.

H is a subgroup of G and closed under the operation of G, so we already know that $\langle x \rangle \subseteq H$. We must show $H \subseteq \langle x \rangle$ to complete the proof. Let $b \in H$, and we need to show that $b = x^n$ for some integer n. As $b \in G$, we know that $b = a^t$ for some integer t. By Theorem 0.18, we can write $t = kq + r$ where $q, r \in \mathbb{Z}$ and $0 \leq r < k$. Since $a^k \in H$, then $(a^k)^{-q} \in H$ as well. But also $a^t \in H$ so by H closed under the operation we have $a^{t-kq} = a^r \in H$. But k was the least positive integer with $a^k \in H$ which tells us that $r = 0$ since $0 \leq r < k$. Thus $t = kq$ so $b = a^t = a^{kq} = (a^k)^q$ and $b = x^q$. Now $b \in \langle x \rangle$ as needed and $H = \langle x \rangle$.

Thus every subgroup of G is cyclic. □

There is always a danger of using theorems incorrectly, so let's consider Theorem 2.13 carefully. We have an implication "G is cyclic $\rightarrow$ G is abelian."

It would be incorrect to automatically assume the implication in the opposite direction without a proof.

Example 2.14 Consider the group $(\mathbb{Q},+)$. We know that this group is abelian, but as seen in Example 2.12 it is NOT cyclic. Thus the converse of the first implication is false.

The converse of the second statement is actually true, because a group is always a subgroup of itself! Thus if we know every subgroup of G is cyclic, then as G is one of those subgroups it is cyclic. But it is still easy to make a mistake using this statement. Consider the group $\mathbb{Z}_2 \times \mathbb{Z}_2 = \{(0,0),(1,0),(0,1),(1,1)\}$ whose operation uses $+_2$ in each coordinate (as in Example 1.16). The proper subgroups (not the whole group), are exactly those shown below.

$$H_0 = \{(0,0)\} \quad H_1 = \{(0,0),(0,1)\} \quad H_2 = \{(0,0),(1,0)\} \quad H_3 = \{(0,0),(1,1)\}$$

These are all cyclic as seen next.

$$H_0 = \langle(0,0)\rangle\,, \quad H_1 = \langle(0,1)\rangle\,, \quad H_2 = \langle(1,0)\rangle\,, \quad H_3 = \langle(1,1)\rangle$$

But $\mathbb{Z}_2 \times \mathbb{Z}_2$ is not a cyclic group since none of its elements will generate the whole group. Thus even if every subgroup other than G is cyclic we cannot assume that G is also cyclic.

The next theorem is a key part of understanding the relationship between the order of elements (Definition 1.24) and cyclic groups.

Theorem 2.15

Let G be a group and $a \in G$.

(i) If $ord(a) = n$ then for every integer m, a^m is equal to exactly one of the distinct elements $a^0,\ a^1,\ a^2,\ a^3, \ldots,\ a^{n-1}$.
(ii) If a has infinite order then no two distinct powers of a are equal, i.e., for any integers $j \neq k$ we must have $a^j \neq a^k$.

Proof Assume G is a group and $a \in G$.

(i) Suppose $ord(a) = n$. Thus n is the least positive integer for which $a^n = e_G$. Let m be any integer, and consider a^m. By Theorem 0.18 there exist integers q and r so that $m = nq + r$ and $0 \leq r < n$. Using $a^n = e_G$ we see the following:

$$a^{nq} = (a^n)^q = (e_G)^q = e_G$$
$$a^m = a^{nq+r} = a^{nq}a^r = e_G a^r = a^r$$

Thus $a^m = a^r$, and r is one of the integers $0, 1, 2, \ldots, n-1$. Hence a^m is equal to one of the elements $a^0,\ a^1,\ a^2,\ a^3, \ldots,\ a^{n-1}$. To see that these elements are distinct, assume we have $a^r = a^s$ with $r \neq s$ and $0 \leq r, s < n$. Without loss of generality we may assume that $r < s$ and notice that

$$a^{s-r} = a^s\,(a^r)^{-1} = a^s\,(a^s)^{-1} = e_G.$$

But $0 < s - r < n$ contradicting that n is the order of a. Thus the powers $a^0,\ a^1,\ a^2,\ a^3, \ldots,\ a^{n-1}$ are all distinct as needed.

(ii) Suppose that a has infinite order. For a contradiction assume there are integers $j \neq k$ with $a^j = a^k$. Without loss of generality assume that $k > j$, so that $k - j > 0$. Now $a^{k-j} = a^k\,(a^j)^{-1} = a^k\,(a^k)^{-1} = e_G$, but the definition of infinite order says no such positive integer can exist. Thus for any integers $j \neq k$ we must have $a^j \neq a^k$. □

Theorem 2.15 gives us tools to prove facts that are frequently used with finite cyclic groups.

Theorem 2.16

Let G be a *finite* cyclic group, with $G = \langle a \rangle$.

(i) $|G| = ord(a)$
(ii) If $b \in G$ then $ord(b)$ evenly divides $ord(a)$.
(iii) Let $b \in G$ where $b \neq e_G$ and $b = a^k$ for some $1 < k < ord(a)$. $\langle b \rangle = \langle a \rangle$ if and only if k is relatively prime to $ord(a)$.

Proof Suppose that G is a finite cyclic group. Let $G = \langle a \rangle$ and assume $|G| = n$ for some positive integer n. Note first that if $|G| = 1$, then $G = \{e_G\}$. But $ord(e_G) = 1$ as needed in (i) and the only choice for b in (ii) is $e_G = a$, making (ii) easily true. Also (iii) does not apply when $G = \{e_G\}$. Thus in the rest of this proof we will assume that $|G| > 1$ and $e_G \neq a$.

(i) We need to show that $ord(a) = n$. Since G is finite we know that $ord(a)$ must be finite as well, or there would be infinitely many distinct powers of a in G by Theorem 2.15. Suppose that $ord(a) = m > n$. By Theorem 2.15 there are exactly m distinct powers of a, each of which is in G. This contradicts that $|G| = n$, so $m \leq n$.

Similarly if $ord(a) < n$ then by Theorem 2.15, every element of G must be equal to one of a^0, a^1, a^2, $a^3, \ldots,$ a^{m-1} and thus G has only m elements in it, a contradiction to our original assumption again.

Therefore $ord(a) = n$ and so $|G| = ord(a)$.

(ii) Left as an exercise at the end of the chapter.

(iii) Suppose that $b \in G$, where $b \neq e_G$, and $b = a^k$ for some $1 < k < ord(a)$. Note that in the proof of (i) it was shown that $ord(a) = n$, thus $1 < k < n$. Our statement requires two proofs (if and only if) so we will include one here and leave the other as an exercise at the end of the chapter.

($\rightarrow$) Assume that $\langle b \rangle = \langle a \rangle$. We need to show that k is relatively prime to $ord(a)$, i.e., to n. Suppose instead that n and k are not relatively prime, but have a common prime factor, $p > 1$. Thus $n = px$ and $k = py$ for some positive integers x and y with $x < n$ and $y < k$. By Theorem 1.23 we have:

$$b^x = \left(a^k\right)^x = (a^{py})^x = (a^{px})^y$$

But $a^{px} = a^n = e_G$ and so $b^x = e_G$. This gives us $ord(b) < n$, which tells us that $|\langle b \rangle| \neq n$ by (i). However, this contradicts our assumption of $\langle b \rangle = \langle a \rangle$, therefore k is relatively prime to n. $\square$

Example 2.17 Consider the group $(\mathbb{Z}_8, +_8)$. $\mathbb{Z}_8$ is cyclic and $\mathbb{Z}_8 = \langle 3 \rangle$. Now $5 \in \mathbb{Z}_8$, and $5 = 3^7$ under $+_8$. Thus we have the conditions in (iii) of Theorem 2.16, using $a = 3$, $b = 5$, $n = 8$, and $k = 7$. Since $k = 7$ is relatively prime to $ord(a) = 8$, the theorem tells us we should have $\langle 3 \rangle = \langle 5 \rangle$, which is verified below.

$$\langle 5 \rangle = \{5^0, 5^1, 5^2, 5^3, 5^4, 5^5, 5^6, 5^7\} = \{0, 5, 2, 7, 4, 1, 6, 3\}$$

Theorem 2.16 also reinforces why we found $\langle 5 \rangle = \mathbb{Z}_6$ in Example 2.9 since $5 = 1^5$, $ord(1) = 6$, and 5 is relatively prime to 6.

Finally, we have an important classification of cyclic groups, which makes the groups discussed in section 1.3 more critical to understand.

Theorem 2.18

Let G be a cyclic group, and $n > 1$.

(i) If $|G| = n$ then G is isomorphic to the group $(\mathbb{Z}_n, +_n)$.
(ii) If G is an infinite group then G is isomorphic to the group $(\mathbb{Z}, +)$.

Proof Let G be a cyclic group and $n > 1$. We will prove (ii) here and leave (i) as an exercise at the end of the chapter. Assume G is an infinite group. As G is cyclic there must exist $a \in G$ with $G = \langle a \rangle$. By Theorem 2.15, a must have infinite order or G would be finite. Hence for any integers $j \neq k$ we must have $a^j \neq a^k$. Define $f : \mathbb{Z} \to G$ by $f(n) = a^n$, and we will show this function is an isomorphism. From Definition 1.34 we need to show that f is a bijection and that $f(m + n) = f(m)f(n)$ (we write the operation on G as multiplication here).

Assume that $m, n \in \mathbb{Z}$ and that $f(n) = f(m)$. Thus $a^m = a^n$ by the definition of f. However, by Theorem 2.15 this can only happen if $n = m$, thus f is injective. Let $y \in G$. Then by $G = \langle a \rangle$, $y = a^n$ for some integer n. Thus $f(n) = y$ and f is surjective. Hence f is a bijection. Finally, assume again that $m, n \in \mathbb{Z}$. $f(m+n) = a^{m+n} = a^m a^n$ by Theorem 1.23. Thus $f(m + n) = a^m a^n = f(m)f(n)$ and f is a homomorphism.

Therefore G is isomorphic to $\mathbb{Z}$ as we wanted to prove. □

2.3 Permutation Groups

Another class of groups, useful as examples but also critical to the study of Modern Algebra (as we shall discover), is *permutation* groups. Before looking at permutation groups however, we must define permutations.

Definition 2.19 Let A be any nonempty set. A ***permutation*** of A is a bijection from A to A.

Example 2.20 Consider the set $\mathbb{Z}$ of integers. The function $f : \mathbb{Z} \to \mathbb{Z}$ defined by $f(n) = n+1$ (usual addition) is a permutation of $\mathbb{Z}$. To verify this we must show that f is a bijection. Suppose we have $a, b \in \mathbb{Z}$ with $f(a) = f(b)$. By the definition of f we thus have $a+1 = b+1$. This is an equation of integers so subtracting one from both sides gives us $a = b$. Thus f is one to one. Suppose we have any $z \in \mathbb{Z}$, and we must find an element $x \in \mathbb{Z}$ with $f(x) = z$. Since $z \in \mathbb{Z}$ then also $z - 1 \in \mathbb{Z}$. But $f(z-1) = (z-1)+1 = z$, so the element $x = z - 1$ has $f(x) = z$. Hence f is onto and therefore is a bijection. Thus f is a permutation of $\mathbb{Z}$.

But the function $g : \mathbb{Z} \to \mathbb{Z}$ defined by $g(n) = 2^n$ is not a permutation as it is not surjective (what element of $\mathbb{Z}$ is not in $g(\mathbb{Z})$?). Is the function $h : \mathbb{Z} \to \mathbb{Z}$ defined by $h(n) = n^2$ a permutation of $\mathbb{Z}$? (Why or why not?)

Example 2.21 For a more interesting example, consider the set $A = \mathbb{R} - \{0, 1, -1\}$ (all real numbers except for 0, 1, and -1), and consider the function below. Is f a permutation of A?

$$f : A \to A \text{ by } f(x) = \frac{x-1}{x+1}$$

Suppose $x \in A$. We must be sure that $f(x) \in A$ as well. As $x \neq -1$ then $x + 1 \neq 0$ so we know that $f(x)$ is a real number, but $f(x)$ must not be equal to 1, -1, or 0. A fraction can only equal 0 if the numerator is equal to 0 so since $x \neq 1$, then $f(x) \neq 0$. Suppose that $f(x) = 1$. Thus we have $x - 1 = x + 1$, which cannot be true for any real number as it is equivalent to $-1 = 1$. Similarly, suppose we have $f(x) = -1$. This gives us $x - 1 = -(x+1)$, or in other words $2x = 0$, which again cannot happen in $\mathbb{R}$ unless $x = 0$. Thus we can be sure that $f(x) \in A$, and since $f(x)$ is uniquely defined for each $x \in A$, f is a function from A to A.

We still need to determine if f is a bijection. Suppose we have $x, y \in A$ with $f(x) = f(y)$.

$$\frac{x-1}{x+1} = \frac{y-1}{y+1}$$

As $x \neq -1$ and $y \neq -1$ it follows that $(x-1)(y+1) = (x+1)(y-1)$. Using basic algebra (do the steps) this equation reduces to $2(x-y) = 0$, but in the real numbers this means $x = y$. Hence f is injective.

Finally, suppose we have any $a \in A$. Can we find $x \in A$ for which $f(x) = a$? Consider the number $x = \frac{a+1}{-a+1}$. As $a \neq 1$, x is a real number, and since $a \neq -1$, we know $x \neq 0$. Also

$x = 1$ would require $a = 0$, and $x = -1$ is equivalent to $-1 = 1$. Thus $x \in A$. Is it the value we hoped to find?

$$f(x) = \frac{x-1}{x+1} = \frac{\left(\frac{a+1}{-a+1}\right) - 1}{\left(\frac{a+1}{-a+1}\right) + 1} = \frac{\frac{a+1+a-1}{-a+1}}{\frac{a+1-a+1}{-a+1}} = \frac{2a}{2} = a$$

Thus $f(x) = a$ and f is surjective. Therefore f is a permutation of A.

An important fact to notice is that if G is a group and f is a permutation of G, we cannot conclude that f is an isomorphism. A permutation only requires that f be a bijection, not have the homomorphism property described in Definition 1.34. To see this consider the permutation from Example 2.20. For $n = 2$ and $m = 3$ we have $f(2+3) = f(5) = 6$ but $f(2) + f(3) = 3 + 4 = 7$. Thus f is not an isomorphism.

For a given nonempty set A we can look at all possible permutations of A, gathered into a new set that we will denote by S_A. The letter S is for the word *symmetry*, which is often used in place of permutation. For any nonempty set A, there will always be at least one permutation of A, namely the identity function. The ***identity permutation*** is traditionally denoted by ε and defined by $\varepsilon(x) = x$ for all $x \in A$.

Theorem 2.22

For any nonempty set A the set $S_A = \{f : A \to A : f \text{ is a permutation of } A\}$ is a group under the operation of composition of functions. The group S_A is called the ***Symmetric Group*** on A.

Proof Let A be a nonempty set, and S_A the set of all possible permutations of A. Recall that the elements of S_A are functions, so let $f, g \in S_A$. Since both f and g are permutations of A, $g(A) \subseteq dom(f)$ and so $f \circ g$ is uniquely defined as a function from A to A. Also the fact that both f and g are bijections guarantees that $f \circ g$ is a bijection by Theorem 0.17, and so $f \circ g \in S_A$. Composition is also uniquely defined so is an operation on S_A according to Definition 1.1.

Now we must verify the properties in Definition 1.10 to show that S_A is a group under the operation of composition. Proof of associativity is left as an exercise at the end of the chapter, and we have (i) of Definition 1.10. We must find a specific function that will act as the identity under composition, and terminology gives us an advantage here. For any nonempty set A, the identity function on A, ε, is the function with $\varepsilon(x) = x$ for all $x \in A$. Notice that for any $f \in S_A$:

$$(f \circ \varepsilon)(x) = f(\varepsilon(x)) = f(x) \text{ and } (\varepsilon \circ f)(x) = \varepsilon(f(x)) = f(x).$$

Thus we have $f \circ \varepsilon = f = \varepsilon \circ f$. Clearly, $\varepsilon \in S_A$ and so ε is our identity under composition.

Finally, consider any $f \in S_A$. We need to find an inverse for f in S_A, and again the name tells us what to do. Define the function $g : A \to A$ by $g(a) = b$ for the unique $b \in A$ with $f(b) = a$. To see that g is uniquely defined recall that f is onto so for any $a \in A$ there is $b \in A$ with $a = f(b)$. If there were two elements $b, c \in A$ for which $g(a) = b$ and $g(a) = c$, this would tell us that $f(b) = a = f(c)$. However, f is one to one so we must have $b = c$. Thus $g(a)$ is a uniquely defined element of A and so g is a function.

If there were $x, y \in A$ with $g(x) = g(y) = z$ then $f(z) = x$ and $f(z) = y$, which violates that f is a function unless $x = y$. Hence $x = y$ and g is one to one. The last piece is actually much easier. Suppose we have $y \in A$, then $x = f(y) \in A$ has $g(x) = y$. Thus g is onto and is a bijection.

To see that g is the inverse of f, consider how $f \circ g$ and $g \circ f$ act as functions. Let $x \in A$. There is $a \in A$ with $f(a) = x$ and $g(x) = a$. But also if $y = f(x)$ then $g(y) = x$ as well.

$$(f \circ g)(x) = f(g(x)) = f(a) = x \qquad (g \circ f)(x) = g(f(x)) = g(y) = x$$

Thus $f \circ g = \varepsilon = g \circ f$ and so f has an inverse that is also a permutation of A, which we will call f^{-1}. Hence every element of S_A has an inverse in S_A and S_A is a group under composition.

□

Example 2.23 Consider the group $S_{\mathbb{Q}}$ where $\mathbb{Q}$ is the set of rational numbers. The functions $f, g : \mathbb{Q} \to \mathbb{Q}$ defined by $f(x) = x+1$ and $g(x) = 2x+3$ are both elements of $S_{\mathbb{Q}}$ (you should verify this). Computing both $(f \circ g)(1)$ and $(g \circ f)(1)$ we see the results below.

$$(f \circ g)(1) = f(g(1)) = f(5) = 6$$
$$(g \circ f)(1) = g(f(1)) = g(2) = 7$$

Thus we have a counterexample showing us $f \circ g \neq g \circ f$ and so $S_{\mathbb{Q}}$ is not an abelian group. In general S_A is not abelian!

Definition 2.24 A group G is a ***permutation group*** (or a group of permutations) if it is a subgroup of the Symmetric Group S_A for some nonempty set A.

Example 2.25 Consider $\mathbb{R}$, the set of real numbers, and $f : \mathbb{R} \to \mathbb{R}$ defined by $f(x) = -x$. You should be able to easily verify that $f \in S_{\mathbb{R}}$. Let $G = \{\varepsilon, f\}$, with ε as the identity function for $\mathbb{R}$, then $G \subseteq S_{\mathbb{R}}$. To see that G is a subgroup of $S_{\mathbb{R}}$ we look at the Cayley table for G:

$\circ$	ε	f
ε	ε	f
f	f	ε

Clearly, G is nonempty, and the table shows it is closed under composition and inverses. Thus G is a subgroup of $S_{\mathbb{R}}$ and so G is a permutation group.

The importance of permutation groups is seen in the next theorem.

Theorem 2.26

Every group is isomorphic to a group of permutations.

Proof Let G be a group. Consider the group S_G under composition of functions. We will show that G is isomorphic to a subgroup of S_G. For each element $a \in G$ define the function $f_a : G \to G$ by $f_a(x) = ax$ for every $x \in G$. As G is a group we know that $f_a(x) \in G$ for every $x \in G$, and f_a is uniquely defined.

Claim: For every $a \in G$, $f_a \in S_G$.

Let $a \in G$. Suppose that $x, y \in G$ and that $f_a(x) = f_a(y)$. Then by definition of f_a we have $ax = ay$. However, by G a group, the Cancellation Law (Theorem 1.20) tells us that $x = y$. Hence f_a is injective. Also for any $y \in G$, we know that $a^{-1}y \in G$ and $f_a(a^{-1}y) = a(a^{-1}y)$. By associativity and $aa^{-1} = e_G$ we have $f_a(a^{-1}y) = y$ and f_a is surjective. Therefore f_a is a bijection and $f_a \in S_G$.

Define: $H = \{f_a : a \in G\}$

We will show that H is a subgroup of S_G and that G is isomorphic to H. We know that G is nonempty, and thus there is at least one $a \in G$. Thus for this a, $f_a \in H$ and H is nonempty. Suppose that $f_a, f_b \in H$. We know $(f_a \circ f_b)(x) = f_a(f_b(x))$ so for all $x \in G$:

$$(f_a \circ f_b)(x) = a(bx) = (ab)x = f_{ab}(x).$$

Thus $f_a \circ f_b = f_{ab}$ and so $f_a \circ f_b \in H$ or H is closed under composition. Now for any $a \in G$ we have $a^{-1} \in G$ so for all $x \in G$:

$$(f_a \circ f_{a^{-1}})(x) = a(a^{-1}x) = x.$$

Thus $f_a \circ f_{a^{-1}} = \varepsilon$. Similarly, $f_{a^{-1}} \circ f_a = \varepsilon$ so we have $(f_a)^{-1} = f_{a^{-1}} \in H$ and H is closed under inverses. Therefore H is a subgroup of S_G by Theorem 1.30.

Define the function $\varphi : G \to H$ by $\varphi(a) = f_a$.

It is clear that φ is a function from G to H, so we need to show that φ is an isomorphism to complete this proof. Suppose $a, b \in G$ for which $\varphi(a) = \varphi(b)$, i.e., $f_a = f_b$. Then for every $x \in G$, $f_a(x) = f_b(x)$. Since $e_G \in G$ we must have $f_a(e_G) = f_b(e_G)$ and so $a = b$, or φ is injective. Also, for any element of H, there is $a \in G$ defining it as f_a, so it is clear that φ is surjective and thus bijective. Finally, for $a, b \in G$:

$$\varphi(ab) = f_{ab} = f_a \circ f_b = \varphi(a) \circ \varphi(b).$$

Thus φ is also a homomorphism.

Hence φ is an isomorphism and G is isomorphic to the permutation group H as needed. □

2.4 Permutations of a Finite Set

Groups of permutations on finite sets are as useful as the groups $\mathbb{Z}_n$ and have more interesting properties. They give simple examples of nonabelian groups and will have many parts to play later in the text.

Consider a set $A = \{a, b\}$. It is not hard to see that there are exactly two different permutations of A. Call them f and g:

$$\begin{array}{ccc} f(a) = a & \text{and} & f(b) = b \\ g(a) = b & \text{and} & g(b) = a \end{array}$$

Another notation we can use to describe these functions is seen below, with the top row simply the list of elements of A and the bottom row showing the result of the function.

$$f = \begin{pmatrix} a & b \\ a & b \end{pmatrix} \quad g = \begin{pmatrix} a & b \\ b & a \end{pmatrix}$$

Written in this form it is clear that f and g are bijections, and that no other options exist. Thus $S_A = \{f, g\}$.

One key to understanding permutations of a finite set is to notice that the names of the elements a and b used above are not important, i.e., any set of two elements will have the same two permutations with the names of a and b changed. For example, if we had $B = \{u, v\}$ then $S_B = \{h, k\}$ where:

$$h = \begin{pmatrix} u & v \\ u & v \end{pmatrix} \quad k = \begin{pmatrix} u & v \\ v & u \end{pmatrix}$$

Clearly by replacing a with u and b with v we would have $f = h$ and $g = k$. Thus we use the following groups for all permutations of finite sets.

Definition 2.27 For each $k \in \mathbb{N}$ the ***Symmetric Group*** S_k is the group of permutations of the set $\{1, 2, \ldots, k\}$ under the operation of composition.

We will often refer to a "product" of permutations. This always means we are using the operation of composition. Our observation before this definition can be stated as a theorem.

Theorem 2.28

For any $k \in \mathbb{N}$ and set A with exactly k elements, S_A is isomorphic to S_k.

Proof Let A be a set with exactly k elements, say $A = \{a_1, a_2, \ldots, a_k\}$. Define the function $\varphi : A \to \{1, 2, \ldots, k\}$ by $\varphi(a_i) = i$. It should be clear (or you should verify) that φ is a bijection and thus has an inverse $\varphi^{-1} : \{a_1, a_2, \ldots, a_k\} \to A$ where $\varphi^{-1}(i) = a_i$.

$$\text{Define } \sigma : S_A \to S_k \text{ by } \sigma(f) = \varphi \circ f \circ \varphi^{-1}.$$

We will show that σ is an isomorphism to complete the proof. First, we must show that $\sigma(f) \in S_k$ for any $f \in S_A$. Let $j \in \{1, 2, 3, \ldots, k\}$. By definition:

$$\sigma(f)(j) = \left(\phi \circ f \circ \phi^{-1}\right)(j) = \varphi\left(f\left(\varphi^{-1}(j)\right)\right) = \varphi\left(f\left(a_j\right)\right)$$

and since $f \in S_A$ we have $f(a_j) = a_m$ for some $m \in \{1, 2, 3, \ldots, k\}$. Thus $\sigma(f(a_j)) = \varphi(a_m) = m$, so $\sigma(f)$ maps from and to $\{1, 2, 3, \ldots, k\}$. Also since φ and f are bijections it is clear that $\sigma(f)$ is a bijection by Theorem 0.17. Therefore $\sigma(f) \in S_k$.

To prove σ is one to one, suppose that $s, t \in S_A$ and $\sigma(s) = \sigma(t)$. To see that $s = t$, we must show that $s(u) = t(u)$ for every $u \in A$. Let $u \in A$ and $m \in \{1, 2, \ldots, k\}$ with $\varphi(u) = m$ (so $\varphi^{-1}(m) = u$). Note the following equations:

$$\begin{aligned} \sigma(s)(m) &= \left(\varphi \circ s \circ \varphi^{-1}\right)(m) = \left(\varphi \circ s\right)(u) = \varphi(s(u)) \\ \sigma(t)(m) &= \left(\varphi \circ t \circ \varphi^{-1}\right)(m) = \left(\varphi \circ t\right)(u) = \varphi(t(u)) \end{aligned}$$

Hence $\varphi(s(u)) = \varphi(t(u))$, but φ is a bijection so $s(u) = t(u)$ for every $u \in A$ and $s = t$.

Thus σ is injective. Also for any $r \in S_k$ it is easy to show that $\varphi^{-1} \circ r \circ \varphi \in S_A$ and $\sigma(\varphi^{-1} \circ r \circ \varphi) = r$ (be sure to verify the details). Hence σ is surjective as well and thus is a bijection.

Finally, we need to verify that σ is a homomorphism. Suppose $s, t \in S_A$, and we will show $\sigma(s \circ t) = \sigma(s) \circ \sigma(t)$. Using the fact that $\varphi^{-1} \circ \varphi = \varepsilon$ we see that

$$\sigma(s \circ t) = \varphi \circ s \circ \left(\varphi^{-1} \circ \varphi\right) \circ t \circ \varphi^{-1} .$$

Associativity of composition gives us the following result:

$$\sigma(s \circ t) = \left(\varphi \circ s \circ \varphi^{-1}\right) \circ \left(\varphi \circ t \circ \varphi^{-1}\right) = \sigma(s) \circ \sigma(t).$$

Thus σ is a homomorphism. Therefore σ is an isomorphism and S_A is isomorphic to S_k. □

From now on when looking at permutations of a finite set, we will simply consider them as belonging to the appropriate symmetric group S_k.

Example 2.29 The Symmetric Group S_3 is the set of permutations of $\{1, 2, 3\}$ (traditionally denoted by lowercase Greek letters):

$$\varepsilon = \begin{pmatrix} 1 & 2 & 3 \\ 1 & 2 & 3 \end{pmatrix} \alpha = \begin{pmatrix} 1 & 2 & 3 \\ 2 & 1 & 3 \end{pmatrix} \beta = \begin{pmatrix} 1 & 2 & 3 \\ 1 & 3 & 2 \end{pmatrix}$$

$$\gamma = \begin{pmatrix} 1 & 2 & 3 \\ 3 & 2 & 1 \end{pmatrix} \delta = \begin{pmatrix} 1 & 2 & 3 \\ 2 & 3 & 1 \end{pmatrix} \kappa = \begin{pmatrix} 1 & 2 & 3 \\ 3 & 1 & 2 \end{pmatrix}$$

The operation of composition gives the Cayley table below:

$\circ$	ε	α	β	γ	δ	κ
ε	ε	α	β	γ	δ	κ
α	α	ε	δ	κ	β	γ
β	β	κ	ε	δ	γ	α
γ	γ	δ	κ	ε	α	β
δ	δ	γ	α	β	κ	ε
κ	κ	β	γ	α	ε	δ

Note that $(\alpha \circ \beta)(1) = \alpha(\beta(1)) = \alpha(1) = 2, (\alpha \circ \beta)(2) = \alpha(\beta(2)) = \alpha(3) = 3$, and $(\alpha \circ \beta)(3) = \alpha(\beta(3)) = \alpha(2) = 1$. Thus $\alpha \circ \beta = \delta$.

You should verify that all of the entries in the Cayley table are correct for practice. Notice that $\alpha \circ \beta = \delta$ and $\beta \circ \alpha = \kappa$ so S_3 is not abelian.

Example 2.30 In the group S_6 define:

$$\alpha = \begin{pmatrix} 1 & 2 & 3 & 4 & 5 & 6 \\ 2 & 1 & 4 & 3 & 5 & 6 \end{pmatrix} \text{ and } \beta = \begin{pmatrix} 1 & 2 & 3 & 4 & 5 & 6 \\ 1 & 3 & 4 & 2 & 6 & 5 \end{pmatrix}$$

Calculating we find:

$$\alpha \circ \beta = \begin{pmatrix} 1 & 2 & 3 & 4 & 5 & 6 \\ 2 & 4 & 3 & 1 & 6 & 5 \end{pmatrix} \qquad \beta \circ \alpha = \begin{pmatrix} 1 & 2 & 3 & 4 & 5 & 6 \\ 3 & 1 & 2 & 4 & 6 & 5 \end{pmatrix}$$

$$\alpha \circ \alpha = \begin{pmatrix} 1 & 2 & 3 & 4 & 5 & 6 \\ 1 & 2 & 3 & 4 & 5 & 6 \end{pmatrix} \qquad \beta \circ \beta = \begin{pmatrix} 1 & 2 & 3 & 4 & 5 & 6 \\ 1 & 4 & 2 & 3 & 5 & 6 \end{pmatrix}$$

We can also easily find β^{-1} with this notation. $\beta(1) = 1$ so we have $\beta^{-1}(1) = 1$, also $\beta(2) = 3$ so we have $\beta^{-1}(3) = 2$. Similarly $\beta^{-1}(2) = 4$, $\beta^{-1}(4) = 3$, $\beta^{-1}(5) = 6$, and $\beta^{-1}(6) = 5$. Thus we have:

$$\beta^{-1} = \begin{pmatrix} 1 & 2 & 3 & 4 & 5 & 6 \\ 1 & 4 & 2 & 3 & 6 & 5 \end{pmatrix}$$

Another notation used for permutations of a finite set is *cycle notation.* This comes from graphing the path an element of $\{1, 2, \ldots, n\}$ takes in a permutation to see a pattern of closed loops, or cycles. For example, consider the element $\phi \in S_7$ where $\phi = \begin{pmatrix} 1 & 2 & 3 & 4 & 5 & 6 & 7 \\ 2 & 5 & 7 & 6 & 1 & 3 & 4 \end{pmatrix}$.

$$\begin{array}{ccccccccc} 1 & & \rightarrow & & 2 & & 3 & \rightarrow & 7 \\ & \nwarrow & & \swarrow & & & \uparrow & & \downarrow \\ & & 5 & & & & 6 & \leftarrow & 4 \end{array}$$

There are two cycles if we follow the action of the permutation ϕ: 1 maps to 2, then 2 maps to 5, and 5 maps back to 1 to complete a cycle; 3 maps to 7, 7 maps to 4, 4 maps to 6, and finally, 6 maps back to 3 completing the second cycle. We write these cycles as (125) and (3746) where the cycle moves from left to right within the parentheses, and the final parenthesis tells us to map back to the beginning element of that cycle. We write $\phi = (125)(3746)$ in this notation and say that ϕ is a product (under composition) of the two cycles (125) and (3746).

Definition 2.31 Let $k > 1$ and $\alpha \in S_k$. α is a ***cycle*** if it is of the form $\alpha = (a_1 a_2 \ldots a_n)$ for some $1 < n \leq k$ where:

- the a_i are distinct elements of $\{1, 2, \ldots, k\}$,
- $\alpha(a_i) = a_{i+1}$ for $i < n$, $\alpha(a_n) = a_1$, and
- elements of $\{1, 2, \ldots, k\}$ not shown in the cycle are mapped to themselves.

A cycle is called a ***transposition*** if it is of the form $\alpha = (a_1 a_2)$. Cycles $\alpha = (a_1 a_2 \ldots a_n)$ and $\beta = (b_1 b_2 \ldots b_r)$ are ***disjoint*** in S_k if $\{a_1, a_2, \ldots, a_n\} \cap \{b_1, b_2, \ldots, b_r\} = \emptyset$.

An interesting fact to notice is that for any transposition α we have $\alpha^{-1} = \alpha$.

Theorem 2.32

For $k \in \mathbb{N}$ with $k > 1$, every nonidentity permutation in S_k can be written as a product (under the operation of composition) of disjoint cycles.

Proof Let $k \in \mathbb{N}$ with $k > 1$. Suppose $\alpha \in S_k$ with $\alpha \neq \varepsilon$ (the identity). Thus there is some i in $\{1, 2, 3, \ldots, k\}$ so that $\alpha(i) \neq i$. Choose the least (in numerical order) i with this property. Now compute $\alpha(i), \alpha(\alpha(i)), \ldots, \alpha^r(i)$ until the first time i is the answer (as we are in S_k this cannot happen more than $k-1$ times without repeating). This gives us the first cycle in the decomposition of α. If every element of $\{1, 2, 3, \ldots, k\}$ not mapping to itself has appeared in our cycle we are done; otherwise, find the least such j that has not yet been used and repeat the process. Continuing this way we will have a product of at most $\frac{k}{2}$ disjoint cycles. □

Notice that in the process of writing a permutation as a product of disjoint cycles, it was not actually necessary to begin with the least i among all that have $\alpha(i) \neq i$. Thus in S_5 the cycles (123) and (312) are the same permutation. We began with the least i only to see the cycles in a more standardized way. Another an important fact (whose proof is an exercise at the end of the chapter) is that under the operation of composition, *disjoint cycles will always commute.* Thus the permutation $(123)(45)$ in S_5 is exactly the same as the permutation $(45)(123)$.

Example 2.33 The permutations from Example 2.30 can be written in cycle notation as $\alpha = (12)(34)$ and $\beta = (234)(56)$. Write $\beta \circ \beta$, $\alpha \circ \beta$, and $\beta \circ \alpha$ as products of disjoint cycles for practice.

For many reasons, including the space required to write them, cycles have become the standard notation for permutations of a finite set. Unfortunately, the process of composing permutations is trickier when they are written as cycles, so we will look at an example to help illustrate the process.

Example 2.34 Consider the permutations in S_7, $\alpha = (135)(27)$ and $\beta = (234)(156)$, each written as a product of disjoint cycles.

To calculate $\alpha\circ\beta$, we first write the cycles in the correct order, $\alpha\circ\beta = (135)(27)(234)(156)$. Now beginning with 1 we follow where it maps through the cycles from the *rightmost* cycle to the *leftmost*. Remember that if the number does not appear in the cycle it maps back to itself.

1 maps to 5, 5 maps to 5, 5 maps to 5, 5 maps to 1

Thus in $\alpha \circ \beta$, 1 maps to 1 and so 1 will not appear in a cycle. Doing the same with 2 we see:

2 maps to 2, 2 maps to 3, 3 maps to 3, 3 maps to 5

Thus $\alpha \circ \beta$ maps 2 to 5. Now when writing $\alpha \circ \beta$ as a product of cycles we continue where 2 ended. Notice that 5 maps to 6, 6 maps to 3, 3 maps to 4, and 4 maps to 7. But since 7 maps to 2 the cycle is completed so $\alpha \circ \beta = (256347)$ and 1 maps to itself. You should try this process again to see why $\beta \circ \alpha =(142736)$.

Consider now $\beta \circ \beta = (234)(156)(234)(156)$. To find $(\beta \circ \beta)(1)$ notice that

$$1 \to 5 \to 5 \to 6 \to 6.$$

giving us $(\beta \circ \beta)(1) = 6$. Following through we should find 6 mapping to 5 and 5 mapping to 1 giving the first cycle (165). Then we begin a new cycle at 2 to get (243). Thus $\beta \circ \beta = (165)(243)$.

It is easy to find the inverse of a single cycle. For $\alpha = (a_1 a_2 \cdots a_k)$ its inverse is found by writing the elements in the cycle in the reverse order, $\alpha^{-1} = (a_k a_{k-1} \cdots a_1)$. Using Theorem 1.17 we can see that for $\beta = (123)(567)$ in S_7 we have $\beta^{-1} = ((123)(567))^{-1} = (567)^{-1}(123)^{-1} = (765)(321)$. A useful and interesting fact is seen in the next theorem. Note that it does not mention the word disjoint.

Theorem 2.35

For $k > 1$, every permutation in S_k can be written as a product (under the operation of composition) of transpositions.

Proof Let $k \in \mathbb{N}$ with $k > 1$ and $\alpha \in S_k$. If α is the identity permutation then $\alpha = (12)(12)$ and thus can be written as a product of transpositions. Assume now that α is not the identity, so can be written as a product of disjoint cycles by Theorem 2.32. For any cycle $\beta = (b_1 b_2 \dots b_r)$ we can write $\beta = (b_1 b_r)(b_1 b_{r-1}) \dots (b_1 b_2)$ as a product of transpositions. Thus writing each <u>cycle</u> in α as a product of transpositions gives us α as a product of transpositions as well. □

There is no unique way to write a permutation as a product of transpositions since when $k > 3$ we can write (1234) in S_k as either $(14)(13)(12)$, $(32)(43)(41)$, $(12)(12)(14)(13)(12)(14)(41)$, or even as $(56)(14)(13)(12)(56)$ if $k > 5$.

What <u>is</u> unique for (1234) is that when we write it as a product of transpositions we will always use an *odd* number of them. This gives rise to the definition of an even or odd permutation.

> **Definition 2.36** Let $k > 1$. A permutation in S_k is ***even*** if it can be written as a product (under the operation of composition) of an even <u>number</u> of transpositions. A permutation is ***odd*** if it can be written as the product (under the operation of composition) of an odd <u>number</u> of transpositions.

Theorem 2.37

For $k > 1$, <u>no</u> permutation in S_k can be *both* even and odd.

Proof Let $k > 1$. Clearly, $\varepsilon = (12)(12)$ so ε can be written with an even number of transpositions. We will show by induction that for all $t \geq 0$, ε cannot be written as a product of $2t + 1$ transpositions, i.e., ε must be <u>even</u>.

The base case for the induction has $t = 0$, so $2t + 1 = 1$. If we have $\varepsilon = (ab)$ then we see that $\varepsilon(a) = b$, so we do not actually have the identity function. Hence ε is not a product of $2(0)+1$ transpositions.

Assume (for the induction hypothesis) we cannot write ε as a product of $2t + 1$ many transpositions for some $t \geq 0$, and we will prove that we cannot write ε as a product of $2(t+1)+1 = 2t+3$ many transpositions. Suppose we had $\varepsilon = s_1 s_2 \cdots s_{2t+3}$ where each s_i is a transposition. Let b be some number from $1, 2, \ldots, k$ that appears in one of the transpositions above. Consider the last (rightmost in the product) transposition that contains b, say s_m. We will show how to rewrite ε to have the last occurrence of b move to the left. Then repeating this we can move b to only appear in the first (leftmost) transposition. Let $s_{m-1}s_m = (ac)(bx)$.

- If (ac) and (bx) are *disjoint* cycles then they commute so we can rewrite $s_{m-1}s_m = (bx)(ac)$.
- If $b = a$ and $x \neq c$ then $(ac)(bx) = (bc)(bx) = (xb)(xc)$.
- If $b \neq a$ and $x = c$ then $(ac)(bx) = (ax)(bx) = (ba)(ax)$.

In each of these cases, the occurrence of b has moved left by one transposition. If this continued we would eventually have b only in the first transpositions, a contradiction since b must map to itself in ε. The only other possibility that can occur is:

- If $b = a$, and $x = c$, they can be eliminated from the list as $(ac)(ac) = \varepsilon$.

This allows us to reduce the number of transpositions used to write ϵ by 2, and we have written ε with $2t+1$ transposition, contradicting our induction hypothesis. Hence ε cannot be written with $2t+3$ transpositions. Thus we can say that ε is even, and not odd.

Finally, we look at an arbitrary permutation α in S_k. Suppose α is both even and odd. Then we can write $\alpha = \beta_1\beta_2\cdots\beta_{2t}$ and $\alpha = \sigma_1\sigma_2\cdots\sigma_{2m+1}$ as products of transpositions. Now using the fact that $(ab)^{-1} = (ab)$ for a transposition and Theorem 1.17:

$$\begin{aligned}
\varepsilon = \alpha\alpha^{-1} &= (\beta_1\beta_2\cdots\beta_{2t})\,(\sigma_1\sigma_2\cdots\sigma_{2m+1})^{-1} \\
&= (\beta_1\beta_2\cdots\beta_{2t})\,(\sigma_1)^{-1}\,(\sigma_2)^{-1}\cdots(\sigma_{2m+1})^{-1} \\
&= (\beta_1\beta_2\cdots\beta_{2t})\,(\sigma_1\sigma_2\cdots\sigma_{2m+1})
\end{aligned}$$

This contradicts that ε is even since we have it written as a product of $2t+2m+1$ many transpositions and $2t+2m+1$ is odd. Thus no permutation can be both even and odd. □

Example 2.38 Look again at the group S_3 described in Example 2.29. We can write its nonidentity elements with cycle notation as:

$$\alpha = (12),\ \beta = (23),\ \gamma = (13),\ \delta = (123),\ \kappa = (132)$$

So $S_3 = \{\varepsilon, (12), (13), (23), (123), (132)\}$. Notice that (12), (13), and (23) are already written as transpositions and are *odd* since the number 1 is odd. However, $(123) = (13)(12)$ and $(132) = (12)(13)$ and thus are *even*.

Write the permutation $\omega = (1234)(5678) \in S_8$ as a product of transpositions to determine if ω is even or odd.

Consider the set $H = \{\varepsilon, (123), (132)\}$ of the even permutations in S_3. Is H a subgroup of S_3? Clearly, $H \neq \emptyset$ since $(123) \in H$. The Cayley table shows us that H is closed under composition.

$\circ$	ε	(123)	(132)
ε	ε	(123)	(132)
(123)	(123)	(132)	ε
(132)	(132)	ε	(123)

Also from the table we see that $\varepsilon^{-1} = \varepsilon$, $(123)^{-1} = (132)$ and $(132)^{-1} = (123)$. Thus H is a subgroup of S_3. Will this always happen?

Theorem 2.39

For $k > 1$, the set of all even permutations in S_k is a subgroup of S_k, called the ***Alternating Group*** on $\{1, 2, \ldots, k\}$ and denoted by A_k.

Proof Let $k > 1$ and define $A_k = \{\alpha \in S_k : \alpha \text{ is } \textit{even}\}$. Since the identity permutation ε is always an even permutation, then $\varepsilon \in A_k$ and so $A_k \neq \emptyset$. Suppose we have $\alpha, \beta \in A_k$. Thus α and β are even permutations, say α is a product of $2n$ transpositions and β is a product of $2m$ transpositions. Now $\alpha \circ \beta$ is a product of $2n+2m$ many transpositions, and $2n+2m$ is an even number. Thus $\alpha \circ \beta \in A_k$ and A_k is closed under composition. Also for any $\alpha \in A_k$, if $\alpha = t_1 t_2 \cdots t_{2n}$ is written as a product of transpositions, then by Theorem 1.17 we see:

$$\alpha^{-1} = (t_1 t_2 \cdots t_{2n})^{-1} = (t_{2n})^{-1}(t_{2n-1})^{-1} \cdots t_1^{-1} = t_{2n} t_{2n-1} \cdots t_1$$

Thus we have α^{-1} written with $2n$ many transpositions. So $\alpha^{-1} \in A_k$ and A_k is closed under inverses. Therefore A_k is a subgroup of S_k by Theorem 1.30. □

From Example 2.38 we find that $A_3 = \{\varepsilon, (123), (132)\}$. But if we instead used all of the *odd* permutations of S_3, the set $K = \{(12), (13), (23)\}$, it should clear that K is not a subgroup of S_3. What is interesting to notice, however, is that $|K| = |A_3|$. The final theorem of the chapter generalizes this idea.

Theorem 2.40

For any $k > 1$, $|S_k| = k!$ and $|A_k| = \frac{1}{2}k!$.

Proof Let $k > 1$. The elements of S_k are defined as functions by their action permuting the numbers $1, 2, \ldots, k$. To see how many there are, notice that for any $\alpha \in S_k$, $\alpha(1)$ has k choices, but once $\alpha(1)$ is chosen there are only $k-1$ choices left for $\alpha(2)$. For each choice of $\alpha(1)$ any of the $k-1$ choices can be part of a permutation so there are $k(k-1)$ many ways to choose the answers to $\alpha(1)$ and $\alpha(2)$. Similarly there are $k-2$ choices for $\alpha(3)$, $\ldots$, 1 choice for $\alpha(k)$. Thus we have $k(k-1)(k-2)\cdots(1) = k!$ possible permutations. Thus $|S_k| = k!$.

To see why $|A_k| = \frac{1}{2}k!$ we will show that the set of even permutations in S_k and the set of odd permutations in S_k have the same cardinality. Let B_k denote the set of all *odd* permutations from S_k. Define $f : A_k \to B_k$ by $f(\alpha) = (12) \circ \alpha$. If α is even it is clear that $(12) \circ \alpha$ is odd, so we have a function that does map from A_k to B_k. We must show that f is a bijection to complete the proof.

Let $\alpha, \beta \in A_k$ so that $f(\alpha) = f(\beta)$. Thus we have $(12) \circ \alpha = (12) \circ \beta$, but as S_k is a group the Cancellation Law (Theorem 1.20) tells us that $\alpha = \beta$. Therefore f is one to one. Finally, assume we have $\beta \in B_k$. Then $(12) \circ \beta$ is even, i.e., $(12) \circ \beta \in A_k$ and $f((12) \circ \beta) = (12) \circ (12) \circ \beta$. However, $(12) \circ (12) = \varepsilon$ so we have $f((12) \circ \beta) = \beta$. Thus f is onto, and is a bijection. Since $S_k = A_K \cup B_k$ and $A_k \cap B_k = \emptyset$ then $|S_k| = |A_k| + |B_k| = 2\,|A_k|$. Thus $|A_k| = \frac{1}{2}k!$ as we wished to prove. □

Example 2.41 There are many interesting facts about permutation groups, but one that will be referenced later is: *any subgroup H of S_5 containing a 5-cycle and a transposition must be all of S_5.* To see why this is true, consider any 5-cycle $\alpha = (a_1 a_2 a_3 a_4 a_5)$ (each $a_i \in \{1, 2, 3, 4, 5\}$), and the transposition $(a_1 a_3)$, and suppose they are in our subgroup H. We can calculate the following:

product	*result*
$\alpha(a_1 a_3)\alpha^{-1}$	$(a_2 a_4)$
$\alpha(a_2 a_4)\alpha^{-1}$	$(a_3 a_5)$
$\alpha(a_3 a_5)\alpha^{-1}$	$(a_1 a_4)$
$\alpha(a_1 a_4)\alpha^{-1}$	$(a_2 a_5)$

Now we have both $(a_2 a_4)$ and $(a_1 a_4)$ in H and thus we know $(a_2 a_4)(a_1 a_4)(a_2 a_4) = (a_1 a_2)$ is in H as well. Repeating the process with α and $(a_1 a_2)$ we find:

product	*result*
$\alpha(a_1a_2)\alpha^{-1}$	(a_2a_3)
$\alpha(a_2a_3)\alpha^{-1}$	(a_3a_4)
$\alpha(a_3a_4)i\alpha^{-1}$	(a_4a_5)
$\alpha(a_4a_5)\alpha^{-1}$	(a_1a_5)

We have found all 10 of the transpositions in S_5 as elements of H. But *every* permutation in S_5 can be found as a product of these 10 transpositions, so $H = S_5$. This same process can be repeated for any transposition instead of (a_1a_3) giving us the result we want.

Exercises for Chapter 2

Section 2.1 Groups of Units

1. Find the elements of $(U(20), \cdot_{20})$ and show its Cayley table.
2. Find the elements of $(U(12), \cdot_{12})$ and show its Cayley table.
3. Find the elements of $(U(14), \cdot_{14})$ and show its Cayley table.
4. Find the elements of $(U(15), \cdot_{15})$ and show its Cayley table.
5. Find the elements of $(U(16), \cdot_{16})$ and show its Cayley table.
6. Find the elements of $(U(21), \cdot_{21})$ and show its Cayley table.
7. Find the elements of $(U(7), \cdot_{7})$ and show its Cayley table.
8. Find the elements of $(U(11), \cdot_{11})$ and show its Cayley table.
9. Find the elements of $(U(24), \cdot_{24})$ and show its Cayley table.
10. Find the elements of $(U(30), \cdot_{30})$ and show its Cayley table.
11. In the group $(U(21), \cdot_{21})$, find the order of each element.
12. In the group $(U(20), \cdot_{20})$, find the order of each element.
13. In the group $(U(7), \cdot_{7})$, find the order of each element.
14. In the group $(U(12), \cdot_{12})$, find the order of each element.
15. In the group $(U(14), \cdot_{14})$, find the order of each element.
16. In the group $(U(15), \cdot_{15})$, find the order of each element.
17. If we have positive integers $n < m$, is $U(n) \subseteq U(m)$? Give a proof or counterexample.
18. Determine if the set $\{1, 3, 5, 7, 9\}$ is a group under the operation $\cdot_{10}$.
19. Use Euler's function to determine the size of the set $U(p)$ for an arbitrary prime p.
20. Prove or disprove: $U(p) = \mathbb{Z}_p - \{0\}$ for p prime.
21. Suppose we have $n = p^2$ when p is prime. Is it true that $U(n) = \mathbb{Z}_n - \{0, p\}$? Be sure to explain why or why not.
22. Using Euler's function, explain how to find the cardinality of $U(pq)$ when p and q are distinct primes.
23. Using Euler's function, what is the cardinality of $U(35)$?

24. Using Euler's function, what is the cardinality of $U(50)$?
25. Using Euler's function, what is the cardinality of $U(105)$?

Section 2.2 Cyclic Groups

26. Find all of the cyclic subgroups in $(U(20), \cdot_{20})$. Is $U(20)$ a cyclic group?
27. Find all of the cyclic subgroups in $(U(15), \cdot_{15})$. Is $U(15)$ a cyclic group?
28. Find all of the cyclic subgroups in $(U(7), \cdot_{7})$. Is $U(7)$ a cyclic group?
29. Find all of the cyclic subgroups in $(U(12), \cdot_{12})$. Is $U(12)$ a cyclic group?
30. Find all of the cyclic subgroups in $(U(30), \cdot_{30})$. Is $U(30)$ a cyclic group?
31. Find all of the cyclic subgroups in $(U(10), \cdot_{10})$. Is $U(10)$ a cyclic group?
32. Find all of the cyclic subgroups in $(U(9), \cdot_{9})$. Is $U(9)$ a cyclic group?
33. Let G denote an arbitrary group with $a, b \in G$. Prove: If $a = b^k$ for some integer k, and $ord(a) = ord(b) = n$ for some $n \in \mathbb{N}$, then $\langle a \rangle = \langle b \rangle$.
34. We needed **finite order** elements in the previous exercise since infinite order elements can make it fail! In the group $\mathbb{Z}$ under usual addition, find nonzero integers a and b with $a = b^k$ for some nonzero integer k, both a and b of infinite order, but $\langle a \rangle \neq \langle b \rangle$.
35. Find two integers $1 \leq a \leq 5$, $1 \leq b \leq 5$ for which $\langle a \rangle = \langle b \rangle$ is true in one of the groups $(U(7), \cdot_7)$ or $(\mathbb{Z}_6, +_6)$ but is false in the other group.
36. Prove: If G is a cyclic group then G is abelian.
37. Let G and K be groups and $f : G \to K$ an isomorphism. Prove: G is cyclic if and only if K is cyclic.
38. Let G and K be groups and $f : G \to K$ a homomorphism. Prove or disprove: If G is cyclic then K is cyclic.
39. Let G and K be groups. Prove: If $G \times K$ is cyclic then both G and K must be cyclic.
40. Prove (ii) of Theorem 2.16.
41. Prove (i) of Theorem 2.18.
42. Consider the group of functions $\Im(\mathbb{R})$ under addition of functions. Determine the elements of the cyclic subgroup $\langle f \rangle$ where $f(x) = x - 1$.

Sections 2.3 and 2.4 Permutation Groups

43. Prove: If $f{:}A \to B$, $g{:}B \to C$, and $h{:}C \to D$ are functions, then $h \circ (g \circ f) = (h \circ g) \circ f$.
44. Let $A = \{x \in \mathbb{R} : x \neq 0, 1\}$. Determine if $f{:}A \to A$ defined by $f(x) = 1 - \frac{1}{x}$ is a permutation of A.
45. Let $A = \{x \in \mathbb{R} : x \neq 0, 1\}$. Determine if $f{:}A \to A$ defined by $f(x) = \frac{x}{x-1}$ is a permutation of A.
46. Determine if $f{:}\{0, 1, 2, 3, 4\} \to \{0, 1, 2, 3, 4\}$ defined by $f(x) = x^3(\text{mod} 5)$, where x^3 denotes usual integer multiplication, is a permutation of $\{0, 1, 2, 3, 4\}$.
47. Determine if $g{:}\{0, 1, 2, 3, 4, 5\} \to \{0, 1, 2, 3, 4, 5\}$ defined by $g(x) = (x+2)(\text{mod } 6)$, where $x + 2$ denotes usual integer addition, is a permutation of $\{0, 1, 2, 3, 4, 5\}$.

48. Determine if $h: \{0,1,2,3,4,5,6\} \to \{0,1,2,3,4,5,6\}$ defined by $h(x) = x^2 (\text{mod} 7)$, where x^2 denotes usual integer multiplication, is a permutation of $\{0,1,2,3,4,5,6\}$.
49. Write $\alpha \circ \alpha$, $\beta \circ \beta$, $\alpha \circ \beta$, and $\beta \circ \alpha$ as products of disjoint cycles for the $\alpha, \beta \in S_7$ defined in Example 2.33.
50. For the permutations $\alpha = (127)(38)(456)$ and $\beta = (23568)(147)$ in S_8, write $\alpha \circ \alpha$, $\beta \circ \beta$, $\alpha \circ \beta$, and $\beta \circ \alpha$ as products of disjoint cycles.
51. For the permutations $\alpha = (1254)(36)$ and $\beta = (14678)$ in S_8, write $\alpha \circ \alpha$, $\beta \circ \beta$, $\alpha \circ \beta$, and $\beta \circ \alpha$ as products of disjoint cycles.
52. For the permutations $\alpha = (234)(567)$ and $\beta = (2358)(147)$ in S_8, write α^2, α^{-1}, $\alpha \circ \beta$, and $\beta \circ \alpha^{-1}$ as products of disjoint cycles.
53. For the permutations $\alpha = (1356)$ and $\beta = (124)(35)$ in S_6, write α^2, β^{-1}, $\alpha \circ \beta$, and $\beta^{-1} \circ \alpha$ as products of disjoint cycles.
54. For the permutations $\alpha = (15)(278)(34)$ and $\beta = (12)(368)$ in S_8, write $\alpha^2 \circ \beta$, $\beta^{-1} \circ \alpha$, $\alpha^{-2} \circ \beta$, and β^{-3} as products of disjoint cycles.
55. For the permutations $\alpha = (468)$ and $\beta = (1234)(58)$ in S_8, write $\alpha^2 \circ \beta$, $\beta^{-1} \circ \alpha$, $\alpha^{-2} \circ \beta$, and β^{-3} as products of disjoint cycles.
56. For the permutations $\alpha = (2367)(15)$ and $\beta = (1467)(23)$ in S_7, write α^3, $\beta \circ \alpha^2$, $\alpha^{-1} \circ \beta^2$, and $\beta \circ \alpha$ as products of disjoint cycles.
57. Show that in S_k $(k > 1)$ two disjoint cycles will commute. That is, if $\alpha = (a_1 a_2 \dots a_n)$ and $\beta = (b_1 b_2 \dots b_m)$ are cycles with $\{a_1, a_2, \dots, a_n\} \cap \{b_1, b_2, \dots, b_m\} = \emptyset$, then $\alpha \circ \beta = \beta \circ \alpha$.

In exercises 58-64, write ω as a product of transpositions to determine if ω is even or odd.

58. $\omega = (1234)(5678) \in S_8$.
59. $\omega = (23456)(17) \in S_8$.
60. $\omega = (135)(2678) \in S_8$.
61. $\omega = (123)(456)(789) \in S_9$.
62. $\omega = (2345678) \in S_8$.
63. $\omega = (13)(2468) \in S_8$.
64. $\omega = (14)(356) \in S_7$.
65. Find an odd permutation of S_8 which (when written as a product of disjoint cycles) includes a cycle of length 5. What is its order?
66. Write the permutation $\omega = (1234)(678) \in S_8$ as a product of transpositions two different ways. Be sure to check that the product of the transpositions really does give you the same permutation.
67. Write the permutation $\omega = (127)(348)(56) \in S_8$ as a product of transpositions two different ways. Be sure to check that the product of the transpositions really does give you the same permutation.
68. Write the permutation $\omega = (13)(456)(27) \in S_7$ as a product of transpositions two

different ways. Be sure to check that the product of the transpositions really does give you the same permutation.

69. Prove: For any $k > 1$, a cycle in S_k of length $m > 1$ is even (as a permutation) if and only if m is odd (as an integer).
70. Prove: For any $k > 1$ and any permutation $\alpha \in S_k$, α^2 must be an even permutation.
71. Prove: If $k \geq 3$ and $\alpha \in S_k$ is a cycle of length 3, $\alpha = (a_1a_2a_3)$, then $ord(\alpha) = 3$.
72. Let $k > 1$ and $\alpha \in S_k$ a cycle of length $m > 1$. Prove: α^2 is a cycle if and only if m is odd.

Projects for Chapter 2

Project 2.1

Consider the group $(U(15), \cdot_{15})$ for this project.

1. Find the elements of $U(15)$ and create the Cayley table for $U(15)$.
2. For each element $a \in U(15)$ find the cyclic subgroup generated by a, $\langle a \rangle$.
3. Is $U(15)$ a cyclic group? (Why or why not?)
4. Calculate powers to find $ord(2)$, $ord(11)$, $ord(7)$ in $U(15)$. Do you notice anything interesting relating the cyclic subgroups and orders of the elements?
5. Many students assume that they can find all subgroups of a group when they compute the cyclic subgroups. To see this is incorrect, consider the set $H = \{1, 4, 11, 14\}$. Create the Cayley table for H and explain why it is a subgroup of $U(15)$ that is **not cyclic**.
6. Can you find any other subgroup of $U(15)$ that is not cyclic and different from H?

Project 2.2

(This was originally published in PRIMUS; see [6] for more.) An important part of the definition for a direct product of groups (see Definition 1.15) is that the operation must act component-wise, with no interaction between the components. To illustrate this we will look at the two groups $\mathbb{Z} \times \mathbb{Z}_3$ and $\mathbb{Z} \circ \mathbb{Z}_3$. Both of these groups have exactly the same elements, $\{(a, b) : a \in \mathbb{Z}, b \in \mathbb{Z}_3\}$. We use the component-wise operation on $\mathbb{Z} \times \mathbb{Z}_3$ with $(a, b) * (x, y) = (a + x, b +_3 y)$.

The group $\mathbb{Z} \circ \mathbb{Z}_3$ has a special operation defined below. See Projects 1.3 and 1.4 for more about such operations.

$$(a, b) \circ (x, y) = (a + x + c, \ \ b +_3 y) \qquad \text{with } c = \begin{cases} 0 & if\ b + y < 3 \\ 1 & if\ b + y \geqslant 3 \end{cases}$$

1. Find a nonidentity element of $\mathbb{Z} \times \mathbb{Z}_3$ with finite order and an element of infinite order. Be sure to explain why they have the correct orders.
2. Explain why an infinite cyclic group cannot have a nonidentity element of finite order. What can we conclude from question 1?
3. Consider the element $x = (-1, 2) \in \mathbb{Z} \circ \mathbb{Z}_3$. Using the correct operation compute the elements x^2, x^3, x^4, x^5, x^6. Do you see a pattern? (If not, calculate more powers of x.)
4. For any positive integer m, determine what power n will give:
   $(-1, 2)^n = (-m, 0)$
   $(-1, 2)^n = (-m, 1)$
   $(-1, 2)^n = (-m, 2)$
5. Find the inverse of $x = (-1, 2)$ in $\mathbb{Z} \circ \mathbb{Z}_3$, i.e., find x^{-1}. The identity in both groups is (0,0).
6. Continue with the element $x = (-1, 2)$ in $\mathbb{Z} \circ \mathbb{Z}_3$. Using the correct operation compute the elements $x^{-2}, x^{-3}, x^{-4}, x^{-5}, x^{-6}$. (Remember that $x^{-n} = (x^{-1})^n$.) Do you see a pattern?
7. For any positive integer m, determine what power n will give:
   $(-1, 2)^n = (m, 0)$
   $(-1, 2)^n = (m, 1)$
   $(-1, 2)^n = (m, 2)$
8. Explain why questions 4 and 7 tell us that $\mathbb{Z} \circ \mathbb{Z}_3$ is a cyclic group.
9. Explain why $\mathbb{Z} \times \mathbb{Z}_3$ and $\mathbb{Z} \circ \mathbb{Z}_3$ cannot be different ways of denoting the same group.

Project 2.3

The group $\mathbb{Z}_2 \times \mathbb{Z}_5$ is the set of ordered pairs (m, k) with $m \in \mathbb{Z}_2$ and $k \in \mathbb{Z}_5$. Its operation is $(m, k) * (u, v) = (m +_2 u, k +_5 v)$.

1. Circle all of the logical flaws in the following "proof," and explain why any circled step is incorrect. Remember to look at the logic of each step to see if it follows from the previous steps, but a mistake does not make each step following it incorrect.

   Proposition: If G is a cyclic group and $|G| = 10$ then $G \cong \mathbb{Z}_2 \times \mathbb{Z}_5$.

   Proof: Assume that G is a cyclic group with exactly 10 elements, and let $G = \langle a \rangle$, so $G = \{e_G,\ a,\ a^2,\ a^3,\ a^4,\ a^5,\ a^6,\ a^7,\ a^8,\ a^9,\ a^{10}\}$.

   Define the function $f : \mathbb{Z}_2 \times \mathbb{Z}_5 \to G$ by $f(m, k) = a^{5m+k}$. The number $5m + k$ is found with usual addition and multiplication on $\mathbb{Z}$.

   Claim: f is one to one.
   Suppose we have $(m, k), (u, v) \in \mathbb{Z}_2 \times \mathbb{Z}_5$ with $(m, k) = (u, v)$, then $m = u$ and $k = v$. Thus $5m = 5u$, so $5m + k = 5u + v$. Therefore $a^{5m+k} = a^{5u+v}$ and $f(m, k) = f(u, v)$.

Claim: f is onto.
Suppose $x \in G$. We know $x = a^n$ for some integer $0 \leq n \leq 9$. Using the division algorithm we can write $n = 5m + k$ with $m, k \in \mathbb{Z}$ and $0 \leq k < 5$. Since the distinct powers of a have $n < 10$, we must have $m = 0$ or $m = 1$. Thus $x = a^{5m+k}$ with $m \in \mathbb{Z}_2$ and $k \in \mathbb{Z}_5$. Thus we have found $(m, k) \in \mathbb{Z}_2 \times \mathbb{Z}_5$ with $f(m, k) = x$ as needed.

Claim f is a homomorphism.
Let $(m, k), (u, v) \in \mathbb{Z}_2 \times \mathbb{Z}_5$. Then $f((m, k)*(u, v)) = f(m+_2u, k+_5v) = a^{5(m+_2u)+(k+_5v)}$, and $f(m, k)f(u, v) = a^{5m+k}a^{5u+v} = a^{5(m+u)+(k+v)}$, so we have $f((m, k) * (u, v)) = f(m, k)f(u, v)$.
Therefore f is an isomorphism and the groups are isomorphic.

2. Is the Proposition true? If so find a correct proof, but if not find a counterexample.

Project 2.4

Consider an equilateral triangle.

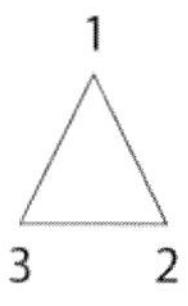

There are six ways we can move the triangle that will not effect its appearance.

- a: Rotate the corners clockwise one position.
- b: Rotate the corners clockwise two positions (or counter-clockwise one position).
- c: Flip the triangle over the vertical line through the top vertex (1).
- d: Flip the triangle over the diagonal line through the vertex to the left (3) and which bisects the opposite side.
- e: Don't move it at all.
- f: Flip the triangle over the diagonal line through the vertex to the right (2) and which bisects the opposite side.

Consider the vertices labeled clockwise from the top, 1, 2, 3 as shown. Each motion described above defines a permutation of the set $A = \{1, 2, 3\}$, denoted by $\begin{pmatrix} 1 & 2 & 3 \\ a & b & c \end{pmatrix}$ with the bottom row giving the position the vertex ended in. For example, $b = \begin{pmatrix} 1 & 2 & 3 \\ 3 & 1 & 2 \end{pmatrix}$.

1. Write each of the motions in permutation notation.

$$a = \qquad b = \qquad c =$$
$$d = \qquad e = \qquad f =$$

2. Create the Cayley table for S_A under the operation of composition.
3. Determine the inverse for each element.

$$a^{-1} = \qquad b^{-1} = \qquad c^{-1} =$$
$$d^{-1} = \qquad e^{-1} = \qquad f^{-1} =$$

4. Find each of the following elements in S_A.

$a^{-1}bc^{-1} =$ $\qquad$ $f^{-1}db =$

$(bcd)^{-1} =$ $\qquad$ $(c^{-1}d)(cf^{-1}) =$

Project 2.5

Consider a square as below:

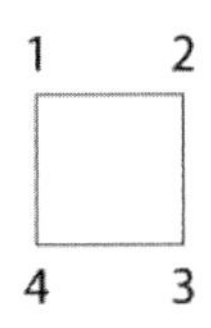

There are eight possible motions of the square. (Three of them will involve rotating.) We will denote the set of motions by $G = \{a, b, d, c, e, f, g, h\}$.

1. Describe each motion of the square in permutation notation as we did for the triangle in Project 2.4. Also describe each motion in words.

$$a = \qquad e =$$
$$b = \qquad f =$$
$$c = \qquad g =$$
$$d = \qquad h =$$

2. Create the Cayley table for this group under composition.
3. Is this the same as S_4?
4. Now consider the symmetries of a rectangle (which is not a square). Describe each motion both in words, and in permutation notation as in #1. Then find the Cayley table for the symmetries as was done for the square.
5. How does this group of symmetries relate to the symmetries of a square? Why does this correspondence occur?

Project 2.6

Consider the following permutations in $(S_7, \circ)$.

$$\alpha = \begin{pmatrix} 1 & 2 & 3 & 4 & 5 & 6 & 7 \\ 4 & 7 & 1 & 3 & 2 & 5 & 6 \end{pmatrix} \beta = \begin{pmatrix} 1 & 2 & 3 & 4 & 5 & 6 & 7 \\ 7 & 3 & 2 & 4 & 6 & 5 & 1 \end{pmatrix}$$

$$\gamma = \begin{pmatrix} 1 & 2 & 3 & 4 & 5 & 6 & 7 \\ 3 & 4 & 5 & 7 & 1 & 6 & 2 \end{pmatrix}$$

1. Write each of the permutations below as a product of disjoint cycles.

   $\alpha =$ $\qquad$ $\alpha^{-1} =$
   $\beta =$ $\qquad$ $\beta^{-1} =$
   $\gamma =$ $\qquad$ $\gamma^{-1} =$

2. Write each of the following permutations as a product of disjoint cycles.

   $\alpha\beta\gamma =$
   $\alpha^2\gamma =$
   $\gamma\alpha^{-1}\beta =$

3. Write each of the permutations α, β, and γ, as a product of transpositions and determine if it is even or odd.

   $\alpha =$
   $\beta =$
   $\gamma =$

4. Let ε denote the identity permutation in S_7, and determine if $G = \{\varepsilon, \alpha, \beta, \gamma\}$ is a subgroup of S_7. Either prove that it is or show how one of the properties fails to hold.

Project 2.7

Given a subset A of a group G, we say A generates G if every element of G can be written as a product of powers of members of A.

For example, consider the group $\mathbb{Z}_2 \times \mathbb{Z}_2 = \{(0,0), (0,1), (1,0), (1,1)\}$ with the usual operation $(a,b) * (u,v) = (a +_2 u, b +_2 v)$. This group is not cyclic as seen after Example 2.14. But if we look at the set $A = \{(1,0), (0,1)\}$, then (using the given operation when it refers to "product") we see the following:

$$\begin{aligned} (0,0) &= (1,0) * (1,0) \\ (1,0) &= (1,0)^1 \\ (0,1) &= (0,1)^1 \\ (1,1) &= (1,0) * (0,1) \end{aligned}$$

Thus each element can be written as a product of powers of elements of A. Thus the set $A = \{(1,0),(0,1)\}$ generates $\mathbb{Z}_2 \times \mathbb{Z}_2$.

Now consider as our group, $(S_k, \circ)$ for the rest of this project and assume $k > 3$.

1. Explain why the set of all transpositions generates S_k.
2. Show that the set $T = \{(1m) : 2 \leq m \leq k\}$ generates S_k.
3. Find a permutation in S_5 that cannot be found as a composition of cycles of length 3, showing that we cannot use the set of all 3-cycles to generate S_5.

Chapter 3

Quotient Groups

The idea of creating new groups out of previously defined ones is very natural, as we saw in Definition 1.15, but Quotient Groups do not always seem natural when first introduced. Quotient Groups are not only useful for understanding group structures, but with them we can describe all possible groups K for which an onto homomorphism $f : G \to K$ exists.

3.1 Cosets

For all that we do here, G denotes a group and H a subgroup of G.

Definition 3.1 For each $a \in G$ define $aH = \{ax : x \in H\}$ where ax is calculated using the operation in G. The set aH is called a ***left coset of H*** in G. If the operation in G is written additively the notation $a + H$ will be used for a left coset.

A ***right coset of H*** in G is a set of the form $Ha = \{xa : x \in H\}$. (With $H + a$ used if the operation is additive.)

Example 3.2 Let G denote the group $(\mathbb{Z}_{12}, +_{12})$. We know $H = \{0, 4, 8\}$ is a subgroup of G since $H = \langle 4 \rangle$. The left cosets of H in G are:

$$\begin{array}{llll} 0 + H = \{0, 4, 8\} & 1 + H = \{1, 5, 9\} & 2 + H = \{2, 6, 10\} & 3 + H = \{3, 7, 11\} \\ 4 + H = \{4, 8, 0\} & 5 + H = \{5, 9, 1\} & 6 + H = \{6, 10, 2\} & 7 + H = \{7, 11, 3\} \\ 8 + H = \{8, 0, 4\} & 9 + H = \{9, 1, 5\} & 10 + H = \{10, 2, 6\} & 11 + H = \{11, 3, 7\}. \end{array}$$

Example 3.3 Let G denote the group $(S_3, \circ)$, and $H = \{\varepsilon, (12)\}$. Since $(12)(12) = \varepsilon$ and $(12)^{-1} = (12)$, then H is closed under composition and inverses. Thus H is a subgroup of G. The left cosets of H in G are:

$$\begin{array}{lcllcl} \varepsilon H & = & \{\varepsilon, (12)\} & (13)H & = & \{(13), (123)\} \\ (23)H & = & \{(23), (132)\} & (12)H & = & \{(12), \varepsilon\} \\ (123)H & = & \{(123), (13)\} & (132)H & = & \{(132), (23)\}. \end{array}$$

In each of the previous examples we used **left** cosets. We could have used right cosets instead, but must choose to use only left or only right cosets in a problem. Most of the theorems dealing with cosets (in this chapter) will be stated using left cosets.

There are several interesting properties to notice about the cosets found in the previous examples. There are repeated cosets, two cosets are either disjoint or equal, all have the same cardinality, and more.

Theorem 3.4

Let G be a group and H a subgroup of G.

(i) For all $a \in G$, $|aH| = |H|$.
(ii) For all $a, b \in G$, either $aH = bH$ or $aH \cap bH = \emptyset$.
(iii) $\bigcup_{a \in G} aH = G$.

This theorem tells us the left cosets of H form a **partition of** G as in Definition 0.10.

Proof Let G be a group and H a subgroup of G.

(i) Let $a \in G$. We want to show that the coset aH has the same cardinality as H, thus by Theorem 0.15 we need to find a bijective function from H to aH.

$$\text{Define } f : H \to aH \text{ by } f(x) = ax \text{ for every } x \in H.$$

To show that f is one to one, let $x, y \in G$ with $f(x) = f(y)$; in other words, $ax = ay$. Since $a, x, y \in G$ by the Cancellation Law (Theorem 1.20), $x = y$, and so f is one to one.

To show that f is onto, let $z \in aH$. Thus $z = ah$ for some $h \in H$. Now we see that $f(h) = ah = z$ so f is onto. As f is a bijection, $|aH| = |H|$.

(ii) Let $a, b \in G$. We must show that either $aH = bH$ or $aH \cap bH = \emptyset$. To prove a statement of the form '*u or v*' it is equivalent to prove '*if not v then u*'. Hence we will assume that $aH \cap bH \neq \emptyset$ and we need to prove that $aH = bH$, which will be done with subsets.

($\subseteq$) Let $y \in aH$, thus $y = aq$ for some $q \in H$. Since $aH \cap bH \neq \emptyset$, there is some element $x \in aH \cap bH$. Now $x = ah$ and $x = bk$ for some $h, k \in H$. As $ah = bk$, we have $a = bkh^{-1}$ so $y = aq = (bkh^{-1})q = b(kh^{-1}q)$. Since H is a subgroup of G with $h, k, q \in H$, $kh^{-1}q \in H$, thus we have $y \in bH$ and $aH \subseteq bH$.

($\supseteq$) You should verify $aH \supseteq bH$ for practice. Thus $aH = bH$ as we needed to show.

(iii) This part of the theorem follows very quickly from the fact that the identity e_G is in H. For any $x \in G$, $x = xe_G$ and so $x \in xH$. Thus $x \in \bigcup_{a \in G} aH$, so $G \subseteq \bigcup_{a \in G} aH$. Since each coset aH is a subset of the group G it is clear that $\bigcup_{a \in G} aH \subseteq G$, therefore $\bigcup_{a \in G} aH = G$. □

Looking back at Example 3.3 we notice that every coset has exactly two elements, the same as H, following (i) of the previous theorem. Also notice $(132) \in (23)H$ and $(23)H = (132)H$, which illustrates (ii) of Theorem 3.4.

Example 3.5 The fact that H is a subgroup of G when we define cosets is critical for (ii) in Theorem 3.4. Again, consider G to be $(\mathbb{Z}_{12}, +_{12})$ and the set $H = \{2, 9, 11\}$. H is not a subgroup of G so consider what "cosets" it might make:

$$\begin{array}{llll} 0+H=\{2,9,11\} & 1+H=\{3,10,0\} & 2+H=\{4,11,1\} & 3+H=\{5,0,2\} \\ 4+H=\{6,1,3\} & 5+H=\{7,2,4\} & 6+H=\{8,3,5\} & 7+H=\{9,4,6\} \\ 8+H=\{10,5,7\} & 9+H=\{11,6,8\} & 10+H=\{0,7,9\} & 11+H=\{1,8,10\}. \end{array}$$

Clearly, (ii) of Theorem 3.4 fails. We see that $0 \in 1+H$ and $0 \in 3+H$ so $(1+H) \cap (3+H) \neq \emptyset$, but $(1+H) \neq (3+H)$.

The properties in Theorem 3.4, as well as those of the next theorem, will help us when we need to calculate cosets. Consider the group G to be $(\mathbb{Z}_{10}, +_{10})$ and the subgroup $H = \langle 2 \rangle$. Notice that $0 + H = \{0, 2, 4, 6, 8\}$. Thus property (ii) in Theorem 3.4 tells us that $2 + H = 0 + H$ since 2 is a member of both cosets! Thus we need not calculate $2 + H$, $4 + H$, $6 + H$, or $8 + H$ since they must be the same as $0 + H$. Now $1 + H = \{1, 3, 5, 7, 9\}$ so in the same way $1 + H = 3 + H = 5 + H = 7 + H = 9 + H$. Thus to find the cosets of H in $\mathbb{Z}_{10}$, we only had to calculate two cosets, namely $0 + H$ and $1 + H$.

Theorem 3.6

Let G be a group and H a subgroup of G. Let $a, b \in G$.

(i) $aH = bH$ if and only if $a \in bH$.
(ii) $aH = bH$ if and only if $b^{-1}a \in H$.
(iii) $aH = H$ if and only if $a \in H$.

Proof Assume H is a subgroup of G and $a, b \in G$. Recall that in the proof of Theorem 3.4 we showed $x \in xH$ for all $x \in G$.

(i) Assume first that $aH = bH$. We know that $a \in aH$, but we assumed that $aH = bH$, so clearly $a \in bH$. Conversely, assume that $a \in bH$. As $a \in aH$ as well we have $a \in aH \cap bH$. Thus by Theorem 3.4 $aH = bH$.

(ii) Suppose $aH = bH$ again. From (i) we know that $a \in bH$, so there is $h \in H$ with $a = bh$. As $a, b, h \in G$ we multiply by b^{-1} to get $b^{-1}a = h$. Thus $b^{-1}a \in H$. Conversely, assume that $b^{-1}a \in H$. Then $b(b^{-1}a) \in bH$, but by associativity we have $a \in bH$. Hence $aH = bH$ by (i).

(iii) This follows immediately from (i) if we assume that $b = e_G$. It is clear that $e_G H = H$ so by (i) we have $aH = e_G H$ if and only if $a \in e_G H$, or $a \in H$. $\square$

Example 3.7 Consider the group $(\mathbb{Q}^*, \cdot)$, the nonzero rational numbers under multiplication. The set $H = \{2^n : n \in \mathbb{Z}\} = \langle 2 \rangle$ is a subgroup.

To decide if two elements determine the same coset we will use Theorem 3.6. Let $a = \frac{3}{8}$ and $b = 6$.

$$b^{-1}a = \tfrac{1}{6} \cdot \tfrac{3}{8} = \tfrac{1}{16} = 2^{-4}$$

Thus we have $b^{-1}a \in H$ so $\frac{3}{8}H = 6H$. Consider instead $c = \frac{1}{6}$.

$$c^{-1}a = 6 \cdot \tfrac{3}{8} = \tfrac{9}{4}$$

But $\frac{9}{4} \notin H$, so $\frac{3}{8}H \neq \frac{1}{6}H$. Notice that since $c = b^{-1}$, this also shows us $bH \neq b^{-1}H$!

Example 3.8 Consider the group $(U(21), \cdot_{21})$. A subgroup is $H = \langle 4 \rangle = \{1, 4, 16\}$ and so $1H = 4H = 16H$. We don't yet know $2H$, so we calculate $2H = \{2, 8, 11\} = 8H = 11H$. Again, looking for one not yet defined, we get $5H = \{5, 20, 17\} = 20H = 17H$. We still need to find cosets with 10, 13, and 19 as elements. Of course, $10H = \{10, 19, 13\} = 13H = 19H$. Thus we only need to compute $1H, 2H, 5H$, and $10H$ to be sure all cosets are found.

The most important theorem of this section relates the cardinality of a subgroup, $|H|$, in a *finite* group to the cardinality of the group, $|G|$, and uses cosets in the proof.

Theorem 3.9

Lagrange's Theorem Let G be a finite group, and H a subgroup of G. Then $|H|$ must evenly divide $|G|$. The quotient of these numbers is called the *index* of H in G, written $[G : H] = \frac{|G|}{|H|}$.

Proof Assume that G is a finite group and that H is a subgroup of G. By Theorem 3.4 we know that $\bigcup_{a \in G} aH = G$. But since G is finite this can be written as a finite union.

$$G = a_1 H \cup a_2 H \cup \cdots \cup a_n H$$

Again, by Theorem 3.4, any two cosets are either equal or disjoint. Hence we eliminate those that are repeats of other cosets and we have a disjoint union.

$$G = a_1 H \cup a_2 H \cup \cdots \cup a_r H$$

Rules about finite disjoint sets discussed in Chapter 0 help us calculate $|G|$.

$$|G| = |a_1 H| + |a_2 H| + \cdots + |a_r H|$$

By Theorem 3.4 we know $|a_i H| = |H|$ for each i so we can write $|G| = r\,|H|$. Since r is a positive integer, $|H|$ evenly divides $|G|$. □

Look back at Example 3.3. We know that $|S_3| = 6$ and for our subgroup $|H| = 2$. Thus we see that $|H|$ evenly divides $|S_3|$. Also Lagrange's Theorem tells us that there cannot exist a subgroup of S_3 with exactly four elements since 4 does not evenly divide 6.

Notice in the previous proof the number of *distinct* cosets, r, is equal to the index $[G : H]$. This fact will frequently be useful. Even with an infinite group G we will use (the index) $[G : H]$ to denote the number of cosets created by a subgroup H. But if G is infinite there may be infinitely many cosets. There are many consequences of Lagrange's Theorem, but we will only mention a few. Several parts of the next theorem are left as exercises at the end of the chapter.

Theorem 3.10

Consequences of Lagrange's Theorem Let G denote a finite group.

(i) If $a \in G$ then $ord(a)$ evenly divides $|G|$.
(ii) If $a \in G$ and $|G| = n$ then $a^n = e_G$.
(iii) If $|G|$ is prime then G is a cyclic group.
(iv) If $|G|$ is prime then $\{e_G\}$ and G are the only subgroups of G.

Proof Let G denote a finite group.

(i) Suppose $a \in G$ and let $H = \langle a \rangle$. Then H is a subgroup of G by Theorem 2.8. By Lagrange's Theorem $|H|$ must evenly divide $|G|$, but by Theorem 2.16 $|H| = ord(a)$. Thus $ord(a)$ evenly divides $|G|$. Proofs of parts (ii), (iii), and (iv) are left as exercises at the end of the chapter and each part uses (i) in the proof. □

Example 3.11 The previous theorem tells us that if we have a group G with $|G| = p$ for a prime number p, we can say by (iii) that G must be cyclic. But then by Theorem 2.18 we know that $G \cong \mathbb{Z}_p$ (using the operation of $+_p$ on $\mathbb{Z}_p$).

Also by (i), for any element $a \in G$ with $a \neq e_G$, $ord(a)$ must evenly divide p (which is prime). The only orders possible for a are 1 and p, but since $a \neq e_G$ then $ord(a) = p$. Thus by Theorem 0.15, $G = \langle a \rangle$. But a was **any** nonidentity element of G, which tells us that every nonidentity element can be used to generate G.

Consider $(\mathbb{Z}_5, +_5)$. Clearly, 5 is prime so we should find that all of $\langle 1 \rangle$, $\langle 2 \rangle$, $\langle 3 \rangle$, $\langle 4 \rangle$ will equal $\mathbb{Z}_5$. We can easily check this by computing the cyclic subgroups.

$$\langle 1 \rangle = \{1, 2, 3, 4, 0\} \qquad \langle 2 \rangle = \{2, 4, 1, 3, 0\}$$
$$\langle 3 \rangle = \{3, 1, 4, 2, 0\} \qquad \langle 4 \rangle = \{4, 3, 2, 1, 0\}$$

Example 3.12 We can use Theorem 3.10 to show that only two nonisomorphic groups of order 4 actually exist. Suppose we consider a group G with exactly 4 elements, $G = \{e_G, a, b, c\}$. For each $x \in G$ we know $ord(x)$ must be 1, 2, or 4 by Theorem 3.10 (i). The only element of order 1 is e_G, so we must look more closely at the other three elements, a, b, and c.

If $ord(a) = 4$ then by Theorem 2.16, $|\langle a \rangle| = 4$. But we know $\langle a \rangle \subseteq G$ and these finite sets have the same size so by Theorem 0.15, $G = \langle a \rangle$. Thus G is a cyclic group and by Theorem 2.18 G is isomorphic to $\mathbb{Z}_4$. Similarly, if either b or c has order 4 then G is isomorphic to $\mathbb{Z}_4$.

Now consider the only other possibility, that each of a, b, c has order 2. This tells us that $aa = e_G$, $bb = e_G$, and $cc = e_G$. According to Theorem 1.20 $ab \neq a$, $ab \neq b$, $ab \neq e_G$, $ba \neq a$, $ba \neq b$, and $ba \neq e_G$ so we must have $ab = c$ and $ba = c$. Similarly, we must have $ac = b$, $ca = b$, $bc = a$, and $cb = a$. Thus we have an abelian group with four elements that is not cyclic. The Cayley table will verify every property except associativity. You should check associativity, for example: $a(bc) = a(a) = e_G$ and $(ab)c = (c)c = e_G$, or $(ab)a = ca = b$ and $a(ba) = ac = b$.

	e_G	a	b	c
e_G	e_G	a	b	c
a	a	e_G	c	b
b	b	c	e_G	a
c	c	b	a	e_G

Consider the group $\mathbb{Z}_2 \times \mathbb{Z}_2 = \{(0,0), (0,1), (1,0), (1,1)\}$ described in Example 1.16, and define the function $f : G \to \mathbb{Z}_2 \times \mathbb{Z}_2$ below.

$$f = \begin{pmatrix} e_G & a & b & c \\ (0,0) & (0,1) & (1,0) & (1,1) \end{pmatrix}$$

This function is one to one and onto by inspection, thus we need only verify that f is a homomorphism to show that G is isomorphic to $\mathbb{Z}_2 \times \mathbb{Z}_2$. We begin with the Cayley table we constructed for G and when its entries are replaced according to f we find:

	$(0,0)$	$(0,1)$	$(1,0)$	$(1,1)$
$(0,0)$	$(0,0)$	$(0,1)$	$(1,0)$	$(1,1)$
$(0,1)$	$(0,1)$	$(0,0)$	$(1,1)$	$(1,0)$
$(1,0)$	$(1,0)$	$(1,1)$	$(0,0)$	$(0,1)$
$(1,1)$	$(1,1)$	$(1,0)$	$(0,1)$	$(0,0)$

This is the correct table for the group $\mathbb{Z}_2 \times \mathbb{Z}_2$ which was shown in Example 1.16, thus f is a homomorphism and G is isomorphic to $\mathbb{Z}_2 \times \mathbb{Z}_2$. Thus every group of order four is either isomorphic to $\mathbb{Z}_4$ or $\mathbb{Z}_2 \times \mathbb{Z}_2$ and there are only two possible groups with four elements.

Since the groups $\mathbb{Z}_2 \times \mathbb{Z}_2$ and $\mathbb{Z}_4$ are both abelian, we can also conclude that every group with exactly 4 elements is abelian.

3.2 Normal Subgroups

Cosets will form the elements of our quotient groups. Unfortunately, the operation we will define requires that the original subgroup used to create the cosets be "special" as described in the next definition.

Definition 3.13 Let G be a group and H a subgroup of G. H is ***closed under conjugates*** from G if for every $a \in H$ and $x \in G$ we have $x^{-1}ax \in H$. The element $x^{-1}ax$ is called a ***conjugate*** of a in G. (Note: xax^{-1} is also conjugate of a.)

Example 3.14 Consider the group of integers under usual addition $(\mathbb{Z}, +)$ and the subgroup $H = \langle 2 \rangle$, i.e., the set of all even integers. Is H closed under conjugates? Let $m \in \mathbb{Z}$ and $n \in H$. The inverse of m is written in additive notation as $-m$, so we look at the conjugate $(-m) + n + m$. As $\mathbb{Z}$ is abelian we know:

$$(-m) + n + m = -m + m + n = n.$$

Since $n \in H$ then $(-m) + n + m \in H$ and H is closed under conjugates from $\mathbb{Z}$.

Example 3.15 Continuing with the group in Example 3.3, $H = \{\varepsilon, (12)\}$ is a subgroup of S_3. Consider the elements $x = (123)$ and $a = (12)$. Then $x \in S_3$ and $a \in H$. As $x^{-1} = (321)$ then $x^{-1}ax = (321)(12)(123) = (13)$ which is not in H. Thus H is *not* closed under conjugates.

Definition 3.16 Let G be a group and H a subgroup of G. H is a ***normal subgroup*** of G if and only if H is closed under conjugates from G. We write $H \triangleleft G$ to denote that H is a normal subgroup of G.

In Example 3.14, we found that $\langle 2 \rangle \triangleleft \mathbb{Z}$. In an exercise at the end of the chapter you will show that for any group G, $\{e_G\} \triangleleft G$ and $G \triangleleft G$.

Checking every possible conjugate of elements of H in G can be very time consuming, especially with an infinite group. The next theorem helps reduce the workload by looking at cosets instead of individual conjugates.

Theorem 3.17

Let G be a group and H a subgroup of G. $H \triangleleft G$ if and only if $aH = Ha$ for every $a \in G$.

Proof Suppose that G is a group and H is a subgroup of G.

($\rightarrow$) First, assume that $H \triangleleft G$ and that $a \in G$. We will show that $aH = Ha$, using subsets. Let $x \in aH$, then $x = ah$ for some $h \in H$. Thus $xa^{-1} = aha^{-1}$, but since $H \triangleleft G$ we know $aha^{-1} \in H$ and so $xa^{-1} \in H$. If we let $xa^{-1} = k$ then $x = ka$ shows $x \in Ha$ and $aH \subseteq Ha$. You should verify that $Ha \subseteq aH$ as well, and thus $aH = Ha$ as needed.

($\leftarrow$) This time assume $aH = Ha$ for *every* $a \in G$. We need to show that $H \triangleleft G$, i.e., that H is closed under conjugates. Let $h \in H$ and $b \in G$. We need to show that $b^{-1}hb \in H$. We know $hb \in Hb$, but we assumed that $aH = Ha$ for every $a \in G$, so we must have $bH = Hb$. Thus $hb \in bH$ as well. Thus there is some $k \in H$ with $hb = bk$. Multiplying by b^{-1} we see that $b^{-1}hb = k$ so $b^{-1}hb \in H$. Thus H is closed under conjugates, and $H \triangleleft G$. □

Recall that in Example 3.15 we found $H = \{\varepsilon, (12)\}$ is not closed under conjugates from S_3. With cosets we find $(13)H = \{(13), (123)\}$ but $H(13) = \{(13), (132)\}$. Thus $(13)H \neq H(13)$, consistent with the fact that H is not closed under conjugates.

For some groups the question of whether a subgroup is normal is easily solved. The proof of the next theorem is left as an exercise at the end of the chapter.

Theorem 3.18

If G is an abelian group then every subgroup of G is a normal subgroup.

The theorem above applies to many of our familiar groups including the integers, real numbers, and rational numbers under addition, as well as $(\mathbb{Z}_n, +_n)$ and $(U(n), \cdot_n)$ for $n > 1$.

Example 3.19 Let G denote the group $U(10) = \{1, 3, 7, 9\}$ using $\cdot_{10}$ and $H = \{1, 9\}$. Since $1 \cdot_{10} 1 = 1$ and $9 \cdot_{10} 9 = 1$ then H is closed under $\cdot_{10}$. Also each of 1, 9 is its own inverse, so H is a subgroup. As $U(10)$ is abelian we can conclude $H \triangleleft U(10)$ by Theorem 3.18.

Sometimes just knowing the size, or cardinality, of a subgroup in a finite group can guarantee it is normal as seen below. This proof easily generalizes to the situation where G is infinite, H is a subgroup of G, and $[G : H] = 2$.

Theorem 3.20

Let G be a finite group and H a subgroup of G for which $|H| = \frac{1}{2}|G|$. Then $H \triangleleft G$.

Proof Suppose that G is a finite group with $|G| = n$, and H is a subgroup of G with $|H| = \frac{1}{2}n$. We will show that $H \triangleleft G$ using Theorem 3.17. Suppose $a \in G$, then we need to show $aH = Ha$. If $a \in H$ then clearly $aH = H = Ha$. So assume instead that $a \notin H$. By Lagrange's Theorem (3.9), $[G : H] = 2$ and thus there are exactly two distinct right cosets, and exactly two distinct left cosets. As $a \notin H$ we know $aH \neq H$ but $G = aH \cup H$, so $aH = G - H$. Similarly, $Ha \neq H$ so that $Ha = G - H = aH$. Therefore for every $a \in G, aH = Ha$ and by Theorem 3.17 we can conclude $H \triangleleft G$. $\square$

The theorem above together with Theorem 2.40 tells us for each $k > 1$, $A_k \triangleleft S_k$, a fact that will be useful in the last chapters of the text.

3.3 Quotient Groups

Given a group G and subgroup H we already know how to find the set of left cosets of H, which we denote by ${}^{G}/_{H}$ and in words say G ***mod*** H. We now define the operation that will create a group of cosets, a *quotient group.*

Definition 3.21 Let G be a group and H a subgroup of G. Define the mapping $*: {}^{G}/_{H} \times {}^{G}/_{H} \to {}^{G}/_{H}$ by $(aH) * (bH) = (ab)H$.

Note that $(ab)H$ means that G's operation has been used on a and b, then the resulting element is used to create a coset of H.

Example 3.22 Let G denote the group $(\mathbb{Z}_{12}, +_{12})$. Also consider the subgroup $H = \{0, 4, 8\}$. We found the distinct cosets of H in Example 3.2, so ${}^{G}/_{H} = \{0 + H, 1 + H, 2 + H, 3 + H\}$. Consider the Cayley table using the mapping $*$ defined above:

$*$	$0+H$	$1+H$	$2+H$	$3+H$
$0+H$	$0+H$	$1+H$	$2+H$	$3+H$
$1+H$	$1+H$	$2+H$	$3+H$	$0+H$
$2+H$	$2+H$	$3+H$	$0+H$	$1+H$
$3+H$	$3+H$	$0+H$	$1+H$	$2+H$

This table should look familiar; if you cover every "+H" it appears to be $\mathbb{Z}_4$.

Example 3.23 Let G denote the group S_3 and $H = \{\varepsilon, (12)\}$. We found the distinct cosets of H in Example 3.3, ${}^{G}/_{H} = \{\varepsilon H, (13)H, (23)H\}$. Again, consider the Cayley table the mapping would give:

$*$	εH	$(13)H$	$(23)H$
εH	εH	$(13)H$	$(23)H$
$(13)H$	$(13)H$	εH	$(23)H$
$(23)H$	$(23)H$	$(13)H$	εH

Notice that $\varepsilon H * (13)H = (23)H * (13)H$, which would cause the Cancellation Law (Theorem 1.20) to fail. Thus we do not have a group.

Recall that the subgroup H in the previous example is not closed under conjugates (Example 3.3) and thus is not a normal subgroup of S_3. The mapping $*$ fails a very important part of the definition of an operation; it does not always give a unique answer to $(aH)*(bH)$. This can be made clear by remembering from Example 3.15 that we have (13)H = (123)H, so the answers to $(13)H * (23)H$ and $(123)H * (23)H$ should be the same.

$$(13)H * (23)H = ((13)(23))H = (132)H$$

$$(123)H * (23)H = ((123)(23))H = (12)H$$

Since $(132)H \neq (12)H$ we see that $*$ is not an operation on ${}^{S_3}/_{H}$.

Theorem 3.24

Let G be a group and H a subgroup of G. If $H \lhd G$ then ${}^{G}/_{H}$ is a group under the mapping $*$ from Definition 3.21. ${}^{G}/_{H}$ is called the ***quotient group***, G mod H.

Proof Suppose that G is a group and $H \lhd G$. We know that ${}^{G}/_{H} = \{aH : a \in G\}$ and for any $a, b \in G$, $aH * bH = (ab)H$. To show that we have a group, we must verify that:

(i) $*$ is an operation on ${}^{G}/_{H}$.
(ii) $*$ is associative on ${}^{G}/_{H}$.
(iii) There is an identity in ${}^{G}/_{H}$.
(iv) Every element of ${}^{G}/_{H}$ has an inverse in ${}^{G}/_{H}$.

(i) To be an operation on ${}^{G}/_{H}$, $*$ must satisfy Definition 1.1. It is clear by the definition of $*$ that $aH * bH$ is defined for every pair of elements in ${}^{G}/_{H}$, and that the answer is again in ${}^{G}/_{H}$. The only part that is not immediate is whether or not this answer is uniquely defined, as discussed before this theorem. Thus we need to be sure that if $aH = cH$ and $bH = dH$, then $(ab)H = (cd)H$.

Assume we have $a, b, c, d \in G$ so that $aH = cH$ and $bH = dH$. To prove $(ab)H = (cd)H$ we will use Theorem 3.6, i.e., we will show that $(cd)^{-1}(ab) \in H$, or $d^{-1}c^{-1}ab \in H$.

From our assumptions, and Theorem 3.6, we know that $a \in cH$ so $a = ch$ for some $h \in H$ and

$$d^{-1}c^{-1}ab = d^{-1}c^{-1}(ch)b = d^{-1}hb.$$

Also $b \in dH$, and so $b = dk$ for some $k \in H$. Thus

$$d^{-1}c^{-1}ab = d^{-1}hb = d^{-1}h(dk) = (d^{-1}hd)k.$$

Notice that $d^{-1}hd \in H$ since $H \lhd G$, so by $k \in H$ we have $d^{-1}c^{-1}ab \in H$ as needed. Thus $(ab)H = (cd)H$ and $*$ is an operation on ${}^{G}/_{H}$.

(ii) For any $a, b, c \in G$, the following two equations hold.

$$(aH * bH) * cH = (ab)H * cH = ((ab)c)H$$

$$aH * (bH * cH) = aH * (bc)H = (a(bc))H$$

But the operation in G is associative so $(ab)c = a(bc)$, giving us $((ab)c)H = (a(bc))H$ and $*$ is associative on ${}^{G}/_{H}$.

(iii and iv) The identity of ${}^{G}/_{H}$ is $e_G H$ and $(aH)^{-1} = a^{-1}H$ as you should be able to easily verify.

Thus ${}^{G}/_{H}$ is a group under $*$ when $H \lhd G$. □

The quotient group and original group are closely related, but will not always have identical properties. For example, ${}^{G}/_{H}$ can be abelian even if G is not abelian.

Example 3.25 Consider the permutation group S_4 which we know is *not* abelian since $(123)(124) = (13)(24)$ and $(124)(123) = (14)(23)$. However, $|A_k| = \frac{1}{2}|S_n|$ by Theorem 2.40, and so $A_4 \lhd S_4$ by Theorem 3.20. Thus ${}^{S_4}/_{A_4}$ is a group with exactly two elements. This tells us that ${}^{S_4}/_{A_4}$ is a cyclic group by Theorem 3.10. Every cyclic group is abelian by Theorem 2.13, so even though S_4 is not abelian the quotient group ${}^{S_4}/_{A_4}$ is abelian.

Example 3.26 Consider the group G as $(\mathbb{Z}_{12}, +_{12})$ and subgroup $H = \{0, 3, 6, 9\}$. You should verify that $H = \langle 3 \rangle$. The quotient group ${}^{G}/_{H} = \{0 + H, 1 + H, 2 + H\}$ has the Cayley table shown.

$*$	$0+H$	$1+H$	$2+H$
$0+H$	$0+H$	$1+H$	$2+H$
$1+H$	$1+H$	$2+H$	$0+H$
$2+H$	$2+H$	$0+H$	$1+H$

Using this table we can calculate $(2 + H)^3 = (2 +_{12} 2 +_{12} 2) + H = 0 + H$. Thus since $2 + H \neq 0 + H$, and $(2 + H)^2 = 4 + H \neq 0 + H$, we have $ord(2 + H) = 3$ in ${}^{G}/_{H}$, even though $ord(2) = 6$ in G.

Some properties of the quotient group ${}^{G}/_{H}$ are always inherited from G as seen in the next theorem.

Theorem 3.27

Let G be a group and $H \lhd G$.

(i) If G is abelian then ${}^{G}/_{H}$ is abelian.
(ii) If $x \in G$ has $ord(x) = n$ then, in the group ${}^{G}/_{H}$, $ord(xH)$ must evenly divide n.
(iii) If G is cyclic then ${}^{G}/_{H}$ is cyclic.

Proof Assume that G is a group and $H \lhd G$.

(i) Left as an exercise at the end of the chapter.

(ii) Suppose that $x \in G$ with $ord(x) = n$. As $x^n = e_G$ then by using the definition of $*$ repeatedly we see that $(xH)^n = (x^n)H = (e_G)H$ so $(xH)^n = e_{G/H}$. Thus Theorem 1.26 tells us that $ord(xH)$ must evenly divide n.

(iii) Suppose G is cyclic, say $G = \langle a \rangle$. We will show that ${}^G/_H = \langle aH \rangle$. Clearly, $\langle aH \rangle \subseteq {}^G/_H$, thus we need only show that $\langle aH \rangle \supseteq {}^G/_H$. Let xH be an element of ${}^G/_H$. As x is an element of G, then $x = a^n$ for some integer n. Now we have $xH = (a^n)H = (aH)^n$. Thus $xH \in \langle aH \rangle$ and $\langle aH \rangle \supseteq {}^G/_H$, so ${}^G/_H = \langle aH \rangle$ and ${}^G/_H$ is cyclic. □

Recall from Lagrange's Theorem (3.9) that the size of a normal subgroup of a finite group must evenly divide the cardinality of the group. This gives us the next important fact, whose proof is actually embedded in the proof of Theorem 3.9.

Theorem 3.28

Let G be a <u>finite</u> group, and $H \triangleleft G$. Then $\left|{}^G/_H\right| = \frac{|G|}{|H|}$.

There are many other facts that can be proven about quotient groups and how they relate to the original groups, and some of them will be explored in exercises or seen later in the text. We turn our attention now to using quotient groups and homomorphisms together.

3.4 The Fundamental Homomorphism Theorem

Group homomorphisms were introduced in section 1.6 but were not always easy to find. In this section we will see how quotient groups can help us understand what relationships must occur between two groups for a homomorphism to exist between them.

Definition 3.29 Let G and K be groups with $f : G \to K$, a homomorphism.

(i) The ***kernel*** of f is the set $ker(f) = \{x \in G : f(x) = e_K\}$.
(ii) The ***image*** of f is the set $f(G) = \{y \in K :$ there exists some $x \in G$ with $f(x) = y\}$.

Notice that with these definitions, $ker(f) \subseteq G$, while $f(G) \subseteq K$. It is critical to keep track of what group our elements are in to use the correct operation.

Example 3.30 Consider the groups $(\mathbb{Z}, +)$ and $(\mathbb{Z}_4, +_4)$. Define the function $f : \mathbb{Z} \to \mathbb{Z}_4$ by $f(a) = a(\text{mod}4)$. Thus we take each element of the integers and find the remainder after dividing by 4. This is easily a homomorphism as $f(a+b) = (a+b)(\bmod 4)$ and, as seen after Definition 1.11, $(a+b)(\text{mod}4) = a(\text{mod}4) +_4 b(\text{mod}4) = f(a) +_4 f(b)$.

By definition $ker(f) = \{a \in \mathbb{Z} : f(a) = 0\} = \{a \in \mathbb{Z} : a(\text{mod}4) = 0\}$, thus any integer that is a multiple of 4 will be in the $ker(f)$ and $\langle 4 \rangle \subseteq ker(f)$. Now let $x \in ker(f)$, thus $f(x) = 0$. But then $x(\text{mod}4) = 0$, so x is evenly divisible by 4, or $x = 4n$ for some $n \in \mathbb{Z}$, and $x \in \langle 4 \rangle$. Thus $\langle 4 \rangle = ker(f)$.

Also $f(\mathbb{Z}) = \mathbb{Z}_4$ since $f(0) = 0, f(1) = 1, f(2) = 2$, and $f(3) = 3$.

Example 3.31 Consider $f : \mathbb{Z}_2 \to \mathbb{Z}_6$ defined by $f(x) = 3x$ ($3x$ means usual multiplication in $\mathbb{Z}$). We can show that f is a homomorphism using Cayley tables as described before in Example 1.39. Replacing each element a in the Cayley table for $\mathbb{Z}_2$ by $f(a)$, we find the second table below. This new table is correct for the operation $+_6$ on $\mathbb{Z}_6$, so f is a homomorphism. By inspection we can see $ker(f) = \{0\}$ and $f(\mathbb{Z}_2) = \{0, 3\}$.

$+_2$	0	1
0	0	1
1	1	0

$\to$

$+_6$	0	3
0	0	3
3	3	0

The sets $ker(f)$ and $f(G)$ can help us understand the properties of a homomorphism as well, as seen in the next theorem.

Theorem 3.32

Let G and K be groups with $f : G \to K$, a homomorphism.

(i) f is one to one if and only if $ker(f) = \{e_G\}$.
(ii) f is onto if and only if $f(G) = K$.

Proof Part (ii) is really just the definition of onto so we will look at (i).

($\to$) Suppose that f is one to one. We need to show that $ker(f) = \{e_G\}$. By Theorem 1.37 we know $f(e_G) = e_K$, thus $\{e_G\} \subseteq ker(f)$. Now let $x \in ker(f)$, thus $f(x) = e_K$. Since $f(e_G) = e_K$ then $f(x) = f(e_G)$. However, we assumed f is one to one, so $x = e_G$ and $ker(f) \subseteq \{e_G\}$ as well. Thus $ker(f) = \{e_G\}$.

($\leftarrow$) This is an exercise at the end of the chapter. □

In Examples 3.30 and 3.31, the images and kernels we found were also *subgroups* of the appropriate groups, which leads us to the following theorem.

Theorem 3.33

Let G and K be groups with $f : G \to K$, a homomorphism.

(i) $ker(f) \triangleleft G$.
(ii) $f(G)$ is a subgroup of K.

Proof Assume G and K are groups with $f : G \to K$, a homomorphism.

(i) We need to show that $ker(f) \triangleleft G$. We use Theorem 1.30 to show that $ker(f)$ is a subgroup of G, then Definition 3.13 helps prove that the subgroup is normal.

As $f(e_G) = e_K$ by Theorem 1.37 then $e_G \in ker(f)$ and so $ker(f) \neq \emptyset$. Suppose $x, y \in ker(f)$. Thus $f(x) = f(y) = e_K$. Now $f(xy) = f(x)f(y)$ since f is a homomorphism, so $f(xy) = e_K$ and $xy \in ker(f)$. Finally, if $x \in ker(f)$ then $f(x) = e_K$, so by Theorem 1.37 $f(x^{-1}) = (e_K)^{-1} = e_K$ and so $x^{-1} \in ker(f)$. By Theorem 1.30 we see that $ker(f)$ is a subgroup of G. It is left as an exercise to show that $ker(f)$ is closed under conjugates and thus $ker(f) \triangleleft G$.

(ii) An exercise at the end of the chapter. □

Notice that in Example 3.31, $ker(f) = \{0\}$, and it was clear that f was one to one. But $f(\mathbb{Z}_2) = \{0, 3\}$, which is not $\mathbb{Z}_6$, just as f is not onto. The key that relates homomorphisms and cosets is next, the theorem this section is named after. This theorem is often referred to as the First Isomorphism Theorem; see [9], for example.

Theorem 3.34

The Fundamental Homomorphism Theorem Let G and K be groups, and assume $f : G \to K$ is a homomorphism. Then the image of f, $f(G)$, is isomorphic to the quotient group ${}^{G}/_{ker(f)}$, i.e., $f(G) \cong {}^{G}/_{ker(f)}$.

Proof Assume G and K are groups, and $f : G \to K$ is a homomorphism. To show that $f(G) \cong {}^{G}/_{ker(f)}$ we need to find an isomorphism between the two groups.

Define $\varphi : {}^{G}/_{ker(f)} \to f(G)$ by $\varphi(xker(f)) = f(x)$ for each $x \in G$.

Showing that φ is well defined is not automatic here, since the domain of φ is a set of cosets. We need to be sure that if $aker(f) = bker(f)$ then φ maps the cosets to the same answer, i.e., that $f(a) = f(b)$. Assume we have $a, b \in G$ with $aker(f) = bker(f)$. Thus $a \in bker(f)$ by Theorem 3.6 and $a = by$ for some $y \in ker(f)$. Now $f(a) = f(by) = f(b)f(y)$ since f is a homomorphism. However, $y \in ker(f)$ so $f(y) = e_K$, and we have $f(a) = f(b)$ as needed. Thus $\varphi(aker(f)) = \varphi(bker(f))$ and φ is well defined.

Suppose the converse, that $\varphi(aker(f)) = \varphi(bker(f))$ for $a, b \in G$. We need to show that $aker(f) = bker(f)$ which will guarantee us that φ is one to one. Since $\varphi(aker(f)) = \varphi(bker(f))$, then $f(a) = f(b)$. Using the properties of a homomorphism we have $f(a^{-1}b) = (f(a)^{-1})f(b) = e_K$. Hence $a^{-1}b \in ker(f)$, so by Theorem 3.6 $aker(f) = bker(f)$. Thus φ is one to one.

To show that φ is onto let $y \in f(G)$. By definition of $f(G)$, $y = f(x)$ for some $x \in G$. Thus $xker(f) \in {}^{G}/_{ker(f)}$ and $\varphi(xker(f)) = f(x) = y$. Hence φ is onto.

Finally, we need to show that φ is a homomorphism. Suppose that $xker(f), yker(f) \in {}^{G}/_{ker(f)}$. Using the definition of φ and that f is a homomorphism, the following equation holds.

$$\begin{aligned} \varphi(xker(f) * yker(f)) &= \varphi((xy)ker(f)) \\ &= f(xy) \\ &= f(x)f(y) \\ &= \varphi(xker(f))\varphi(yker(f)) \end{aligned}$$

Hence φ is a homomorphism, so φ is an isomorphism and $f(G) \cong {}^{G}/_{ker(f)}$. □

Example 3.35 Consider the set of complex numbers, $\mathbb{C} = \{a + bi : a, b \in \mathbb{R}\}$, where we use i to denote $\sqrt{-1}$ which is *not* a real number. We can define addition $+_{\mathbb{C}}$ on $\mathbb{C}$ by $(a + bi) +_{\mathbb{C}} (c + di) = (a +_{\mathbb{R}} c) + (b +_{\mathbb{R}} d)i$ for any real numbers a, b, c, d. This is clearly an operation on $\mathbb{C}$ since the usual operation of addition in the real numbers is used to calculate $a +_{\mathbb{R}} c$ and $b +_{\mathbb{R}} d$. Associativity is easy to verify (you should try it), and the identity element is $0 + 0i$, and $(a + bi)^{-1} = (-a) + (-b)i$. Thus we have a group under the operation $+_{\mathbb{C}}$.

Define the function $f : \mathbb{C} \to \mathbb{R}$ by $f(x + yi) = y$, for all $x + yi \in \mathbb{C}$. This function is a homomorphism since for $a + bi, c + di \in \mathbb{C}$,

$$\begin{aligned} f((a + bi) +_{\mathbb{C}} (c + di)) &= f((a +_{\mathbb{R}} c) + (b +_{\mathbb{R}} d)i) \\ &= b +_{\mathbb{R}} d \\ &= f(a + bi) +_{\mathbb{R}} f(c + di). \end{aligned}$$

To use the Fundamental Homomorphism Theorem (3.34) we need to find the kernel and image of f.

$$\begin{aligned} ker(f) &= \{a + bi \in \mathbb{C} : f(a + bi) = 0\} \\ &= \{a + bi \in \mathbb{C} : b = 0\} \\ &= \{a : a \in \mathbb{R}\} \\ &= \mathbb{R} \end{aligned}$$

Also $f(\mathbb{C}) = \{b \in \mathbb{R} : f(x+yi) = b \text{ for some } x+yi \in \mathbb{C}\}$, but for any $b \in \mathbb{R}$ we know $0 + bi \in \mathbb{C}$ and $f(0+bi) = b$. Thus $f(\mathbb{C}) = \mathbb{R}$. Now by Theorem 3.34 we conclude that $\mathbb{R} \cong {}^{\mathbb{C}}/_{\mathbb{R}}$.

Example 3.36 For any group G, we can show that ${}^{G}/_{\{e_G\}} \cong G$. Define the identity function $f : G \to G$ where $f(x) = x$ for all $x \in G$. Then $f(G) = G$ as f is onto, and $ker(f) = \{e_G\}$ since f is one to one (Theorem 3.32). Hence by the Fundamental Homomorphism Theorem 3.34, ${}^{G}/_{\{e_G\}} \cong G$. An exercise at the end of the chapter asks you to verify that ${}^{G}/_{G} \cong \{e_G\}$ for any group G as well.

3.5 Homomorphic Images of a Group

When working with a homomorphism $f : G \to K$, between groups, we must be given the group K ahead of time. If the function is onto it is then easy to know $f(G) = K$, and use the Fundamental Homomorphism Theorem. But how do you know *which* groups K will actually allow you to create such an onto homomorphism?

Definition 3.37 Let G and K be groups. If there exists an onto homomorphism from G to K, then K is called a ***homomorphic image*** of G.

For a finite group G, and occasionally an infinite group, it is possible to determine *exactly* what groups can be found as homomorphic images of G. Before classifying all of them, the next theorem gives us a first clue to the answer.

Theorem 3.38

Let G be a group. If $H \triangleleft G$ then ${}^{G}/_{H}$ is a homomorphic image of G.

Proof Let G be a group and $H \triangleleft G$. We need to find an onto homomorphism from G to ${}^{G}/_{H}$ to complete the proof.

Define $f : G \to {}^{G}/_{H}$ by $f(x) = xH$ for all $x \in G$.

Let $y \in {}^{G}/_{H}$, then by definition $y = xH$ for some $x \in G$. Thus $f(x) = xH = y$, and f is onto. Now assume we have $x, y \in G$.

$$f(xy) = (xy)H = xH * yH = f(x) * f(y)$$

Thus f is also a homomorphism and ${}^{G}/_{H}$ is a homomorphic image of G. $\square$

Example 3.39 Consider the group $(\mathbb{Z}_6, +_6)$. We know that $H = \langle 3 \rangle$ is a cyclic subgroup of $\mathbb{Z}_6$, and as $\mathbb{Z}_6$ is abelian we have $H \lhd \mathbb{Z}_6$ by Theorem 3.18. Now the group ${}^{\mathbb{Z}_6}/_{H}$ is a homomorphic image of $\mathbb{Z}_6$, but what does it look like? ${}^{\mathbb{Z}_6}/_{H} = \{0 + H, 1 + H, 2 + H\}$ so our quotient group has exactly three elements. By Theorem 3.10 we know that ${}^{\mathbb{Z}_6}/_{H} \cong \mathbb{Z}_3$ so $\mathbb{Z}_3$ is a homomorphic image of $\mathbb{Z}_6$.

The previous theorem actually tells us what *all* homomorphic images must look like.

Theorem 3.40

Let G and K be groups. K is a homomorphic image of G if and only if ${}^{G}/_{H} \cong K$ for some normal subgroup H of G.

Proof Let G and K be groups. For one half $(\rightarrow)$ we simply recognize that if an onto homomorphism $f : G \rightarrow K$ exists then $ker(f) \lhd G$ by Theorem 3.33, and by Theorem 3.34 ${}^{G}/_{ker(f)} \cong K$.

$(\leftarrow)$ Now assume we have $H \lhd G$ with ${}^{G}/_{H} \cong K$. Then there exists an isomorphism $\varphi : {}^{G}/_{H} \rightarrow K$. Also ${}^{G}/_{H}$ is a homomorphic image of G by Theorem 3.38, so by Definition 3.37 there is an onto homomorphism $f : G \rightarrow {}^{G}/_{H}$. Consider the function $\varphi \circ f : G \rightarrow K$ which is well defined since $f(G) \subseteq dom(\varphi)$. As both φ and f are onto, we know that $\varphi \circ f$ will also be onto by Theorem 0.17. Let $x, y \in G$. As f and φ are both homomorphisms, the next equation follows immediately.

$$\begin{aligned}
(\varphi \circ f)(xy) &= \varphi(f(xy)) \\
&= \varphi(f(x)f(y)) \\
&= \varphi(f(x))\varphi(f(y)) \\
&= ((\varphi \circ f)(x))((\varphi \circ f)(y))
\end{aligned}$$

Thus $\varphi \circ f$ is a homomorphism as well, and so by Definition 3.37 K is a homomorphic image of G. $\square$

Theorem 3.40 tells us we only have to list every normal subgroup of G, and determine each quotient group, to know *every* homomorphic image of G.

Example 3.41 Let G be the group $(\mathbb{Z}_{18}, +_{18})$. To find all of its homomorphic images we need to list every normal subgroup of G. As G is abelian every subgroup of G is normal by Theorem 3.18, so we just need to list every subgroup. Also G is cyclic so we only have to look for cyclic subgroups by Theorem 2.13.

Using Theorem 2.16 we see that the distinct subgroups of $\mathbb{Z}_{18}$ are:

$$\begin{aligned}
&\langle 0\rangle = \{0\}\\
&\langle 1\rangle = \langle 5\rangle = \langle 7\rangle = \langle 11\rangle = \langle 13\rangle = \langle 17\rangle = \mathbb{Z}_{18}\\
&\langle 2\rangle = \langle 4\rangle = \langle 8\rangle = \langle 10\rangle = \langle 14\rangle = \langle 16\rangle = \{0,2,4,6,8,10,12,14,16\}\\
&\langle 3\rangle = \langle 15\rangle = \{0,3,6,9,12,15\}\\
&\langle 6\rangle = \langle 12\rangle = \{0,6,12\}\\
&\langle 9\rangle = \{0,9\}.
\end{aligned}$$

As these subgroups have sizes 1, 18, 9, 6, 3, and 2, respectively, then the sizes of the quotient groups they give us are 18, 1, 2, 3, 6, and 9, respectively, by Theorem 3.28.

When $H = \mathbb{Z}_{18}$, then as in Example 3.36 we know ${}^{\mathbb{Z}}/_{H} \cong \{0\}$. For the other subgroups, Theorem 3.27 tells us that ${}^{\mathbb{Z}_{18}}/_{H}$ is cyclic, so isomorphic to some $\mathbb{Z}_n$ by Theorem 2.18. Hence we know that:

$$\begin{array}{lll}
{}^{\mathbb{Z}_{18}}/_{\langle 0\rangle} \cong \mathbb{Z}_{18} & {}^{\mathbb{Z}_{18}}/_{\langle 1\rangle} \cong \{0\} & {}^{\mathbb{Z}_{18}}/_{\langle 2\rangle} \cong \mathbb{Z}_2\\
{}^{\mathbb{Z}_{18}}/_{\langle 3\rangle} \cong \mathbb{Z}_3 & {}^{\mathbb{Z}_{18}}/_{\langle 6\rangle} \cong \mathbb{Z}_6 & {}^{\mathbb{Z}_{18}}/_{\langle 9\rangle} \cong \mathbb{Z}_9
\end{array}$$

Thus, we know that the only possible homomorphic images for the group $\mathbb{Z}_{18}$ are $\{0\}$, $\mathbb{Z}_2$, $\mathbb{Z}_3$, $\mathbb{Z}_6$, $\mathbb{Z}_9$, and $\mathbb{Z}_{18}$ (each using the appropriate $+_n$ as operation).

When the group is not cyclic, or not abelian, this process can be more complicated since subgroups are found in different ways, and not all of them will be normal.

Example 3.42 Consider the group $S_3 = \{\varepsilon, (12), (13), (23), (123), (132)\}$. Its subgroups are shown below.

$$\begin{array}{lll}
H_1 = \{\varepsilon\} & H_2 = \{\varepsilon, (12)\} & H_3 = \{\varepsilon, (13)\}\\
H_4 = \{\varepsilon, (23)\} & H_5 = \{\varepsilon, (123), (132)\} & H_6 = S_3
\end{array}$$

Only H_1, H_5, and H_6 are normal. H_1 and H_6 are trivially normal, and H_5 is normal by Theorem 3.20, since $|H_5| = \frac{1}{2}|S_3|$. You should verify that the other subgroups are *not* normal.

Thus *all* of the homomorphic images of S_3 are ${}^{S_3}/_{H_1}$, ${}^{S_3}/_{H_5}$, and ${}^{S_3}/_{H_6}$. Since $[S_3 : H_5] = 2$ is prime the quotient group is isomorphic to $\mathbb{Z}_2$. Thus we know all of the quotient groups.

$${}^{S_3}/_{\{\varepsilon\}} \cong S_3,\ {}^{S_3}/_{S_3} \cong \{\varepsilon\},\text{ and } {}^{S_3}/_{H_5} \cong \mathbb{Z}_2$$

Hence, the only homomorphic images of S_3 are S_3, $\{\varepsilon\}$, and $\mathbb{Z}_2$.

More examples will be investigated in the exercises.

3.6 Theorems of Cauchy and Sylow ($*$ optional)

The theorems presented in this section are beautiful, but more difficult than many others in the text and thus optional. We will not prove these theorems since most of them require topics not covered here, but they will be mentioned in the final topics of the book. (See [1] for the details of these theorems.)

Lagrange's Theorem says that the order of a subgroup must divide the size of a finite group, and ultimately that every element has an order that divides the size of the group. Conversely Cauchy's Theorem looks at the order of the group and tells us there are elements of certain orders that must exist.

Theorem 3.43

Cauchy's Theorem If a prime p divides $|G|$ in the finite group G then G must contain an element whose order is p.

Example 3.44 Suppose H is a subgroup of the group S_5 and 5 divides $|H|$. Then Cauchy's Theorem tells us H must contain an element of order 5. But in S_5, what elements can have order 5? We know that a 5-cycle will have order five, such as (12345), but can any other element also have order 5? Remember that a single cycle of length $t > 1$ always has order t.

Suppose we have a permutation $\alpha \in S_5$ of order 5. Then α can be written as a product of disjoint cycles, $\alpha = \beta_1\beta_2\cdots\beta_k$ where $k < 5$. Since the β_i are disjoint then for all n, the permutations $(\beta_1)^n, (\beta_2)^n, \ldots, (\beta_k)^n$ are also disjoint (or equal to ε) and so we have:

$$\alpha^n = (\beta_1\beta_2\cdots\beta_k)^n = \beta_1^n\beta_2^n\cdots\beta_k^n.$$

Since $\alpha^5 = \varepsilon$ and the permutations $(\beta_1)^5, (\beta_2)^5, \ldots, (\beta_k)^5$ are disjoint (or equal to ε) then for each j we must have $(\beta_j)^5 = \varepsilon$. Thus the order of each cycle β_j is either 1 or 5. But as a t-cycle has order t, only a 5-cycle has order 5. Thus the subgroup H of S_5 must contain a 5-cycle if 5 divides $|H|$.

Another collection of theorems, the *Sylow Theorems*, relate the existence of subgroups to the cardinality of the group.

Definition 3.45 Suppose G is a finite group and a prime p divides $|G|$. A ***p-subgroup*** of G is a subgroup H in which every $a \in H$ has $ord(a) = p^k$ for some nonnegative integer k.

Example 3.46 Consider the group S_4 which has 24 elements. We know that 2 is a prime dividing 24, and we found before the subgroup $H = \{\varepsilon, (12)\}$ which is a 2-subgroup. There are other 2-subgroups as well such as $\{\varepsilon, (12)\}, \{\varepsilon, (13)\}$, and $\{\varepsilon, (1234), (13)(24), (1432)\}$. The size of a 2-subgroup is not unique here.

Definition 3.47 Suppose G is a finite group and a prime p divides $|G|$. A ***Sylow p-subgroup*** H is a maximal p-subgroup, i.e., there is no p-subgroup of G, K, with $H \subset K \subseteq G$.

Note: H is a Sylow p-subgroup of G if and only if H is a p-subgroup of G for which $[G : H]$ is not divisible by p.

Example 3.48 In S_4 (as seen in Example 3.46) we have $H_1 = \{\varepsilon, (12)\}$ as a 2-subgroup. However, $[G : H_1] = 12$ which is divisible by 2 so H_1 is *not* a Sylow 2-subgroup. Similarly, $H_2 = \{\varepsilon, (12)(34), (13)(24), (14)(23)\}$ has $[G : H_2] = 6$ which is still divisible by 2 so is *not* a Sylow 2-subgroup. A Sylow 2-subgroup must have order 8 in this case, such as the subgroup $H_3 = \{\varepsilon, (1234), (13)(24), (1432), (14)(23), (12)(34), (13), (24)\}$. Verify that H_3 is a subgroup by creating its Cayley table.

If we have a subgroup H of a group G, we can define the *conjugate* of H by g to be the set $g^{-1}Hg = \{g^{-1}hg : h \in H\}$. Also, we can define the normalizer of H as $N(H) = \{g \in G : gH = Hg\}$. The verification that $g^{-1}Hg$ and $N(H)$ are subgroups of G is left as an exercise at the end of the chapter.

Theorem 3.49

The Sylow Theorems Let G be a finite group with $|G| = p^n m$ where p does not divide m.

(i) There exists a Sylow p-subgroup of G of order p^n. Also every p-subgroup of G is contained in a Sylow p-subgroup.

(ii) If H and K are two Sylow p-subgroups of G then there exists an element $g \in G$ with $K = g^{-1}Hg$, i.e., H and K are conjugates (and thus isomorphic).

(iii) The number r of Sylow p-subgroups of G must satisfy

r divides m,

$r = 1(\text{mod} p)$, and

$r = [G : N(P)]$ where P is a Sylow p-subgroup of G.

There are many applications of this theorem, and we will look at only a simple one.

Example 3.50 Let G be a group with $|G| = 15$. Since $15 = 3 \cdot 5$, we know from Theorem 3.49 that a Sylow 3-subgroup has exactly three elements, and the number of Sylow 3-subgroups is a number r dividing 5 that has $r = 1(\text{mod} 3)$. But the only divisors of 5 are 1 and 5, and $5 = 2(\text{mod} 3)$, thus there can only exist one Sylow 3-subgroup we will call P_3. Similarly, there is only one Sylow 5-subgroup we call P_5.

But $|P_5 \cup P_3| = 7$ so there must be an element of G not in either of these subgroups. Suppose we have $a \in G$, where $a \notin P_3$ and $a \notin P_5$. Since $ord(a)$ must divide 15 it is either 1, 3, 5, or 15. We know $ord(a)$ cannot be 1 since $e_G \in P_5 \cup P_3$. If $ord(a) = 3$, it must be contained in a Sylow 3-subgroup, and thus $a \in P_3$ which is impossible. Similarly, if $ord(a) = 5$ we have $a \in P_5$ which cannot be true. Thus $ord(a) = 15$, and G must be a cyclic group, thus isomorphic to $\mathbb{Z}_{15}$.

The previous example can give the impression that a group of order pq for distinct primes p and q will always be cyclic, which is not actually true. The next theorem tells us how this can occur.

Theorem 3.51

Let G be a finite group with $|G| = pq$ for distinct primes $p < q$. If G is not cyclic then:

(i) p divides $q - 1$.
(ii) G is a generated by elements a and b which satisfy $a^p = e_G, b^q = e_G$, and $ba = ab^n$ for a positive integer n with $n \neq 1(\text{mod} q)$ and $n^p = 1(\text{mod} q)$.

Example 3.52 The previous theorem can be illustrated with the group S_3 which we know is not cyclic (since it is not abelian). We know $|S_3| = 6 = 2 \cdot 3$. Using $p = 2$ and $q = 3$ notice that p divides $q - 1$. Let $a = (12)$ and $b = (123)$ which are in S_3. We have $a^2 = \varepsilon$ and $b^3 = \varepsilon$. Now $ab = (23)$, $ba = (13)$, and $ab^2 = (23)$. Since every non-identity element of S_3 can be found as a product using only (12), (13), and (23), i.e., is a product of transpositions, then a and b generate all of S_3. Finally, $ba = (13) = ab^2$, where $2 \neq 1(\bmod 3)$ and $2^2 = 1(\bmod 3)$.

One of the most well known theorems, which relates to the prime factorization of $|G|$, is the *Fundamental Theorem of Finite Abelian Groups.* Although we will not prove this theorem, the key is writing the cardinality of the group in its prime factorization and finding p-subgroups.

Theorem 3.53

The Fundamental Theorem of Finite Abelian Groups Suppose G is a *finite abelian* group. Then G is isomorphic to a direct product of cyclic groups of (not necessarily distinct) prime power orders, $\mathbb{Z}_{p_1^{n_1}} \times \mathbb{Z}_{p_2^{n_2}} \times \cdots \times \mathbb{Z}_{p_k^{n_k}}$. This product is unique up to the order of the factors and $|G| = p_1^{n_1} p_2^{n_2} \cdots p_k^{n_k}$.

Example 3.54 If we have an abelian group with $|G| = 20$, then as $20 = 2^2 5$, the Fundamental Theorem of Finite Abelian Groups tells us there are only two different options for G, $G \cong \mathbb{Z}_{2^2} \times \mathbb{Z}_5$, or $G \cong \mathbb{Z}_2 \times \mathbb{Z}_2 \times \mathbb{Z}_5$.

Exercises for Chapter 3

Section 3.1 Cosets

1. Compute the left cosets of $\langle 9 \rangle$ in $(\mathbb{Z}_{12}, +_{12})$.
2. Compute the left cosets of $\langle 6 \rangle$ in $(\mathbb{Z}_{12}, +_{12})$.
3. Compute the left cosets of $\langle 3 \rangle$ in $(\mathbb{Z}_{15}, +_{15})$.
4. Compute the left cosets of $\langle 6 \rangle$ in $(\mathbb{Z}_{20}, +_{20})$.
5. Compute the left cosets of $\langle 3 \rangle$ in $(\mathbb{Z}_{36}, +_{36})$.
6. Compute the left cosets of $\langle (123) \rangle$ in $(S_4, \circ)$.
7. Compute the left cosets of $\langle (1234) \rangle$ in $(S_4, \circ)$.
8. Compute the left cosets of $H = \{\varepsilon, (12)(34), (13)(24), (14)(23)\}$ in $(S_4, \circ)$.
9. Compute the left cosets of $H = \{(0,0), (2,0), (0,2), (2,2)\}$ in $\mathbb{Z}_4 \times \mathbb{Z}_4$. Remember that the operation uses $+_4$ in each coordinate.
10. Compute the left cosets of $\langle 7 \rangle$ in $(U(16), \cdot_{16})$.
11. Compute the left cosets of $\langle 11 \rangle$ in $(U(12), \cdot_{12})$.
12. Compute the left cosets of $\langle 2 \rangle$ in $(U(15), \cdot_{15})$.
13. Prove (ii) of Theorem 3.10.
14. Prove (iii) of Theorem 3.10.
15. Prove (iv) of Theorem 3.10.
16. Without listing the elements of the left cosets, determine if $(123)H = (132)H$ for the subgroup $H = \{\varepsilon, (12)\}$ in $(S_3, \circ)$.
17. Without listing the elements of the left cosets, determine if $4 + H = 7 + H$ for the subgroup $H = \{0, 3, 6, 9\}$ in $(\mathbb{Z}_{12}, +_{12})$.
18. Without listing the elements of the left cosets, determine if $7H = 5H$ for the subgroup $H = \{1, 8\}$ in $(U(9), \cdot_9)$.
19. Without listing the elements of the left cosets, determine if $3 + H = 9 + H$ for the subgroup $H = \{0, 5, 10, 15\}$ in $(\mathbb{Z}_{20}, +_{20})$.
20. Use Lagrange's Theorem to determine the possible subgroup sizes in a group with exactly 40 elements.
21. Use Lagrange's Theorem to determine the possible subgroup sizes in a group with exactly 16 elements.
22. Use Lagrange's Theorem to determine the possible subgroup sizes in a group with exactly 30 elements.

23. Use Lagrange's Theorem to determine the possible subgroup sizes in a group with exactly 17 elements.
24. Use Lagrange's Theorem to determine the possible subgroup sizes in a group with exactly 10 elements.
25. Without checking subgroup properties, how do we know that $H = \{0, 2, 4\}$ cannot be a subgroup of $(\mathbb{Z}_{16}, +_{16})$?
26. Prove or find a counterexample: For any group G and subgroup H, if $a, b, c \in G$ and (ba)H$=(bc)H$ then $aH = cH$.
27. Prove or find a counterexample: For any group G and subgroup H, if $a, b, c \in G$ and $(ab)H = (cb)H$ then $aH = cH$.

Section 3.2 Normal Subgroups

28. Prove: If G is a group then $\{e_G\} \triangleleft G$ and $G \triangleleft G$.
29. Prove: If G is an abelian group then every subgroup of G is a normal subgroup.
30. Determine if the subgroup $\{\varepsilon, (123), (132)\}$ is a normal subgroup of $(S_4, \circ)$.
31. Determine if the subgroup $H = \{\varepsilon, (1234), (13)(24), (1432)\}$ is a normal subgroup of $(S_4, \circ)$.
32. Determine if the subgroup $H = \{\varepsilon, (12)(34), (13)(24), (14)(23)\}$ is a normal subgroup of $(S_4, \circ)$.
33. Using the group defined in Exercise 21 of Chapter 1 formed by moving a coin, determine if $\langle H \rangle \triangleleft G$.
34. In Chapter 1 we saw the group (using matrix multiplication):

$$A = \left\{ \begin{bmatrix} 1 & 0 \\ 0 & -1 \end{bmatrix}, \begin{bmatrix} 1 & 0 \\ 0 & 1 \end{bmatrix}, \begin{bmatrix} -1 & 0 \\ 0 & 1 \end{bmatrix}, \begin{bmatrix} -1 & 0 \\ 0 & -1 \end{bmatrix} \right\}.$$

Determine if $H = \left\{ \begin{bmatrix} 1 & 0 \\ 0 & -1 \end{bmatrix}, \begin{bmatrix} 1 & 0 \\ 0 & 1 \end{bmatrix} \right\}$ is a normal subgroup of A.

35. In $(S_3, \circ)$, determine if the subgroups $H_2 = \{\varepsilon, (12)\}$, $H_3 = \{\varepsilon, (13)\}$, and $H_4 = \{\varepsilon, (23)\}$ are normal subgroups.
36. Consider the group $G = \{e, a, b, c, d, i, j, k\}$ whose Cayley table is shown. Let $H = \{e, a\}$. Show that $H \triangleleft G$.

*	e	a	b	c	d	i	j	k
e	e	a	b	c	d	i	j	k
a	a	e	i	j	k	b	c	d
b	b	i	a	k	c	e	d	j
c	c	j	d	a	i	k	e	b
d	d	k	j	b	a	c	i	e
i	i	b	e	d	j	a	k	c
j	j	c	k	e	b	d	a	i
k	k	d	c	i	e	j	b	a

37. Using the same group G from the previous exercise and $K = \{e, a, k, d\}$ determine if $K \triangleleft G$.
38. Suppose G is a group and H is a subgroup of G. For $a \in G$ define the set $a^{-1}Ha = \{a^{-1}ha : h \in H\}$. Prove: $a^{-1}Ha$ is a subgroup of G. Is $a^{-1}Ha = H$ for each $a \in G$?
39. Let G be a group, H a subgroup of G, and $a \in G$. Prove: $a^{-1}Ha \cong H$. (The set $a^{-1}Ha$ is defined in the previous problem.)
40. Suppose G is a group and H is a subgroup of G. Define $K = \{a \in G : aH = Ha\}$. Prove: K is a subgroup of G and $H \triangleleft K$.
41. Let G be a group with $H \triangleleft G$ and K a subgroup of G. Prove that the set $HK = \{xy : x \in H, y \in K\}$ is a subgroup of G.
42. Let G be a group with $H \triangleleft G$ and K a subgroup of G. Prove that $H \triangleleft HK$.

Section 3.3 Quotient Groups

43. Find the Cayley table for the group ${}^{\mathbb{Z}_{12}}/_{\langle 4 \rangle}$. Use $+_{12}$ on $\mathbb{Z}_{12}$.
44. Find the Cayley table for the group ${}^{\mathbb{Z}_{15}}/_{\langle 5 \rangle}$. Use $+_{15}$ on $\mathbb{Z}_{15}$.
45. Find the Cayley table for the group ${}^{U(15)}/_{\langle 2 \rangle}$. Use $\cdot_{15}$ on $U(15)$.
46. Find the Cayley table for the group ${}^{U(20)}/_{\langle 11 \rangle}$. Use $\cdot_{20}$ on $U(20)$.
47. For the group in Exercise 36, find the Cayley table for ${}^{G}/_{H}$.
48. Find an example of a group G, normal subgroup H, and an element $x \in G$ so that $ord(xH) \neq ord(x)$.
49. In the group ${}^{\mathbb{Z}_{10}}/_{\langle 5 \rangle}$ find the order of each element. Use $+_{10}$ on $\mathbb{Z}_{10}$.
50. In the group ${}^{\mathbb{Z}_{16}}/_{\langle 4 \rangle}$ find the order of each element. Use $+_{16}$ on $\mathbb{Z}_{16}$.
51. In the group ${}^{\mathbb{Z}_{18}}/_{\langle 6 \rangle}$ find the order of each element. Use $+_{18}$ on $\mathbb{Z}_{18}$.
52. In the group ${}^{U(9)}/_{\langle 8 \rangle}$ find the order of each element. Use $\cdot_{9}$ on $U(9)$.
53. In the group ${}^{U(15)}/_{\langle 2 \rangle}$ find the order of each element. Use $\cdot_{15}$ on $U(15)$.
54. Prove (i) of Theorem 3.27.
55. Suppose G is an abelian group and $H = \{a^2 : a \in G\}$. Prove: For every $xH \in {}^{G}/H$ we have $(xH)^{-1} = xH$.
56. Suppose G is an abelian group and define $H = \{a^3 : a \in G\}$. Prove: For every $x \in G$ with $x \notin H$ we must have $ord(xH) = 3$.
57. Let G be an abelian group with subgroups H, K so that $K \subseteq H \subseteq G$. Prove that ${}^{H}/_{K}$ is a subgroup of ${}^{G}/_{K}$.
58. Let G be a group with $H \triangleleft G$ and K a subgroup of G. Prove that for every left coset xH of the quotient group ${}^{HK}/_{H}$ there is $k \in K$ for which $xH = Hk$

Sections 3.4 and 3.5 The Fundamental Homomorphism Theorem and Homomorphic Images

59. Complete the proof of Theorem 3.32 by proving the second implication, $(\leftarrow)$.
60. Complete the proof of (i) in Theorem 3.33 by showing $ker(f)$ is closed under conjugates so that $ker(f) \triangleleft G$.

61. Prove (ii) in Theorem 3.33.
62. Prove: For any group G, ${}^{G}/_{G} \cong \{e_G\}$.
63. Consider the group of functions $\Im(\mathbb{R})$ defined in Example 1.13. Let $\varphi : \Im(\mathbb{R}) \to \mathbb{R}$ be defined by $\varphi(f) = f(3)$. Show that φ is an onto homomorphism, and describe $ker(\varphi)$.
64. Define $K = \{(0,0),(1,1),(2,2)\}$ (a subgroup of $\mathbb{Z}_3 \times \mathbb{Z}_3$), and prove that $\mathbb{Z}_3 \cong {}^{\mathbb{Z}_3 \times \mathbb{Z}_3}/_{K}$. As usual $\mathbb{Z}_3$ a group under $+_3$.
65. Let G be an abelian group. Define $H = \{x^2 : x \in G\}$ and $K = \{x \in G : x^2 = e_G\}$. Prove $H \cong {}^{G}/_{K}$.
66. Without creating homomorphisms, explain why we know that $\mathbb{Z}_4$ cannot be a homomorphic image for a group of order 30.
67. Find all of the possible homomorphic images of the group $(\mathbb{Z}_8, +_8)$.
68. Find all of the possible homomorphic images of the group $(\mathbb{Z}_9, +_9)$.
69. Find all of the possible homomorphic images of the group $(U(9), \cdot_9)$. [Show that $U(9)$ is cyclic to help.]
70. Let G be a group with $H \lhd G$ and K a subgroup of G. Prove ${}^{K}/_{H \cap K} \cong {}^{HK}/_{H}$.

Projects for Chapter 3

Project 3.1

(This was originally published in PRIMUS; see [6] for more). In Project 1.3 we considered a group made up from other groups but with a special operation involving the "carry" in each coordinate. For the rest of this project we will use:

$$G = \mathbb{Z} \times \mathbb{Z}_2 \times \mathbb{Z}_4 = \{(u,v,w) \ : \ u \in \mathbb{Z},\ v \in \mathbb{Z}_2,\ w \in \mathbb{Z}_4\}.$$

With operation $(u,v,w) * (r,s,t) = (u + r + c_1 + c_2,\ \ v +_2 s,\ \ w +_4 t)$ where c_1 and c_2 are the "carries", each is either 0 or 1. Note that $c_1 = 1$ when $v + s \geq 2$ and $c_2 = 1$ when $w + t \geq 4$.

Define the set $H = \{(a,\ 0,\ b) :\ a \in \mathbb{Z}$ and either $b = 0$ *or* $b = 2\}$.

1. H is clearly a subSET of G and $(1,0,2) \in H$, so H is nonempty. Use the following steps to show H is a subgroup of G.

   (i) Show that H is closed under the operation of G.
   (ii) Determine the inverse for an arbitrary element of the form $(a,0,0)$ to show it is in H.
   (iii) Determine the inverse for an arbitrary element of the form $(a,0,2)$ to show that H is closed under inverses.

2. Consider the (left) cosets of H in G.

Is $(1,1,3) \in (2,1,1)H$? Be sure to explain why or why not.
Is $(1,1,1)H = (2,1,1)H$? Why or why not?

3. Determine the elements of the cosets (use set builder notation):

   $(0,0,0)H =$

   $(0,1,0)H =$

   $(0,0,1)H =$

   $(0,1,1)H =$

4. Is any element of G missing from these cosets?

Project 3.2

Let G be a group and H and K subgroups of G.

1. Prove: If $a, b \in G$ and $x \in aH \cap bK$ then $aH = xH$ and $bK = xK$.
2. Prove: $yH \cap yK = y(H \cap K)$ for all $y \in G$.
3. Prove: If $a, b \in G$ and $aH \cap bK \neq \emptyset$, then $aH \cap bK = c(H \cap K)$ for some $c \in G$.

Project 3.3

Multiplication of two complex numbers is defined by $(a+bi)*(c+di) = (ac-d)+(ad+bc)i$. For any real number x, define $cis(x) = \cos(x) + \sin(x)i$, and let $T = \{cis(x) : x \in \mathbb{R}\}$.

1. Using the trigonometric identities below for $cos(x+y)$ and $sin(x+y)$, show that $cis(x+y) = cis(x) * cis(y)$.

$$\sin(x+y) = \sin(x)\cos(y) + \sin(y)\cos(x)$$
$$\cos(x+y) = \cos(x)\cos(y) - \sin(y)\sin(x)$$

   Therefore complex multiplication is an operation on T which you may assume to be associative already.

2. What is the identity of T, and does every element have an inverse in T? (In other words, is $(T, *)$ a group?) Remember that the identity and the inverses must again be of the form $cis(x)$.

Define: $f : \mathbb{R} \to T$ by $f(x) = cis(2\pi x)$.

3. Prove that f is an onto homomorphism.
4. Prove that $ker(f) = \mathbb{Z}$. [Hint: When are $\cos(2\pi x) = 1$ and $\sin(2\pi x) = 0$ both true?]
5. How do we conclude that $T \cong {}^{\mathbb{R}}/_{\mathbb{Z}}$?

Project 3.4

Let G be the group $(\mathbb{Z}, +)$, and $H = \{2n : n \in \mathbb{Z}\}$ (the set of even integers). Note that $2n$ means usual multiplication here.

1. Prove that H is a normal subgroup of G.
2. Prove: If $k \in \mathbb{Z}$ then either $0 + H = k + H$ or $1 + H = k + H$.
3. Explain why we know ${}^{G}/_{H} \cong \mathbb{Z}_2$ without finding an isomorphism.

Project 3.5

Consider the group $G = \{e, a, b, c, d, i, j, k\}$ with the operation given in the Cayley table below.

$*$	e	a	b	c	d	i	j	k
e	e	a	b	c	d	i	j	k
a	a	e	i	j	k	b	c	d
b	b	i	a	k	c	e	d	j
c	c	j	d	a	i	k	e	b
d	d	k	j	b	a	c	i	e
i	i	b	e	d	j	a	k	c
j	j	c	k	e	b	d	a	i
k	k	d	c	i	e	j	b	a

1. Find the subgroups $M = \langle a \rangle$ and $N = \langle i \rangle$, and show they are both normal using their cosets.
2. Find the Cayley tables for the groups ${}^{G}/_{M}$ and ${}^{G}/_{N}$, with the usual operation for quotient groups.
3. Find onto homomorphisms f :G$\rightarrow \mathbb{Z}_2 \times \mathbb{Z}_2$ with $ker(f) = M$ and $g : G \rightarrow \mathbb{Z}_2$ with $ker(g) = N$. Verify that one of them is a homomorphism.

Project 3.6

Let G be a group and $H \triangleleft G$.

Theorem: If $a, c \in G$ and $H(ac^{-1}) = H(ac)$ then $c^2 \in H$.

1. Circle all of the logical flaws in the following "proof," and explain why any circled step

is incorrect. Remember to look at the logic of each step to see if it follows from the previous steps; a mistake does not make each step following it incorrect.

Proof: Assume that $a, c \in G$ and that $H(ac^{-1}) = H(ac)$. We need to show that $c^2 \in H$. Since we know $ac^{-1} \in H(ac)$, there is $h \in H$ for which $ac^{-1} = ach$. Thus by the Cancellation Law $c^{-1} = ch$. Multiplying by c^{-1} gives us $(c^{-1})^2 = c^{-1}hc^{-1}$. But H is normal and thus closed under conjugates so $c^{-1}hc^{-1} \in H$ and hence $(c^{-1})^2 \in H$. Since $(c^{-1})^2 = (c^2)^{-1}$, we now have $c^2 \in H$ as needed.

2. Give a correct proof of the theorem above.

Project 3.7

Let G be a group, H a subgroup of G, and $x, y, z \in G$. Do **not** assume that H is a normal subgroup.

1. Prove: If $xH = y^2H$ then $x^{-1}y \in Hy^{-1}$.
2. Prove: If $(x^{-1}y)H = Hz$ then $z^{-1}H \subseteq H(y^{-1}x)$.
3. If we still assume that $(x^{-1}y)H = Hz$, is it also true that $z^{-1}H \supseteq H(y^{-1}x)$? Give a proof or counterexample.

Chapter 4

Rings

After the definition of a group in Chapter 1, the first example of a group was $(\mathbb{Z}, +)$. However, even elementary school children learn that we can both add and multiply integers, thus having two operations on a set is also a familiar concept. Having two operations that satisfy specific rules will define a *Ring*.

4.1 Rings

Definition 4.1 Let A be a nonempty set. Suppose there are two operations on A, $\oplus$ and $\otimes$. Then $(A, \oplus, \otimes)$ is a ***ring*** if the following hold:

(i) $(A, \oplus)$ forms an abelian group.
(ii) $\otimes$ is associative on A.
(iii) The distributive laws hold, i.e., for any $a, b, c \in A$:
$$a \otimes (b \oplus c) = (a \otimes b) \oplus (a \otimes c)$$
$$(a \oplus b) \otimes c = (a \otimes c) \oplus (b \otimes c).$$

The operation $\oplus$ will be referred to as the *addition* of A, and $\otimes$ the *multiplication* of A, even though they may not be usual addition or multiplication. As with groups, $|A|$ denotes the cardinality of A.

Example 4.2 As the inspiration for the definition, it should be clear that $(\mathbb{Z}, +, \cdot)$ forms a ring. You should easily be able to verify that the usual operations of addition and multiplication on the sets $\mathbb{Q}$ and $\mathbb{R}$ form rings as well.

One important thing to notice, which is again inspired by $\mathbb{Z}$, is that the operations of addition and multiplication have different properties. While $(\mathbb{Z}, +)$ is a group, $(\mathbb{Z}, \cdot)$ is *not* a group. Thus it is critical to know which of the two given operations is considered the addition and which is the multiplication.

Example 4.3 Another group we considered in Chapter 1 is the set $M_2(\mathbb{R})$ under usual addition of matrices. To create a ring we also include multiplication of 2×2 matrices which is an operation on $M_2(\mathbb{R})$. In a Linear Algebra course you should have seen (or will see) that this multiplication is associative and the distributive laws hold. Thus $M_2(\mathbb{R})$ forms a ring. However, matrix multiplication is not commutative as seen below.

$$\begin{bmatrix} -1 & 2 \\ 3 & 1 \end{bmatrix}\begin{bmatrix} 1 & -1 \\ 2 & 3 \end{bmatrix} = \begin{bmatrix} 3 & 7 \\ 5 & 0 \end{bmatrix} \qquad \begin{bmatrix} 1 & -1 \\ 2 & 3 \end{bmatrix}\begin{bmatrix} -1 & 2 \\ 3 & 1 \end{bmatrix} = \begin{bmatrix} -4 & 1 \\ 7 & 7 \end{bmatrix}$$

Thus, commutativity of multiplication will be considered a special property and cannot be assumed in an arbitrary ring.

Example 4.4 We will often use a ring whose additive structure was discussed as a group in the previous chapters, but <u>not</u> every group can actually be used to create a ring by adding another operation. The group S_3, the permutations of $\{1, 2, 3\}$, cannot become a ring by adding any operation of "multiplication." Why not? Think about the definition of a ring and what we know about S_3.

So far most of our examples have had "usual" operations on familiar sets, but there may be other operations we are asked to consider. We must carefully determine if the properties of a ring hold in such cases, as seen in the next example.

Example 4.5 Let $A = \{3m : m \in \mathbb{Z}\}$. Define $x \oplus y = x + y - 3$ and $x \otimes y = 3$ for each $x, y \in A$. (Note that the answer to every "multiplication" problem is 3.) These new operations are defined with usual addition or subtraction when $x + y$ or $x - y$ are mentioned. Is $(A, \oplus, \otimes)$ a ring?

For $x = 3m, y = 3n$ in A we have $x \oplus y = 3m + 3n - 3 = 3(m + n - 1)$ and $x \otimes y = 3 = 3(1)$. These answers are uniquely defined and back in A so $\oplus$ and $\otimes$ are operations on A.

Let $x, y, z \in A$. Using the fact that usual addition in $\mathbb{Z}$ is associative and commutative we can calculate $(x \oplus y) \oplus z$ and $(x \oplus y) \oplus z$.

$$\begin{aligned}(x \oplus y) \oplus z &= (x+y-3) \oplus z \\ &= x+y-3+z-3 \\ &= x+y+z-6\end{aligned}$$

$$\begin{aligned}x \oplus (y \oplus z) &= x \oplus (y+z-3) \\ &= x+(y+z-3)-3 \\ &= x+y+z-6\end{aligned}$$

Thus we have $(x \oplus y) \oplus z = (x \oplus y) \oplus z$ and $\oplus$ is associative. Notice that $x \oplus 3 = x+3-3 = x$ and $3 \oplus x = 3+x-3 = x$ for all $x \in A$. Thus the identity for $\oplus$ is 3, which is in A. Also for each $x \in A$, say $x = 3m$ for some $m \in \mathbb{Z}$, we can see that $6-x = 6-3m = 3(2-m)$. Thus $6-x \in A$ with $x \oplus (6-x) = x+(6-x)-3 = 3$ and $(6-x) \oplus x = (6-x)+x-3 = 3$. Hence the inverse under $\oplus$ for x is $6-x$ and $(A, \oplus)$ is a group. Since addition in $\mathbb{Z}$ is commutative it is easy to see that $x \oplus y = y \oplus x$ and so $(A, \oplus)$ is an abelian group.

It is straightforward to see that $\otimes$ is associative as shown next.

$$(x \otimes y) \otimes z = 3 \otimes z = 3 \text{ and } x \otimes (y \otimes z) = x \otimes 3 = 3$$

Since $x \otimes y = 3 = y \otimes x$ for any $x, y \in A$ we also know $\otimes$ is commutative. The last part of the definition of a ring is the Distributive Law, and since we know $\otimes$ is commutative we only need to check one of them to know both hold.

$$(x \oplus y) \otimes z = (x+y-3) \otimes z = 3$$
$$(x \otimes z) \oplus (y \otimes z) = 3 \oplus 3 = 3+3-3 = 3$$

Hence $(x \oplus y) \otimes z = (x \otimes z) \oplus (y \otimes z)$ so the Distributive Laws hold and $(A, \oplus, \otimes)$ is a ring. Defining $x \otimes y = 3$ for every $x, y \in A$, when 3 is identity for $\oplus$, is a useful technique; a generalization of this is an exercise at the end of the chapter.

The rings we discuss do not need to be sets of numbers either as seen in the next, even more unusual, example.

Example 4.6 The *Power Set* of a nonempty set C is $\wp(C) = \{U : U \subseteq C\}$. With the two operations $\oplus$ and $\otimes$ on $\wp(C)$ defined below, is $(\wp(C), \oplus, \otimes)$ a ring?

For $U, V \in \wp(C)$:

$$U \oplus V = (U-V) \cup (V-U) \qquad U \otimes V = U \cap V.$$

Notice that $U \oplus V$ can also be written as $U \oplus V = (U \cup V) - (U \cap V)$ (an exercise in Chapter 0).

The most difficult part of showing we have a ring is associativity of our addition. Suppose we have three elements $U, V, W \in \wp(C)$.

$$(U \oplus V) \oplus W = [(U \oplus V) \cup W] - [(U \oplus V) \cap W]$$
$$U \oplus (V \oplus W) = [U \cup (V \oplus W)] - [U \cap (V \oplus W)]$$

An exercise at the end of the chapter asks you to prove $(U \oplus V) \oplus W = U \oplus (V \oplus W)$.

The definition of $U \oplus V = (U \cup V) - (U \cap V)$ makes it easy to see that $\oplus$ is commutative as union and intersection are commutative operations on sets. The empty set, $\emptyset$, is the additive identity since $U \oplus \emptyset = U$. To see if we have additive inverses for each element consider $U \in \wp(C)$ and define $U^{-1} = U$. Then $U \oplus (U^{-1}) = (U \cup U) - (U \cap U)$ but $U \cup U = U \cap U$ so $U \oplus (U^{-1}) = \emptyset$ as needed. Hence $(\wp(C), \oplus)$ is an abelian group.

You are asked to show that $\otimes$ is associative as an exercise at the end of the chapter, so the final step is to verify the distributive laws $U \otimes (V \oplus W) = (U \otimes V) \oplus (U \otimes W)$ and $(U \oplus V) \otimes W = (U \otimes W) \oplus (V \otimes W)$. However, it is easy to see that $\otimes$ is commutative so we only need to verify one of them.

$$U \otimes (V \oplus W) = U \cap [V \oplus W] = U \cap [(V \cup W) - (V \cap W)]$$
$$(U \otimes V) \oplus (U \otimes W) = (U \cap V) \oplus (U \cap W) = [(U \cap V) \cup (U \cap W) - (U \cap V \cap W)]$$

Assume that $x \in U \otimes (V \oplus W)$. Thus $x \in U$ and $x \in V \oplus W$, so we must have $x \in U$, $x \in V \cup W$, and $x \notin V \cap W$. Thus either $x \in U \cap V$ or $x \in U \cap W$ so $x \in (U \cap V) \cup (U \cap W)$ and $x \notin U \cap V \cap W$. Hence we have $x \in (U \otimes V) \oplus (U \otimes W)$ and $U \otimes (V \oplus W) \subseteq (U \otimes V) \oplus (U \otimes W)$.

For the other half, suppose we have $x \in (U \otimes V) \oplus (U \otimes W)$. Thus $x \in (U \cap V) \cup (U \cap W)$ and $x \notin U \cap V \cap W$. If $x \in U \cap V$ then $x \notin W$, so we have $x \in U$, $x \in V$, and $x \notin W$. Now $x \in U \cap [(V \cup W) - (V \cap W)]$ and $x \in U \otimes (V \oplus W)$. If instead we have $x \in U \cap W$ then $x \notin V$, so we have $x \in U$, $x \in W$, and $x \notin V$. Thus $x \in U \cap [(V \cup W) - (V \cap W)]$ and $x \in U \otimes (V \oplus W)$. Therefore $U \otimes (V \oplus W) \supseteq (U \otimes V) \oplus (U \otimes W)$ so $U \otimes (V \oplus W) = (U \otimes V) \oplus (U \otimes W)$.

Since the distributive laws hold we have a ring.

Recall that when we discussed an arbitrary group we used the letter G to stand for the set. This makes sense for the reminder that G is a *group*, so it would seem natural to use R to denote a ring. However, this can be too easily confused with our notation for the set of real numbers $\mathbb{R}$, so we will continue to use A to denote a ring.

Since there are two operations in a ring, when we talk about an identity, inverses, or even commutativity, it must be clear which operation we are using. Thus we will define terms that are specific to each operation, beginning with addition.

Definition 4.7 Suppose $(A, \oplus, \otimes)$ is a ring.

(i) The identity under addition $\oplus$ in A is called the ***zero*** of A and is denoted by 0_A. The subscript is there to remind us it may not be the number 0.
(ii) The inverse for an element $a \in A$ under the operation of addition $\oplus$ is called the ***negative*** of a and denoted by $-a$.

There is no requirement that the operation of multiplication have any of the properties in the next definition, so each begins with the word "if."

Definition 4.8 Suppose $(A, \oplus, \otimes)$ is a ring.

(i) If $\otimes$ is also commutative we say that A is a ***commutative ring***. (The term *abelian* is only used for groups.)
(ii) If there is an identity for $\otimes$ in A, we call it the ***unity*** of A and denote it by 1_A. (Again, it need not be the number one.) If A contains a unity, we say A is a ***ring with unity***.
(iii) If an element $a \in A$ has an inverse under $\otimes$, then we say that a is a ***unit*** of A and denote its inverse by a^{-1}.

Example 4.9 $(\mathbb{Z}, +, \cdot)$ is a commutative ring and it has 1 as its unity. However, not every element has an inverse under multiplication. Can you identify the two integers that are the only units? However, $\mathbb{Q}$ is a commutative ring with unity 1, *and* every nonzero element is a unit. What is the inverse of a nonzero element $\frac{a}{b}$?

Example 4.10 Recall from Example 1.13 that we had a group of functions, $\Im(\mathbb{R}) = \{f : \mathbb{R} \to \mathbb{R} : f \text{ is a function}\}$. For the operation on $\Im(\mathbb{R})$ we used addition of functions, i.e., for $f, g \in \Im(\mathbb{R})$ the function $f + g$ is defined by $(f + g)(x) = f(x) + g(x)$ for all x in $\mathbb{R}$.

For multiplication we now define the function fg by $fg(x) = f(x)g(x)$ for all x in $\mathbb{R}$. We simply multiply the answers each function would give. The proof that $\Im(\mathbb{R})$ is a ring under these operations is an exercise at the end of the chapter. Is it a commutative ring with unity?

Some critical examples of groups were $(\mathbb{Z}_n, +_n)$. We also defined multiplication $\cdot_n$ so it is natural to ask if $(\mathbb{Z}_n, +_n, \cdot_n)$ is a ring.

Theorem 4.11

If $n > 1$ then $(\mathbb{Z}_n, +_n, \cdot_n)$ is a commutative ring with unity.

Proof For $n > 1$, Theorem 1.12 tells us that $(\mathbb{Z}_n, +_n)$ forms an abelian group. We verified in Chapter 2 that $\cdot_n$ is associative and commutative on $\mathbb{Z}_n$ as well. We only need to verify that for any $a, b, c \in \mathbb{Z}_n$, $a \cdot_n (b +_n c) = (a \cdot_n b) +_n (a \cdot_n c)$ to know we have a ring.

Let $a, b, c \in \mathbb{Z}_n$. There are $x, y \in \mathbb{Z}$ with $b + c = nx + y$ and $0 \leq y < n$. Now we can calculate:

$$\begin{aligned} a \cdot_n (b +_n c) &= ay(\mathrm{mod} n) \\ &= [anx + ay](\mathrm{mod} n) \\ &= [a(nx + y)](\mathrm{mod} n) \\ &= [a(b + c)](\mathrm{mod} n) \end{aligned}$$

Similarly, there are $q, r, u, v \in \mathbb{Z}$, $0 \leq r, v < n$, with $ab = nq + r$ and $ac = nu + v$. Again, we calculate:

$$\begin{aligned} (a \cdot_n b) +_n (a \cdot_n c) &= r +_n v \\ &= (r + v)(\mathrm{mod} n) \\ &= [n(q + u) + (r + v)](\mathrm{mod} n) \\ &= [(nq + r) + (nu + v)](\mathrm{mod} n) \\ &= [(ab + ac)](\mathrm{mod} n) \end{aligned}$$

In the integers with its usual operations $a(b + c) = ab + ac$, so we have $a \cdot_n (b +_n c) = (a \cdot_n b) +_n (a \cdot_n c)$, and the distributive laws hold. Therefore for $n > 1$, $(\mathbb{Z}_n, +_n, \cdot_n)$ forms a commutative ring. Clearly, 1 is also the unity of $(\mathbb{Z}_n, +_n, \cdot_n)$ completing the proof. □

Recall that the trivial group has only one element in it. Similarly, $A = \{a\}$ is the ***trivial ring***. The only way to define addition and multiplication on A is $a \oplus a = a$ and $a \otimes a = a$. You should be able to verify that these operations create a commutative ring with unity.

We also define a direct product of rings, similar to Definition 1.15.

Definition 4.12 Suppose we have rings $(A, +_A, \cdot_A)$ and $(B, +_B, \cdot_B)$. On the set $A \times B = \{(a, b) : a \in A, b \in B\}$, we define the operations by coordinates:

$$(a, b) +_{A \times B} (c, d) = (a +_A c, b +_B d)$$
$$(a, b) \cdot_{A \times B} (c, d) = (a \cdot_A c, b \cdot_B d).$$

Since the coordinates are in sets that satisfy the necessary properties to be rings, it is straightforward to see that $A \times B$ is also a ring with $0_{A\times B} = (0_A, 0_B)$. The steps needed to verify this are exercises at the end of the chapter.

Example 4.13 Let A denote the ring $(\mathbb{Z}_4, +_4, \cdot_4)$ and B denote the ring $(\mathbb{Q}, +, \cdot)$. Then in $A \times B$ we find $\left(2, \frac{1}{3}\right) +_{A\times B} \left(3, \frac{1}{2}\right) = \left(1, \frac{5}{6}\right)$ and $\left(2, \frac{1}{3}\right) \cdot_{A\times B} \left(3, \frac{1}{2}\right) = \left(2, \frac{1}{6}\right)$.

At first the new structure we call a ring may appear to be simply repeating the ideas of a group twice (once for addition and once for multiplication). However, the interaction between addition and multiplication can often have interesting consequences, some of which will be used frequently in later chapters of the text.

Definition 4.14 Suppose we have a nontrivial ring $(A, \oplus, \otimes)$.

(i) An element $a \in A$ is called a ***zero divisor*** if $a \neq 0_A$ and there exists an element $b \in A$ with $b \neq 0_A$ but $a \otimes b = 0_A$.
(ii) If A is a ring with unity and has no zero divisors then we say that A is a ***domain***.
(iii) If A is a commutative ring with unity and has no zero divisors we say that A is an ***integral domain***.
(iv) If A is a commutative ring with unity and every nonzero element of A is a unit, then A is a ***field***.

The requirement (in the previous definition) that A be a nontrivial ring will ensure that we do not consider a ring with one element to be an integral domain or field. The trivial ring $\{a\}$ is easily commutative and would have the unity and zero of the ring to be equal, i.e., $0_A = 1_A$. The ring would vacuously have no zero divisors and the nonzero elements would be units, since there are no nonzero elements. Thus without our restriction the trivial ring would qualify as an integral domain or a field.

Example 4.15 Consider the ring $M_2(\mathbb{R})$. We saw in Example 4.3 that this ring is not commutative. Are there zero divisors or units? Consider the elements

$$a = \begin{bmatrix} -1 & 0 \\ 0 & 1 \end{bmatrix}, \quad b = \begin{bmatrix} 1 & 0 \\ -1 & 0 \end{bmatrix}, \quad c = \begin{bmatrix} 0 & 0 \\ 3 & 2 \end{bmatrix}.$$

As the zero is $\begin{bmatrix} 0 & 0 \\ 0 & 0 \end{bmatrix}$ and the unity is $\begin{bmatrix} 1 & 0 \\ 0 & 1 \end{bmatrix}$, we have b and c as nonzero elements, but b and c are zero divisors as seen below.

$$bc = \begin{bmatrix} 1 & 0 \\ -1 & 0 \end{bmatrix} \begin{bmatrix} 0 & 0 \\ 3 & 2 \end{bmatrix} = \begin{bmatrix} 0 & 0 \\ 0 & 0 \end{bmatrix}$$

Notice that a is a unit since

$$aa = \begin{bmatrix} -1 & 0 \\ 0 & 1 \end{bmatrix} \begin{bmatrix} -1 & 0 \\ 0 & 1 \end{bmatrix} = \begin{bmatrix} 1 & 0 \\ 0 & 1 \end{bmatrix}.$$

Many of our familiar rings are integral domains. The integers, rational, or real numbers with usual addition and multiplication do not have zero divisors. This is one of the properties of the real numbers we use to help us solve polynomial equations such as $3x^2 + 2x = 0$. We write $x(3x + 2) = 0$ and then say that either $x = 0$ or $3x + 2 = 0$ to give the solutions as $x = 0$ and $x = -\frac{2}{3}$. However, this type of reasoning will only be valid when the ring involved is a domain.

Example 4.16 If we look in the ring $(\mathbb{Z}_8, +_8, \cdot_8)$ we notice that $3(2)^2+2(2) = 0$, $3(0)^2+2(0) = 0$, $3(4)^2 + 2(4) = 0$, and $3(6)^2 + 2(6) = 0$ so there are four solutions to the equation $3x^2 + 2x = 0$. Also notice that $3(6) + 2 = 4$ so $x = 6$ does <u>not</u> satisfy either $x = 0$ or $3x + 2 = 0$. This is all because we do not have a domain, as 2, 4, and 6 are zero divisors. Polynomial equations will become critical to our study of Modern Algebra beginning in Chapter 7.

The following theorem will be used frequently, and strengthened later in Theorem 6.13.

Theorem 4.17

For $n > 1$, the ring $(\mathbb{Z}_n, +_n, \cdot_n)$, is an integral domain if and only if n is prime.

Proof Consider the ring $(\mathbb{Z}_n, +_n, \cdot_n)$ which we already know is a commutative ring with unity by Theorem 4.11. We will prove one direction of the implication, and the other will be left as an exercise at the end of the chapter.

($\leftarrow$) Suppose that n is a prime, $n > 1$. We need to show that $\mathbb{Z}_n$ is an integral domain. We only need to show there are no zero divisors in $\mathbb{Z}_n$. Assume instead we have nonzero $a, b \in \mathbb{Z}_n$ so that $a \cdot_n b = 0$. This means that the integers a, b make ab evenly divisible by n, or that n evenly divides ab. As n is prime then by Theorem 0.21 either n evenly divides a or n evenly divides b. As the elements of $\mathbb{Z}_n$ are $0, 1, 2, \ldots, n-1$, the only one evenly divisible by n is 0. Thus one of a or b must have been 0 contradicting our assumptions. Thus $\mathbb{Z}_n$ is an integral domain when n is prime. $\square$

Following the same pattern we used for groups, we will now look at basic algebra in rings. The fact that we have two operations makes the algebra a bit more interesting, but it is critical to remember that we may not be dealing with real numbers.

4.2 Basic Algebra in Rings

In order to make the notation easier to read and write we will stop using the symbols $\oplus$ and $\otimes$. Thus instead of writing $a \oplus b$ or $a \otimes b$ we use the standard notation of $a + b$ and ab to refer to the addition and multiplication in our ring. Instead of saying $(A, \oplus, \otimes)$ is a ring we will simply say A is a ring, but it does not imply the operations are any "usual" addition or multiplication. If there is more than one ring involved we use $+_A$ and $\cdot_A$ to denote the operations in A.

As each ring is an abelian group under its addition, all of the theorems we proved about groups are valid here, but must be translated into additive notation (since we only have a group under addition). The following theorem is a translation of Theorem 1.17, and uses the fact that addition is commutative in the statement of (ii).

Theorem 4.18

Suppose A is a ring.

(i) For every $a \in A, -(-a) = a$.
(ii) For every $a, b \in A, -(a + b) = -b + -a = -a + -b$.

Frequently, we will add a negative of one element to another element. In $\mathbb{Z}$ we call this ***subtraction***, and we will use this notation in a ring as well.

Definition 4.19 Let A be a ring. For any $a, b \in A$ we define $a - b = a +_A (-b)$ where $-b$ is the negative of b in A.

The zero of a ring and negatives of its elements have well-understood properties since $(A, +)$ is a group. Now they must also interact with multiplication, and the next theorem tells us that some familiar properties will hold.

Theorem 4.20

Suppose that A is a ring.

(i) For every $a \in A, a0_A = 0_A = 0_A a$.
(ii) For every $a, b \in A, -(ab) = (-a)b = a(-b)$.

Proof Assume A is a ring.

(i) Let $a \in A$. We know that $0_A + 0_A = 0_A$ as 0_A is the identity under addition. Thus $a0_A = a(0_A + 0_A)$. Using the Distributive Law we see that $a0_A = a0_A + a0_A$. Since $a0_A + 0_A = a0_A$ then $a0_A + 0_A = a0_A + a0_A$. But $(A, +)$ is a group so by the Cancellation Law we can conclude $0_A = a0_A$. Similarly, you should be able to verify that $0_A a = 0_A$.

(ii) Assume now that we have $a, b \in A$. In order to say that $-(ab) = (-a)b$ we must show that $(-a)b$ is the negative of ab. Since addition is commutative in A we only need to show that $ab + (-a)b = 0_A$. Notice that using the Distributive Law we know that $ab + (-a)b = (a - a)b$. But by definition of negatives $a - a = 0_A$. Thus $ab + (-a)b = 0_A b$. By (i) we know $0_A b = 0_A$ so we have $ab + (-a)b = 0_A$ as needed. Thus $-(ab) = (-a)b$. You should show that $-(ab) = a(-b)$. □

It is important to remember that a theorem about groups can only be applied to a ring using the addition. One common mistake occurs when using the Cancellation Law, Theorem 1.20. It has become a familiar step to simplify an equation of the form $abc = xyc$ to $ab = xy$. Unfortunately, the proof of Theorem 1.20 relies on the use of inverses. Thus to use this theorem with multiplication in a ring, the element you wish to "cancel" must be a unit. The following theorem is the most we can say in a ring (unless we have a domain) and is left as an exercise at the end of the chapter.

Theorem 4.21

Suppose A is a ring. If $a, b, c \in A$ with $a \neq 0_A$ and $ab = ac$, then either a is a zero divisor or $b = c$.

In a trivial ring (exactly one element) we saw that $0_A = 1_A$, but implied that this would not be the case for a nontrivial ring. The next theorem finally shows us why.

Theorem 4.22

If a ring with unity A contains more than one element then $0_A \neq 1_A$.

Proof Suppose A is a ring with unity, and has at least two elements. Thus there must exist some element $a \in A$ with $a \neq 0_A$. Suppose instead that we have $0_A = 1_A$. Thus it must be true that $a0_A = a1_A$. Since 1_A is the unity of A we know that $a1_A = a$. Also by Theorem 4.20 we know that $a0_A = 0_A$. But this tells us that $0_A = a$, contradicting our choice of a. Thus we must have $0_A \neq 1_A$. □

This brings up two interesting questions that will be considered in exercises:

- If A is a ring with exactly two elements, must it be a ring with unity?
- If A is a ring with unity we know that $1_A^{-1} = 1_A$, but is $-(1_A) = 1_A$ as well?

Finally, we need to use the notation for exponentiation as we did in groups, but must be careful to keep our notation for $a+a+\cdots+a$ and $aaa\cdots a$ distinct. Most students are used to thinking of $a + a + \cdots + a(n - many)$ with the shorthand na. Unfortunately, this is the notation we would use to multiply two elements of A, and thus seems to imply that $n \in A$. However, this will not generally be the case (think about $M_2(\mathbb{Z})$ or $\wp(C)$, for example) and can be misleading. Thus we introduce a new symbol.

Definition 4.23 Let A denote a ring. For every $n \in \mathbb{Z}$ and $a \in A$, define $n \bullet a$ to be the element of A as follows:

$$n \bullet a = \begin{cases} \underbrace{a + a + \cdots + a}_{n-many} & n > 0 \\ 0_A & n = 0 \\ \underbrace{-a + -a + \cdots + -a}_{-n-many} & n < 0. \end{cases}$$

The previous definition translates Definition 1.21 into additive notation. The symbol is a large "dot" to distinguish it from the multiplication defined in A. We can only use the multiplication of A on two elements of A, as mentioned before the definition.

Example 4.24 Consider the ring $(\mathbb{Z}_8, +_8, \cdot_8)$. Using $a = 3$ we calculate $12 \bullet a$ and $-6 \bullet a$ as follows. (Recall that the negative of 3 is 5.)

$$12 \bullet a = 3 +_8 3 +_8 \cdots +_8 3 = 36(\text{mod}8) = 4$$
$$-6 \bullet a = 5 +_8 5 +_8 5 +_8 5 +_8 5 +_8 5 = 30(\text{mod}8) = 6$$

Definition 4.25 Let A denote a ring and $a \in A$.

(i) For a positive integer n, $a^n = \underbrace{aa \cdots a}_{n-many}$.

(ii) If A is a ring with unity 1_A and $a \neq 0_A$ then $a^0 = 1_A$.

(iii) If a is a unit in A and n is a negative integer then

$$a^n = \underbrace{a^{-1}a^{-1} \cdots a^{-1}}_{-n-many}.$$

Example 4.26 Consider again the ring $(\mathbb{Z}_8, +_8, \cdot_8)$ and $a = 3$. As $3^2 = 9(\text{mod}8) = 1$ then 3 is a unit with $3^{-1} = 3$, so we can calculate $3^{-6} = 3 \cdot_8 3 \cdot_8 3 \cdot_8 3 \cdot_8 3 \cdot_8 3 = 729(\text{mod}8) = 1$. However, we cannot calculate 4^{-6} since 4 is not a unit in this ring (as you can verify).

Example 4.27 Consider the power set of a nonempty set, $\wp(C)$, which is a ring under the operations defined in Example 4.6. If $U \in \wp(C)$ what are $6 \bullet U$ and U^6?

$$\begin{aligned} 6 \bullet U &= U \oplus U \oplus U \oplus U \oplus U \oplus U = \emptyset \\ U^6 &= U \otimes U \otimes U \otimes U \otimes U \otimes U = U \end{aligned}$$

Both the additive notation, $n \bullet a$, and exponentiation, a^n, satisfy the basic power rules we are familiar with. A translation of Theorem 1.23 gives us the first three parts of the next theorem.

Theorem 4.28

Suppose A is a ring, $a \in A$, and $n, m \in \mathbb{Z}$.

(i) $(n +_{\mathbb{Z}} m) \bullet a = (n \bullet a) +_A (m \bullet a)$.
(ii) $(n \cdot_{\mathbb{Z}} m) \bullet a = n \bullet (m \bullet a)$.
(iii) $(-n) \bullet a = n \bullet (-a) = -(n \bullet a)$.
(iv) If A has a unity 1_A then $n \bullet a = (n \bullet 1_A)a$.

Proof Suppose A is a ring, $a \in A$. Parts (i), (i), and (iii) follow directly from Theorem 1.23 so we only need to prove (iv). Assume A has unity 1_A and let $n \in \mathbb{Z}$. If $n = 0$, then by definition $n \bullet a = 0_A$ and $n \bullet 1_A = 0_A$. Thus $(n \bullet 1_A)a = 0_A a = 0_A$, so $n \bullet a = (n \bullet 1_A)a$.

If $n > 0$, then $n \bullet a = a + a + \cdots + a(n - many)$ and since $a = 1_A a$ we have $n \bullet a = 1_A a + 1_A a + \cdots + 1_A a(n - many)$. Using the Distributive Law we factor out the a on the right to get $n \bullet a = (1_A + 1_A + \cdots + 1_A)a = (n \bullet 1_A)a$.

Finally, if $n < 0$ then $(-n) > 0$ so let $m = -n$. Then $m \bullet a = (m \bullet 1_A)a$ as seen above. Thus with help from (iii) and Theorem 4.20 we see can show that $n \bullet a = (n \bullet 1_A)a$.

$$\begin{aligned} n \bullet a &= (-m) \bullet a \\ &= -(m \bullet a) \\ &= -[(m \bullet 1_A)a] \\ &= [-(m \bullet 1_A)]a \\ &= [((-m) \bullet 1_A)]a \\ &= (n \bullet 1_A)a \end{aligned}$$

□

For multiplication we also have the usual power rules.

Theorem 4.29

Suppose A is a ring, $a \in A$, and n, m are integers.

(i) If $n > 0$ and $m > 0$ then $a^n a^m = a^{n+m}$.
(ii) If $n > 0$ and $m > 0$ then $(a^n)^m = a^{nm}$.
(iii) If a is a unit in A (a ring with unity) then $a^{-n} = (a^{-1})^n = (a^n)^{-1}$.

The proof of this is very similar to our proof of Theorem 1.23 for positive integers.

Suppose A is a ring and $a, b \in A$. To calculate the element $(a+b)^2$ we use the Distributive Law.

$$(a+b)^2 = (a+b)(a+b) = a(a+b) + b(a+b) = a^2 + ab + ba + b^2$$

This is not quite the same formula you likely remember, $(a+b)^2 = a^2 + 2ab + b^2$. Why? Can you see what assumption we are making if we write our final step as $a^2 + 2 \bullet ab + b^2$?

A similar formula can be found for any positive integer n, if the ring A is commutative. It is called the ***Binomial Formula*** (see [7] for more information) and involves a special notation used in combinatorics, $\binom{n}{k}$, called n *choose* k, the number of ways to choose k things out of a list of n many.

Theorem 4.30

Suppose that A is a commutative ring with unity and that $a, b \in A$. For any positive integer n,

$$(a+b)^n = a^n + \binom{n}{1} \bullet a^{n-1}b + \binom{n}{2} \bullet a^{n-2}b^2 + \cdots + \binom{n}{n-1} \bullet ab^{n-1} + b^n$$

where for each $k < n$, $\binom{n}{k} = \frac{n!}{k!(n-k)!}$.

For example, when $n = 3$, we have $\binom{3}{1} = \frac{3!}{1!(2)!} = \frac{6}{2} = 3$ and $\binom{3}{2} = \frac{3!}{2!(1)!} = \frac{6}{2} = 3$, giving the familiar formula $(a+b)^3 = a^3 + 3 \bullet a^2b + 3 \bullet ab^2 + b^3$.

4.3 Subrings

Just as we did in Chapter 1 for groups, after the definition of a new structure and some of the properties of that structure, the concept of a substructure logically follows.

Definition 4.31 Suppose A is a ring. A nonempty subset S of A is a ***subring*** of A if S is also a ring under the addition and multiplication of A.

Example 4.32 Consider the subset $S = \{0, 2, 4\}$ in the ring $(\mathbb{Z}_6, +_6, \cdot_6)$. We need to determine if S is a ring under $+_6$ and $\cdot_6$. Since we have a finite set the Cayley tables for the operations will help us verify the needed properties.

$+_6$	0	2	4
0	0	2	4
2	2	4	0
4	4	0	2

$\cdot_6$	0	2	4
0	0	0	0
2	0	4	2
4	0	2	4

Notice from the first table that S is closed under $+_6$. Thus we know $+_6$ is an operation on S. Also it should be clear that as $0 \in S$ there is a zero element (it was the additive identity in $\mathbb{Z}_6$). Since $-2 = 4$, and $-4 = 2$, then each element of S has a negative in S as well. We already knew that $+_6$ is commutative so $(S, +_6)$ forms an abelian group, a subgroup of $(\mathbb{Z}_6, +_6)$. The second table tells us that $\cdot_6$ is an operation on S. As the operations in $\mathbb{Z}_6$ are commutative, associative, and satisfy the distributive laws, we need not check them on S since the elements of S are from $\mathbb{Z}_6$.

Thus S is a subring of $\mathbb{Z}_6$.

There is something important to notice in the previous example. In the original ring $\mathbb{Z}_6$, there is a unity element, namely 1. But in S the unity is equal to 4 since $4 \cdot_6 2 = 2$ and $4 \cdot_6 4 = 4$. Thus if a subring has a unity it is not required to agree with the unity of the original ring. In fact, a subring is not required to have a unity even if the original ring does. In an exercise you will be asked to show that the set of even integers, E, is a subring of the integers, $\mathbb{Z}$, under its usual operations but has no unity element.

Not every part of the definition of a ring needed to be checked in the previous example, giving rise to the following theorem which is analogous to Theorem 1.30.

Theorem 4.33

Suppose A is a ring and S is a subset of A. S is a subring of A if and only if the following hold:

(i) S is nonempty.
(ii) S is closed under subtraction in A.
(iii) S is closed under multiplication in A.

Proof Suppose A is a ring and S is a subset of A.

($\rightarrow$) Suppose we know that S is a subring of A. By Definition 4.31 S is a ring under the operations of A, so $(S, +)$ must be a subgroup of $(A, +)$. Thus by Theorem 1.30 S is nonempty. Also multiplication from A is an operation on S so S is closed under multiplication. To show that S is closed under subtraction, let $a, b \in S$. As $b \in S$ and S is closed under negatives by Theorem 1.30, we have $-b \in S$. But S is also closed under addition by Theorem 1.30 so $a + (-b) \in S$. Thus $a - b \in S$ and S is closed under subtraction completing this half of the proof.

($\leftarrow$) Now suppose that S satisfies (i), (ii), and (iii). We must show that S is a subring of A. From (i) we know S is nonempty. Since S is nonempty there is some $a \in S$. But S is closed under subtraction by (ii) so $a - a \in S$ and thus $0_A \in S$. Using that S is closed under subtraction again, $0_A - a \in S$ and so $-a \in S$ or S is closed under negatives. For addition, suppose $a, b \in S$ then $-b \in S$ as well, so $a - (-b) \in S$ since S is closed under subtraction. But $a - (-b) = a + -(-b) = a + b$ so $a + b \in S$ and S is closed under addition. Thus by Theorem 1.30, $(S, +)$ is a subgroup of $(A, +)$, i.e., is an abelian group.

Finally (iii) tells us multiplication is an operation on S, and the associativity of multiplication and distributive laws hold in S since the elements are from A and A is a ring. Thus S is a subring of A. $\square$

Example 4.34 As groups we had $\mathbb{Z}$, a subgroup of both $\mathbb{Q}$ and $\mathbb{R}$ under their usual addition, as well as $\mathbb{Q}$, a subgroup of $\mathbb{R}$. Similarly, under usual addition and multiplication, $\mathbb{Z}$ is a subring of both $\mathbb{Q}$ and $\mathbb{R}$ while $\mathbb{Q}$ is a subring of $\mathbb{R}$.

Example 4.35 Consider the set $M_2(\mathbb{Z})$ of 2×2 matrices whose entries are *integers*. Similar to Example 4.3, under the usual addition and multiplication of matrices $M_2(\mathbb{Z})$ is a ring. Define the set:

$$S = \left\{ \begin{bmatrix} a & b \\ c & 0 \end{bmatrix} : a, b, c \in \mathbb{Z} \right\}.$$

We know that $\begin{bmatrix} 0 & 0 \\ 0 & 0 \end{bmatrix} \in S$ so S is not empty. S is closed under subtraction (you should prove it), but S is not closed under multiplication. For a counterexample, notice that using $x, y \in M_2(\mathbb{Z})$ below we have $xy \notin M_2(\mathbb{Z})$.

$$x = \begin{bmatrix} 2 & 3 \\ 4 & 0 \end{bmatrix} \; y = \begin{bmatrix} 1 & 2 \\ 3 & 0 \end{bmatrix}$$

$$xy = \begin{bmatrix} 11 & 4 \\ 4 & 8 \end{bmatrix}$$

Thus S is not a subring of $M_2(\mathbb{Z})$.

4.4 Ring Homomorphisms

As with groups, we now consider homomorphisms between rings. This time we need to consider how the function interacts with both operations of the ring.

Definition 4.36 Let $(A, +_A, \cdot_A)$ and $(K, +_K, \cdot_K)$ denote two rings. A ***ring homomorphism*** is a function $f : A \to K$ so that for any $a, b \in A$, $f(a +_A b) = f(a) +_K f(b)$ and $f(a \cdot_A b) = f(a) \cdot_K f(b)$.

If the homomorphism f is also a bijection from A to K we say that f is a ***ring isomorphism*** or that A and K are isomorphic.

Notice that the first part of this definition, namely $f(a +_A b) = f(a) +_K f(b)$ for any $a, b \in A$, tells us that the function is a group homomorphism using the addition of our rings. Thus facts from previous chapters about group homomorphisms are valid for rings and addition, including the following translation of Theorem 1.37.

Theorem 4.37

Let A and K be rings, and suppose that $f : A \to K$ is a ring homomorphism.

(i) $f(0_A) = 0_K$.
(ii) $f(-a) = -f(a)$ for each $a \in A$.
(iii) For every $n \in \mathbb{Z}$ and $a \in A$, $n \bullet f(a) = f(n \bullet a)$.

Example 4.38 Consider the rings $(\mathbb{Z}_6, +_6, \cdot_6)$ and $(\mathbb{Z}_{12}, +_{12}, \cdot_{12})$. Define a function $f : \mathbb{Z}_6 \to \mathbb{Z}_{12}$ by $f(x) = (3x)(\text{mod}12)$. Is this a ring homomorphism?

Consider the elements $4, 5 \in \mathbb{Z}_6$. By definition we have $f(4 +_6 5) = f(3) = 9$ since in $\mathbb{Z}_6$ we have $4 +_6 5 = 3$. However, $f(4) +_{12} f(5) = 0 +_{12} 3 = 3$. Thus $f(4 +_6 5) \neq f(4) +_{12} f(5)$ and f is not a ring homomorphism.

Consider instead the function $g : \mathbb{Z}_6 \to \mathbb{Z}_{12}$ defined by $g(x) = (4x)(\text{mod}12)$. As in Chapter 1 we will use the Cayley tables for the rings to help us decide if g is a homomorphism. The two Cayley tables for $\mathbb{Z}_6$ are shown here:

$+_6$	0	1	2	3	4	5
0	0	1	2	3	4	5
1	1	2	3	4	5	0
2	2	3	4	5	0	1
3	3	4	5	0	1	2
4	4	5	0	1	2	3
5	5	0	1	2	3	4

$\cdot_6$	0	1	2	3	4	5
0	0	0	0	0	0	0
1	0	1	2	3	4	5
2	0	2	4	0	2	4
3	0	3	0	3	0	3
4	0	4	2	0	4	2
5	0	5	4	3	2	1

Using our function g, we see that $g(0) = 0, g(1) = 4, g(2) = 8, g(3) = 0, g(4) = 4, g(5) = 8$. Replacing each element in the Cayley tables for $\mathbb{Z}_6$ with its image under g:

$+_{12}$	0	4	8	0	4	8
0	0	4	8	0	4	8
4	4	8	0	4	8	0
8	8	0	4	8	0	4
0	0	4	8	0	4	8
4	4	8	0	4	8	0
8	8	0	4	8	0	4

$\cdot_{12}$	0	4	8	0	4	8
0	0	0	0	0	0	0
4	0	4	8	0	4	8
8	0	8	4	0	8	4
0	0	0	0	0	0	0
4	0	4	8	0	4	8
8	0	8	4	0	8	4

All of these answers are correct in $(\mathbb{Z}_{12}, +_{12}, \cdot_{12})$, so g is a ring homomorphism.

Example 4.39 Consider the rings $M_2(\mathbb{Z})$ and $\mathbb{Z}$, and define $f : \mathbb{Z} \to M_2(\mathbb{Z})$ by $f(a) = \begin{bmatrix} a & 0 \\ 0 & a \end{bmatrix}$. Let $a, b \in \mathbb{Z}$.

$$f(a) = \begin{bmatrix} a & 0 \\ 0 & a \end{bmatrix}, f(b) = \begin{bmatrix} b & 0 \\ 0 & b \end{bmatrix}, f(a+b) = \begin{bmatrix} a+b & 0 \\ 0 & a+b \end{bmatrix}$$

$$f(a) + f(b) = \begin{bmatrix} a & 0 \\ 0 & a \end{bmatrix} + \begin{bmatrix} b & 0 \\ 0 & b \end{bmatrix} = \begin{bmatrix} a+b & 0 \\ 0 & a+b \end{bmatrix}$$

$$f(ab) = \begin{bmatrix} ab & 0 \\ 0 & ab \end{bmatrix}$$

$$f(a)f(b) = \begin{bmatrix} a & 0 \\ 0 & a \end{bmatrix} \begin{bmatrix} b & 0 \\ 0 & b \end{bmatrix} = \begin{bmatrix} ab & 0 \\ 0 & ab \end{bmatrix}$$

Then $f(a+b) = f(a)+f(b)$ and $f(ab) = f(a)f(b)$ so our function is a ring homomorphism.

Notice that in Example 4.39 both of the rings have a unity and the homomorphism f maps the integer 1 to the identity matrix. But in the first function in Example 4.38 we have $f(1) = 3$. Could that be why the function failed to be a ring homomorphism, i.e., is it necessary to have $f(1_A) = 1_K$ to have a ring homomorphism from A to K?

A simple example tells us we cannot prove that $f(1_A) = 1_K$ in general since $f(x) = 0_K$ for all $x \in A$ is a ring homomorphism (be sure you can verify this). One fact that will be used in later chapters is given below, but left as an exercise at the end of the chapter. Remember that $\{0_A\}$ is not a field.

Theorem 4.40

Let A and K be **fields**. If $f : A \to K$ is an *isomorphism*, then $f(1_A) = 1_K$.

There is a collection of ring homomorphisms we will use throughout the book, described in the next theorem.

Theorem 4.41

Using the rings $(\mathbb{Z}, +, \cdot)$ and $(\mathbb{Z}_n, +_n, \cdot_n)$, the function $f : \mathbb{Z} \to \mathbb{Z}_n$ defined by $f(x) = x(\text{mod} n)$ is a ring homomorphism for any integer $n > 1$.

Proof Define $f : \mathbb{Z} \to \mathbb{Z}_n$ by $f(x) = x(\text{mod} n)$. To see that this is a ring homomorphism consider integers x, y. We need to show that $f(x+y) = f(x) +_n f(y)$ and that $f(xy) = f(x) \cdot_n f(y)$.

There exist $q, r, k, s \in \mathbb{Z}$, $0 \leq r, s < n$ with $x = nq + r$ and $y = nk + s$. Thus $f(x) = r$ and $f(y) = s$ so $f(x) +_n f(y) = r +_n s$ and $f(x) \cdot_n f(y) = r \cdot_n s$. The following steps complete the proof.

$$\begin{aligned} f(x+y) &= f((nq+r)+(nk+s)) \\ &= f(n(q+k)+(r+s)) \\ &= (r+s)(\text{mod} n) \\ &= f(x)+_n f(y) \end{aligned} \qquad \begin{aligned} f(xy) &= f((nq+r)(nk+s)) \\ &= f(n(nqk+qs+kr)+rs) \\ &= (rs)(\text{mod} n) \\ &= f(x)\cdot_n f(y) \end{aligned}$$

Therefore f is a ring homomorphism. □

The definitions of image and kernel for a function are identical here to those in Chapter 3, where the identity referred to in $ker(f)$ is the additive identity, i.e., zero of the ring.

Definition 4.42 Let A and K denote rings and $f : A \to K$ a ring homomorphism.

(i) The ***kernel*** of f is defined by $ker(f) = \{x \in A : f(x) = 0_K\}$.
(ii) The ***image*** of f is defined by $f(A) = \{y \in K : y = f(x) \text{ for some } x \in A\}$.

There will be more about the kernel and image of a homomorphism in Chapter 5 where Quotient Rings and another Fundamental Homomorphism Theorem are discussed.

Exercises for Chapter 4

Section 4.1 Rings

In each of the problems 1 through 8 below, a set and operations are given. Determine if $(A, \oplus, \otimes)$ is a ring. If it is not a ring, determine which parts of the definition hold, and which fail by proving or finding a counterexample for each.

1. Consider the set $A = \{5n : n \in \mathbb{Z}\}$. Determine if A is a ring under the operations defined by $a \oplus b = a + b + 5$ and $a \otimes b = ab$. Note that $5n$ and ab mean usual integer multiplication, while $+$ refers to usual integer addition.
2. Define $A = \{0, 2, 4, 6\}$. Using $+_8$ and $\cdot_8$, show that A is a ring. Is it a commutative ring with unity?

3. On the set $\mathbb{Z}$ of integers define the following operations: $a \oplus b = a + b - 1$ and $a \otimes b = ab - (a+b) + 2$. Determine if $(\mathbb{Z}, \oplus, \otimes)$ is a ring. Is it a commutative ring with unity? Note that ab implies usual integer multiplication, and $+$ or $-$ refer to usual integer addition or subtraction.
4. On the set $\mathbb{R} - \{-1\}$ define the operations $a \oplus b = a + b + ab$ and $a \otimes b = 0$. Determine if $(\mathbb{R} - \{-1\}, \oplus, \otimes)$ is a ring. Is it a commutative ring with unity? Note that ab implies usual multiplication, and $+$ refers to usual addition on the real numbers.
5. Consider the set $\mathbb{Q}$ and define the operations $a \oplus b = a + b$ and $a \otimes b = \frac{1}{2}ab$. Determine if $(\mathbb{Q}, \oplus, \otimes)$ is a ring. Note that ab implies usual multiplication, and $a + b$ refers to usual addition on the rational numbers.
6. With the set $\mathbb{Z}$ of integers, using $a \oplus b = a + b$ and $a \otimes b = a(b-1)$, determine if $(\mathbb{Z}, \oplus, \otimes)$ is a ring. Note that ab implies usual integer multiplication, and $+$ or $-$ refer to usual integer addition or subtraction.
7. Determine if $\mathbb{Q}^* = \{\frac{a}{b} : a, b \in \mathbb{Z}, a, b \neq 0\}$ (the nonzero rationals) is a ring under the operations $\frac{a}{b} \oplus \frac{c}{d} = \frac{ac}{bd}$ and $\frac{a}{b} \otimes \frac{c}{d} = \frac{ac}{bd}$ (yes, the same for each operation). Note that xy refers to usual multiplication in the rational numbers.
8. Define the set $A = \{a + b\sqrt{3} : a, b \in \mathbb{Z}\}$ and the operations on A shown below. Determine if $(A, \oplus, \otimes)$ is a ring. Note that $+_{\mathbb{Z}}$ means usual integer addition in the definitions, and ac, $3bd$, ad, bc are found by usual integer multiplication.

$$\begin{aligned}(a + b\sqrt{3}) \oplus (c + d\sqrt{3}) &= (a +_Z c) + (b +_Z d)\sqrt{3} \\ (a + b\sqrt{3}) \otimes (c + d\sqrt{3}) &= (ac +_Z 3bd) + (ad +_Z bc)\sqrt{3}\end{aligned}$$

9. Using the notation of Example 4.6, show that $\oplus$ is associative.
10. Prove: If $(A, \oplus)$ is an abelian group with identity e_A, then the operation $a \otimes b = e_A$ for all $a, b \in A$ makes $(A, \oplus, \otimes)$ a ring.
11. Show the Cayley tables for the ring $(\mathbb{Z}_9, +_9, \cdot_9)$ and identify the zero divisors and units of the ring.
12. Show the Cayley tables for the ring $(\mathbb{Z}_5, +_5, \cdot_5)$ and identify the zero divisors and units of the ring.
13. Show the Cayley tables for the ring $(\mathbb{Z}_{10}, +_{10}, \cdot_{10})$ and identify the zero divisors and units of the ring.
14. Show the Cayley tables for the ring $(\mathbb{Z}_7, +_7, \cdot_7)$ and identify the zero divisors and units of the ring.
15. Suppose $(A, +_A, \cdot_A)$ and $(B, +_B, \cdot_B)$ are rings. Prove that the set $A \times B = \{(a, b) : a \in A, b \in B\}$ is a ring under the operations described in Definition 4.12.
16. Show the Cayley tables for the ring $(\mathbb{Z}_2 \times \mathbb{Z}_2, \oplus, \otimes)$ where $(a, b) \oplus (c, d) = (a +_2 c, b +_2 d)$ and $(a, b) \otimes (c, d) = (a \cdot_2 c, d \cdot_2 d)$. Identify the zero divisors and units of the ring.
17. Show the Cayley tables for the ring $(\mathbb{Z}_2 \times \mathbb{Z}_4, \oplus, \otimes)$ where $(a, b) \oplus (c, d) = (a +_2 c, b +_4 d)$ and $(a, b) \otimes (c, d) = (a \cdot_2 c, b \cdot_4 d)$. Identify the zero divisors and units of the ring.

18. Show the Cayley tables for the ring $(\mathbb{Z}_3 \times \mathbb{Z}_3, \oplus, \otimes)$ where $(a,b) \oplus (c,d) = (a +_3 c, b +_3 d)$ and $(a,b) \otimes (c,d) = (a \cdot_3 c, d \cdot_3 d)$. Identify the zero divisors and units of the ring.
19. If A and B are finite nontrivial rings, show that the ring $A \times B$ (described in Definition 4.12) contains at least $|A| + |B| - 2$ many zero divisors.
20. Let C denote a nonempty set, and $\wp(C) = \{U : U \subseteq C\}$ the ring whose operations were defined in Example 4.6. Prove that C is the unity and determine if any elements in $\wp(C)$ are units. Are there zero divisors in $\wp(C)$?
21. Show that the set of functions $\Im(\mathbb{R}) = \{f : \mathbb{R} \to \mathbb{R} : f \text{ is a function}\}$ forms a ring where $f + g$ is defined by $(f + g)(x) = f(x) +_{\mathbb{R}} g(x)$ for all x in $\mathbb{R}$, and fg is defined by $fg(x) = f(x) \cdot_{\mathbb{R}} g(x)$ for all x in $\mathbb{R}$. Is it a commutative ring with unity?
22. Prove the second half of Theorem 4.17.

Section 4.2 Basic Algebra in Rings

23. Let A be a ring and $a, b \in A$. Prove $0_A a = 0_A$ and $-(ab) = a(-b)$.
24. Suppose A is a nontrivial ring with unity, and $a \in A$. Prove: If a is a unit then a cannot be a zero divisor.
25. Suppose A is a nontrivial ring with unity, and $a, b, c \in A$ with $a \neq 0_A$. Prove: If $ab = ac$ then either a is a zero divisor or $b = c$.
26. Suppose A is a commutative ring with unity. Prove: If $a, b \in A$, $a \neq 0_A$, and $a(b + 1_A) = b(a + a)$ then either a is a zero divisor or $b = 1_A$.
27. Suppose A is a nontrivial ring with unity 1_A, $x, y \in A$ and $y \neq 0_A$. Prove: If $y^2 = y + yx$ then either y is a zero divisor or $y - x = 1_A$.
28. Prove: If A is a ring and $a^2 = a$ for every element $a \in A$ then $-a = a$ for each $a \in A$.
29. Give an example of a ring with more than two elements for which $a^2 = a$ for each $a \in A$.
30. Prove: If A is a ring and $a^2 = a$ for every element $a \in A$ then A is a commutative ring.
31. Suppose A is a nontrivial ring with unity 1_A, and for every element $a \in A$, $a^2 = a$. Prove: If $a \neq 1_A$ and $a \neq 0_A$ then a is a zero divisor.
32. Suppose A is a ring with unity 1_A and for every element $a \in A$, $a^3 = a$. Prove: For each $a \in A$ if $a \neq 0_A$ and a is not a zero divisor then a is a unit.
33. Give an example of a nontrivial ring with more than two elements so that $a^3 = a$ for each $a \in A$.
34. Suppose A is a nontrivial ring with unity. Prove: If for each $a \in A$, $a^2 = a$ then for each $a \in A$, $a^3 = a$. Show that the converse of this statement is not true by finding an example of a nontrivial ring A with $a^3 = a$ for each $a \in A$ but $a^2 \neq a$ for some $a \in A$.
35. If A is a ring with exactly two elements, must it be a ring *with unity*? Prove it is true or find a counterexample.
36. If A is a ring with exactly two elements and no zero divisors, must it be a ring *with unity*? Prove it is true or find a counterexample.
37. If A is a ring with unity we know that $1_A^{-1} = 1_A$, but is $-(1_A) = 1_A$? Use examples to help support your answer.

38. Let A be a nontrivial ring with unity. Prove: If $a, b \in A$ and ab is a zero divisor then at least one of a or b must be a zero divisor.
39. Let A be a nontrivial commutative ring with unity. If $a, b \in A$ and $a+b$ is a zero divisor, can we conclude that one of a or b is a zero divisor? Prove it is true or find a nontrivial counterexample.
40. Let A be a nontrivial commutative ring with unity. If $a, b \in A$ and ab is a unit can we conclude that both a and b must be units? Prove it is true or find a nontrivial counterexample. Would the answer change if the conclusion was that at least one of a or b is a unit?
41. Let A be a nontrivial commutative ring with unity. If $a, b \in A$ and $a + b$ is a unit, can we conclude that both of a or b must be units? Prove it is true or find a nontrivial counterexample. Would the answer change if the conclusion was that at least one of a or b is a unit?

Section 4.3 Subrings

42. Prove that the set of even integers E is a subring of $\mathbb{Z}$ under its usual operations, but has no unity.
43. If A is a nontrivial ring with unity, must $S = \{0_A, 1_A\}$ always be a subring of A? Prove that S is a subring or give an example of a nontrivial commutative ring in which S is not a subring.
44. Let A be a ring. Prove that $\{0_A\}$ and A are subrings of A.
45. Determine if $S = \{0, 3, 6\}$ is a subring of $(\mathbb{Z}_8, +_8, \cdot_8)$. Prove it is a subring or find a counterexample to see why it fails to be a subring.
46. Determine if $S = \{0, 3, 6\}$ is a subring of $(\mathbb{Z}_9, +_9, \cdot_9)$. Prove it is a subring or find a counterexample to see why it fails to be a subring.
47. Determine if $S = \{x^2 : x \in \mathbb{Z}\}$ is a subring of $(\mathbb{Z}, +, \cdot)$. Prove it is a subring or find a counterexample to see why it fails to be a subring.
48. Determine if $S = \{0, 3, 6, 9\}$ is a subring of $(\mathbb{Z}_{12}, +_{12}, \cdot_{12})$. Prove it is a subring or find a counterexample to see why it fails to be a subring.
49. Prove that $S = \left\{ \begin{bmatrix} a & 0 \\ 0 & a \end{bmatrix} : a \in A \right\}$ is a subring of $M_2(\mathbb{Z})$ with its usual addition and multiplication of matrices.
50. Determine if $S = \left\{ \begin{bmatrix} 0 & a \\ a & 0 \end{bmatrix} : a \in \mathbb{Z} \right\}$ is a subring of $M_2(\mathbb{Z})$ with usual addition and multiplication of matrices. Prove it is true or find a nontrivial counterexample.
51. Determine if $S = \left\{ \begin{bmatrix} 0 & 0 \\ a & 0 \end{bmatrix} : a \in \mathbb{Z} \right\}$ is a subring of $M_2(\mathbb{Z})$ with its usual addition and multiplication of matrices. Prove it is true or find a nontrivial counterexample.
52. Determine if $S = \left\{ \begin{bmatrix} a & b \\ 0 & 0 \end{bmatrix} : a, b \in \mathbb{Z} \right\}$ is a subring of $M_2(\mathbb{Z})$ with its usual addition

and multiplication of matrices. Prove it is true or find a nontrivial counterexample.

53. Let A be a commutative ring with unity, and $S = \{a \in A : a \text{ is a unit}\}$. Prove that S is a group under the multiplication of A.
54. Let A be a commutative ring with unity, and $S = \{a \in A : a \text{ is a unit}\}$. Determine if S is always a subring of A. Either prove that S is a subring or give an example of a nontrivial ring with unity in which S is not a subring.
55. Prove that the set of all continuous functions from $\mathbb{R}$ to $\mathbb{R}$, $C(\mathbb{R})$ is a subring of $\Im(\mathbb{R})$. Remember that the operations on $\Im(\mathbb{R})$ are usual addition and multiplication of functions defined in Exercise 21.
56. Determine if the subset $S = \left\{f \in C(\mathbb{R}) : \int_0^1 f dx = 0\right\}$ is a subring of $C(\mathbb{R})$.

Section 4.4 Ring Homomorphisms

57. Let A be a ring and define $f : A \to A$ by $f(x) = x$ for all $x \in A$. Prove that f is a ring homomorphism.
58. Consider two positive integers $m, n > 1$ and the rings $(\mathbb{Z}_n, +_n, \cdot_n)$ and $(\mathbb{Z}_m, +_m, \cdot_m)$. Define the function $f : \mathbb{Z}_n \to \mathbb{Z}_m$ by $f(x) = x(\text{mod} m)$ for all $x \in \mathbb{Z}_n$. Is this function always a ring homomorphism when the rings have their usual operations? Either prove it is always a homomorphism or find specific m, n and a counterexample showing it fails to be a homomorphism.
59. Consider the function $f : \mathbb{Z}_3 \times \mathbb{Z}_2 \to \mathbb{Z}_6$ defined by $f(a, b) = ab(\text{ mod } 6)$ where ab denotes usual integer multiplication. Determine if f is a ring homomorphism when each ring has its usual operations of addition and multiplication. Either prove it is a homomorphism or find a counterexample showing it fails to be a homomorphism.
60. Consider the function $f : \mathbb{Z}_3 \times \mathbb{Z}_2 \to \mathbb{Z}_6$ defined by $f(a, b) = (4a + 3b)(\text{mod} 6)$ where $4a + 3b$ denotes usual integer addition and multiplication. Determine if f is a ring homomorphism when each ring has its usual operations of addition and multiplication. Either prove it is a homomorphism or find a counterexample showing it fails to be a homomorphism.
61. Define $f : \mathbb{Z}_4 \times \mathbb{Z}_6 \to \mathbb{Z}_{12}$ by $f(a, b) = (3a + 2b)(\text{mod} 12)$. Determine if f is a ring homomorphism. Either prove it is a homomorphism or find a counterexample showing it fails to be a homomorphism. Each ring has the usual operations on it.
62. Define $f : M_2(\mathbb{Z}) \to \mathbb{Z}$ by $f\left(\begin{bmatrix} a & b \\ c & d \end{bmatrix}\right) = b$. Determine if f is a ring homomorphism. Either prove it is a homomorphism or find a counterexample showing it fails to be a homomorphism. Each ring has its usual operations.
63. Define $f : M_2(\mathbb{Z}) \to \mathbb{Z}$ by $f\left(\begin{bmatrix} a & b \\ c & d \end{bmatrix}\right) = ad - bc$. Determine if f is a ring homomorphism. Either prove it is a homomorphism or find a counterexample showing it fails to be a homomorphism. Each ring has its usual operations.

64. Consider $S = \left\{ \begin{bmatrix} a & b \\ 0 & 0 \end{bmatrix} : a, b \in \mathbb{Z} \right\}$, the subring of $M_2(\mathbb{Z})$ under usual addition and multiplication of matrices. Define $f : S \to \mathbb{Z}$ by $f\left(\begin{bmatrix} a & b \\ 0 & 0 \end{bmatrix} \right) = a$. Determine if f is a ring homomorphism. Either prove it is a homomorphism or find a counterexample showing it fails to be a homomorphism. Each ring has its usual operations.
65. Let A and K be rings. Prove: The function $f : A \to K$ defined by $f(a) = 0_K$ for all $a \in A$ is a ring homomorphism.
66. Define $g : \mathbb{Z}_3 \times \mathbb{Z}_3 \to \mathbb{Z}_3$ by $g(a,b) = a +_3 b$. With the usual operations of $+_3$ and $\cdot_3$ on $\mathbb{Z}_3$, determine if g is a ring homomorphism. Either prove it is a homomorphism or find a counterexample showing it fails to be a homomorphism. Each ring has its usual operations.
67. Define $g : \mathbb{Z}_3 \times \mathbb{Z}_3 \to \mathbb{Z}_3$ by $g(a,b) = a \cdot_3 b$. With the usual operations of $+_3$ and $\cdot_3$ on $\mathbb{Z}_3$, determine if g is a ring homomorphism. Either prove it is a homomorphism or find a counterexample showing it fails to be a homomorphism. Each ring has its usual operations.
68. Let A be a ring with unity and $a \in A$. Determine if the function $h : A \to A$ defined by $h(x) = ax$ is always a ring homomorphism. Must a have a special property to guarantee it?
69. Prove Theorem 4.40.
70. Suppose we have rings A, B, and C with ring homomorphisms $f : A \to B$ and $g : B \to C$. Prove that the function $g \circ f$ is a ring homomorphism.

Projects for Chapter 4

Project 4.1

Consider the set of ordered pairs of integers, $\mathbb{Z} \times \mathbb{Z} = \{(a,b) : a, b \in \mathbb{Z}\}$ with the operations defined below. Note that $+, -$ refer to usual addition and subtraction, while writing uv means usual multiplication in these definitions. The definitions make it clear that they are operations already.

$$(x,y) \oplus (a,b) = (x+a, y+b) \qquad (x,y) \otimes (a,b) = (xa - yb, xb + ya)$$

1. Prove that $(\mathbb{Z} \times \mathbb{Z}, \oplus, \otimes)$ is a ring.
2. Is this a commutative ring? Either prove that it is true or find a counterexample showing it fails to be commutative.
3. Determine if there is a unity for the ring $(\mathbb{Z} \times \mathbb{Z}, \oplus, \otimes)$. Either find the unity and verify it is the unity or explain why none can exist.

4. Prove that $(-1, 1)$ is *not* a zero divisor. Do zero divisors exist in this ring?
5. Prove that $(-1, 1)$ is also *not* a unit. Can you find a unit (other than the unity) of this ring?

Project 4.2

When checking to see if operations create a ring, associativity and the distributive laws are usually the most difficult to check. Thus we will practice using a variety of operations. Every operation is defined on the set $\mathbb{Z}$.

1. Define the operation $x \oplus y = x - y + xy$. Determine if $\oplus$ is associative. Either prove that it is associative or find a counterexample.
2. Define the operation $x \otimes y = y$. Determine if $\otimes$ is associative. Either prove that it is associative or find a counterexample.
3. Define the operations $x \oplus y = xy$ and $x \otimes y = y$. Determine if the distributive laws hold. For each one either prove it holds or find a counterexample. Remember to use the correct operations.
4. Define the operations $x \oplus y = x + y$ and $x \otimes y = x + y$ (yes, they are the same). Determine if the distributive laws hold. For each one either prove it holds or find a counterexample. Remember to use the correct operations.

Project 4.3

Let G denote an abelian group with operation $*$ and identity e. Define: An *endomorphism of* G is a group homomorphism from G to G. (No assumptions of one to one or onto are included here!)

Let END(G) denote the set of all endomorphisms of G. The operations for END(G) are defined below.

For $f, g \in END(G)$

$$\begin{aligned} (f+g)(x) &= f(x) * g(x) \quad \textit{(the operation in G)} \\ (fg)(x) &= f(g(x)) \quad \textit{(usual composition)} \end{aligned}$$

1. No matter what the group G is, what element will always exist in END(G) so we can guarantee it is nonempty?
2. To prove that END(G) is a ring, let $f, g, h \in END(G)$:

   (i) $f + (g + h) = (f + g) + h$ follows from what property in G?
   (ii) What endomorphism will be the zero of END(G)?
   (iii) How will the negative of f be defined in END(G)?
   (iv) $f + g = g + h$ follows from what property of G?
   (v) $f(gh) = (fg)h$ follows from what property of functions?
   (vi) $f(g+h) = fg + fh$ and $(g+h)f = gf + hf$ follow from what property of f, g, h?

(vii) What endomorphism will be the unity of END(G)?

Thus END(G) is a ring with unity whenever G is an abelian group!

3. Suppose now that G is a cyclic group generated by a, and $f \in END(G)$. Explain why $f(a)$ completely determines f, i.e., for any $b \in G$, $f(b)$ can be found as a power of $f(a)$.

   We can now say that when G is cyclic, the size of END(G) is the same as the size of G since we can use each element of G as our $f(a)$ to determine a different member of END(G).

4. Use $G = \mathbb{Z}_4$ (which is a cyclic group under $+_4$) and find all of the elements of END(G). Then show the two Cayley tables for END(G) using the operations defined at the beginning of the project.

Project 4.4

Consider the ring $\mathbb{Z}_2 \times \mathbb{Z}_6$ with $+_2$ and $\cdot_2$ in the first coordinate and $+_6$ and $\cdot_6$ in the second.

1. Find all of the zero divisors in $\mathbb{Z}_2 \times \mathbb{Z}_6$.
2. Find all of the units in $\mathbb{Z}_2 \times \mathbb{Z}_6$.
3. Give an example of a nontrivial subring A of $\mathbb{Z}_2 \times \mathbb{Z}_6$ which does not have a unity. Be sure to show its addition and multiplication tables, show that it is a subring, and explain why there is no unity.
4. Give an example of a nontrivial subring of $\mathbb{Z}_2 \times \mathbb{Z}_6$ which has a unity, but whose unity is not the same as the unity of $\mathbb{Z}_2 \times \mathbb{Z}_6$. Be sure to show its addition and multiplication tables, that it is a subring, and what the unity is.
5. If a subring S of a ring with unity A contains a unity 1_S that is not the same as 1_A, is it possible for an element $a \in S$ to be a unit in S but a zero divisor in A? If it is possible find an example; otherwise, prove it is not possible.

Project 4.5

For the rings $(\mathbb{Z}, +, \cdot)$ and $(\mathbb{Z}_n, +_n, \cdot_n)$, $n > 1$, we saw that the function $f : \mathbb{Z} \to \mathbb{Z}_n$ defined by $f(x) = x(\text{mod} n)$ is a ring homomorphism. These homomorphisms can help us solve problems about the integers!

Consider the equation $x^2 - 7y^2 - 24 = 0$. We will determine if it has integer solutions.

1. Suppose we had integers a, b with $a^2 - 7b^2 - 24 = 0$. Use the homomorphism $f : \mathbb{Z} \to \mathbb{Z}_7$ to show that $(f(a))^2 - 3 = 0$ in $\mathbb{Z}_7$.
2. Show that no $u \in \mathbb{Z}_7$ will satisfy $u^2 - 3 = 0$. Thus there are no integer solutions to $x^2 - 7y^2 - 24 = 0$.

3. Use $\mathbb{Z}_5$ to see why the infinite sequence of integers $3, 8, 13, 18, 23, \ldots$ does not include a square.
4. Suppose the positive integer n is the product of two consequtive positive integers, i.e., $n = m(m+1)$. Use an appropriate homomorphism to show that the "ones digit" of n must be either 0, 2, or 6.

Project 4.6

Suppose A and B are nontrivial commutative rings with unity, and $f : A \to B$ is a ring homomorphism.

1. Prove: If $f(1_A) = 0_B$ then for every $x \in A$ we must have $f(x) = 0_B$.
2. What does the contrapositive of the previous statement tell us about a homomorphism?
3. Fill in the blanks to complete the proof of the statement below.

   Prove: If B has no zero divisors then either $f(1_A) = 0_B$ or $f(1_A) = 1_B$.

   "Proof": Assume A and B are nontrivial commutative rings with unity and $f : A \to B$ is a __________. Since $1_A \in A$ then $f(1_A) \in$ ______. Now $(1_A)^2 =$ ______ so by f a homomorphism $f($________$) = f(1_A)f(1_A)$. Subtracting ________ from both sides we have ________ $= f(1_A)f(1_A) - f(1_A)$. Using the distributive law we find ________ $= f(1_A)($________$)$. Since B has no zero divisors then either $f(1_A) =$ ________ or ____________ $= 0_B$. However, if $f(1_A) - 1_B = 0_B$, then by adding ________ to both sides we have $f(1_A) =$ ________. Thus either $f(1_A) = 0_B$ or $f(1_A) = 1_B$.

4. Consider the function $f : \mathbb{Z} \to M_2(\mathbb{Z})$ by $f(x) = \begin{bmatrix} 0 & 0 \\ 0 & x \end{bmatrix}$. Prove f is a homomorphism.
5. How do the previous parts of the project guarantee us that $M_2(Z)$ must have zero divisors?

Chapter 5

Quotient Rings

As in Chapter 3, we now want to look at the formation of quotients, this time quotient rings. We continue to use A as our notation for a ring, to avoid confusion with $\mathbb{R}$.

5.1 Cosets

In Chapter 3, for a subgroup H of a group G, we wrote cosets in either an additive $(a + H)$ or multiplicative (aH) form to match the operation on G. Since a ring has two operations we cannot use both of them to mean the same set. Thus all cosets for rings use the addition of the ring to define them.

Definition 5.1 Let A be a ring and S a subring of A. A ***left coset*** of S is a set of the form $a + S = \{a +_A x : x \in A\}$. The set of all left cosets of S in A is denoted by ${}^{A}/_{S}$.

Similarly, a ***right coset*** is a set of the form $S + a = \{x +_A a : x \in A\}$.

Recall that since A is a ring, a subring S must be a subgroup $(S, +)$ of the group $(A, +)$. With a subgroup we defined cosets in Definition 3.1. Compare this definition (translated into addition) with the one above to see that the cosets we use in rings are exactly the same cosets the additive subgroup would create.

The fact that these cosets are the same as those defined for groups can cause one common error. If A is a ring and $(S, +)$ is a subgroup of $(A, +)$, we can define a group coset $a + S$. However, *ring* cosets cannot be discussed unless S is also a **subring** of A. A subgroup

S will not always be a subring as seen with $M_2(\mathbb{Z})$ and $S = \left\{ \begin{bmatrix} a & b \\ c & 0 \end{bmatrix} : a, b, c \in \mathbb{Z} \right\}$ of Example 4.35.

Addition is always commutative in a ring so by Theorem 3.18 in any subring S, the additive subgroup $(S, +)$ is a normal subgroup of $(A, +)$. Thus by Theorem 3.17, the left and right cosets of S will always be the same, but we will generally write left cosets in our theorems and examples for consistency.

Example 5.2 In Example 3.2 we considered the group $(\mathbb{Z}_{12}, +_{12})$ with subgroup $H = \{0, 4, 8\}$. Now consider the ring $(\mathbb{Z}_{12}, +_{12}, \cdot_{12})$. If we use the same set $H = \{0, 4, 8\}$ will it be a subring? Consider the Cayley tables for H:

$+_{12}$	0	4	8
0	0	4	8
4	4	8	0
8	8	0	4

$\cdot_{12}$	0	4	8
0	0	0	0
4	0	4	8
8	0	8	4

H is closed under both addition and multiplication, and since $-4 = 8, -8 = 4$ it is closed under negatives. Thus H is a subring by Theorem 4.33 and the left cosets are:

$$0 + H = \{0, 4, 8\} \quad 1 + H = \{1, 5, 9\} \quad 2 + H = \{2, 6, 10\} \quad 3 + H = \{3, 7, 11\}$$

Example 5.3 Consider the power set ring defined in Example 4.6, with $C = \{1, 2, 3\}$. We have $\wp(C) = \{\emptyset, \{1\}, \{2\}, \{3\}, \{1,2\}, \{1,3\}, \{2,3\}, \{1,2,3\}\}$ in this case. Define the set $S = \{\emptyset, \{1\}, \{2\}, \{1,2\}\}$. The proof that S is a subring of $\wp(C)$ is an exercise at the end of the chapter. The distinct cosets of S in $\wp(C)$ are seen below. Remember $U \oplus V = (U - V) \cup (V - U)$ for $U, V \in \wp(C)$.

$$\emptyset + S = \{\emptyset, \{1\}, \{2\}, \{1,2\}\} \qquad \{1,3\} + S = \{\{1,3\}, \{3\}, \{1,2,3\}, \{2,3\}\}$$

Again, since the cosets are the same as the cosets of the additive groups, all of the coset theorems from Chapter 3 (translated into additive notation) are also valid for rings. When translated, we see Theorem 3.4 gives us the following rules.

Theorem 5.4

Let A be a ring and S a subring of A.

(i) For all $a \in A$, $|a + S| = |S|$.
(ii) For all $a, b \in A$, either $a + S = b + S$ or $(a + S) \cap (b + S) = \emptyset$.
(iii) $\bigcup_{a \in A}(a + S) = A$.

In Theorem 3.6, which is translated below, (ii) should say $-b + a \in S$, but the addition of a ring is commutative so $-b + a = a + -b = a - b$.

Theorem 5.5

Let A be a ring and S a subring of A. Let $a, b \in A$.

(i) $a + S = b + S$ if and only if $a \in b + S$.
(ii) $a + S = b + S$ if and only if $a - b \in S$.

Example 5.6 Consider the ring $M_2(\mathbb{Z})$ and $S = \left\{ \begin{bmatrix} x & 0 \\ 0 & x \end{bmatrix} : x \in \mathbb{Z} \right\}$. S is a subring of $M_2(\mathbb{Z})$ [an exercise in Chapter 4]. Consider the following elements of $M_2(\mathbb{Z})$:

$$a = \begin{bmatrix} -1 & 1 \\ 0 & 2 \end{bmatrix} \quad b = \begin{bmatrix} 1 & 0 \\ -1 & 4 \end{bmatrix}$$

$$a - b = \begin{bmatrix} -1-1 & 1-0 \\ 0+1 & 2-4 \end{bmatrix} = \begin{bmatrix} -2 & 1 \\ 1 & -1 \end{bmatrix}$$

Since $a - b \notin S$ then $a + S \neq b + S$.

In Definition 3.21 we defined an operation on cosets (translated into addition), $(a+S)+(b+S) = (a +_A b) + S$. By Theorem 3.24 we know that using this operation ${}^{A}/_{S}$ forms a **group**.

Example 5.7 We found the distinct cosets of $S = \{0, 4, 8\}$ in the ring $(\mathbb{Z}_{12}, +_{12}, \cdot_{12})$ in Example 5.2. Thus ${}^{\mathbb{Z}_{12}}/_{S} = \{0 + S, 1 + S, 2 + S, 3 + S\}$ and this addition of cosets gives us the Cayley table:

$+$	$0+S$	$1+S$	$2+S$	$3+S$
$0+S$	$0+S$	$1+S$	$2+S$	$3+S$
$1+S$	$1+S$	$2+S$	$3+S$	$0+S$
$2+S$	$2+S$	$3+S$	$0+S$	$1+S$
$3+S$	$3+S$	$0+S$	$1+S$	$2+S$

Another operation is still needed to make ${}^{A}/_{S}$ a ring, but this will not be possible for every subring S of a ring A. Among the collection of subrings of a ring there are special subrings, just as normal subgroups were needed for quotient groups.

5.2 Ideals

According to Theorem 4.33 a subring S of a ring A must be closed under multiplication, meaning that if $a, b \in S$ then $ab \in S$. However, it does not require the product to be in S when multiplying an element of S with any element of A.

> **Definition 5.8** Suppose S is a subring of a ring A. We say S ***absorbs multiplication*** from A if for any $a \in S$ and $x \in A$ we have $ax \in S$ and $xa \in S$. A subring S with this property is called an ***ideal*** of A.

Example 5.9 Consider the subring from Example 4.32, $S = \{0, 2, 4\}$ in the ring $(\mathbb{Z}_6, +_6, \cdot_6)$. Is S also an ideal? We already know that products of two elements of S are back in S. We only need to check products with one element from S and one element not in S. Thus consider the following products:

$$\begin{array}{lll} 1(0) = 0 & 3(0) = 0 & 5(0) = 0 \\ 1(2) = 2 & 3(2) = 0 & 5(2) = 4 \\ 1(4) = 4 & 3(4) = 0 & 5(4) = 2 \end{array}$$

Since multiplication in $\mathbb{Z}_6$ is commutative, these tell us that S absorbs multiplication from $\mathbb{Z}_6$, and thus S is an ideal of $\mathbb{Z}_6$.

If a set S absorbs multiplication from A it must also be closed under multiplication. Thus we have a modification of Theorem 4.33 for an ideal.

Theorem 5.10

Suppose S is a subset of a ring A. S is an ideal of A if and only if

(i) S is nonempty.
(ii) S is closed under subtraction of A.
(iii) S absorbs multiplication from A.

Example 5.11 Consider the ring of rational numbers, $(\mathbb{Q}, +, \cdot)$. The set of all integers, $\mathbb{Z}$, is easily a subring of $\mathbb{Q}$ (as we said before). However, the element $\frac{1}{2} \cdot 3$ is not an integer, so $\mathbb{Z}$ does not absorb multiplication from $\mathbb{Q}$ and so $\mathbb{Z}$ is not an ideal of $\mathbb{Q}$.

This example shows a significant difference between ideals of rings and normal subgroups of groups. If a group is abelian then every subgroup is normal, but even if we have a commutative ring with unity, such as $\mathbb{Q}$, not every subring will be an ideal.

When looking for subrings and ideals there will always be two of them in a nontrivial ring, $\{0_A\}$ and A. The proof that these are ideals of A is left as an exercise at the end of the chapter. There is also one special property of ideals (an exercise at the end of the chapter) that is useful in later chapters.

Theorem 5.12

If A is a ring with unity and S is an ideal of A with $1_A \in S$ then $S = A$.

Example 5.13 Notice that the previous theorem can help us quickly determine that certain subsets of a ring cannot be ideals. In the ring $\mathbb{Z} \times \mathbb{Z}$ (usual addition and multiplication in each coordinate) define the set $S = \{(x, x) : x \in \mathbb{Z}\}$. Since $(1, 1)$ is the unity of $\mathbb{Z} \times \mathbb{Z}$ and $(1, 1) \in S$ then S cannot be an ideal unless $S = \mathbb{Z} \times \mathbb{Z}$. However, $(1, 3) \notin S$ so S is not an ideal of $\mathbb{Z} \times \mathbb{Z}$.

If a ring is not commutative we can define left ideals or right ideals as well. Since in later chapters all of our rings will be assumed commutative, we will not use left or right ideals and thus only mention them here (see [1] for more).

Definition 5.14 Suppose S is a subring of a ring A.

(i) S is a ***left ideal*** of A if for any $a \in S$ and $x \in A$ we have $xa \in S$. (S absorbs multiplication on the left by elements of A.)
(ii) S is a ***right ideal*** of A if for any $a \in S$ and $x \in A$ we have $ax \in S$. (S absorbs multiplication on the right by elements of A.)

It should be clear that if S is both a left ideal of A and a right ideal of A then S is an ideal of A.

In Chapter 2 cyclic groups were introduced. This concept helped us find many subgroups of a given group. In rings the parallel concept is a *principal ideal*, generated by a single element.

Definition 5.15 Let A be a ring and $a \in A$. The ***principal ideal*** generated by a is the set:

$$\langle a \rangle = \{x_1ay_1 + x_2ay_2 + \cdots + x_nay_n : n \in \mathbb{Z}^+ \text{ and } x_i, y_i \in A \text{ for each } i\}.$$

It is easy to see why the set $\langle a \rangle$ is in fact an ideal of S. Since we have $a \in A$, then using $n = 1$, $x_1 = a$, and $y_1 = a$, $aaa \in \langle a \rangle$, making it nonempty. Consider two elements $u, v \in \langle a \rangle$.

$$u = x_1 a y_1 + x_2 a y_2 + \cdots + x_n a y_n \qquad v = w_1 a z_1 + w_2 a z_2 + \cdots + w_m a z_m$$

We can see that $u - v$, ub and bu for $b \in A$ are back in $\langle a \rangle$, showing us that $\langle a \rangle$ is an ideal by Theorem 5.10.

$$\begin{aligned} u - v &= x_1 a y_1 + x_2 a y_2 + \cdots + x_n a y_n + (-w_1) a z_1 + (-w_2) a z_2 + \cdots + (-w_m) a z_m \\ ub &= x_1 a (y_1 b) + x_2 a (y_2 b) + \cdots + x_n a (y_n b) \\ bu &= (b x_1) a y_1 + (b x_2) a y_2 + \cdots + (b x_n) a y_n \end{aligned}$$

The notation in the definition above, $x_1 a y_1 + x_2 a y_2 + \cdots + x_n a y_n$, is very cumbersome and difficult to work with in general. Thus we will only use principal ideals in commutative rings with unity where there is a simpler description.

Theorem 5.16

Suppose that A is a ring and $a \in A$. If A is a commutative ring with unity then $\langle a \rangle = \{xa : x \in A\}$.

Proof Suppose that A is a commutative ring with unity and $a \in A$. $\langle a \rangle$ is the principal ideal generated by a as in Definition 5.15. We must show $\langle a \rangle \subseteq \{xa : x \in A\}$ and $\langle a \rangle \supseteq \{xa : x \in A\}$.

($\subseteq$) Let $u \in \langle a \rangle$ then there are $c_i, y_i \in A$ with

$$u = c_1 a y_1 + c_2 a y_2 + \cdots + c_n a y_n.$$

As A is commutative and using the distributive laws we find:

$$\begin{aligned} u &= (c_1 y_1) a + (c_2 y_2) a + \cdots + (c_n y_n) a \\ &= (c_1 y_1 + c_2 y_2 + \cdots + c_n y_n) a. \end{aligned}$$

Since $c_1 y_1 + c_2 y_2 + \cdots + c_n y_n \in A$ then $u = xa$ for some $x \in A$. Thus $\langle a \rangle \subseteq \{xa : x \in A\}$.

($\supseteq$) Suppose $w \in \{xa : x \in A\}$ then $w = xa$ for some $x \in A$. We assumed A has a unity so we can write $w = xa1_A$ and so $w \in \langle a \rangle$. Thus we have $\langle a \rangle \supseteq \{xa : x \in A\}$ and therefore $\langle a \rangle = \{xa : x \in A\}$. $\square$

Example 5.17 Consider the ring $\mathbb{Z}$ of integers with its usual operations and the element $3 \in \mathbb{Z}$. We can see the elements of $\langle 3 \rangle$ easily:

$$\langle 3 \rangle = \{3n : n \in \mathbb{Z}\} = \{\ldots, -12, -9, -6, -3, 0, 3, 6, 9, 12, \ldots\}$$

Thus $\langle 3 \rangle$ is a proper subset of $\mathbb{Z}$. However, in the ring $\mathbb{Q}$ of rational numbers we in fact have $\langle 3 \rangle = \mathbb{Q}$. Note $\frac{1}{3} \in \mathbb{Q}$ and so by $\langle 3 \rangle$ an ideal we have $\frac{1}{3}(3) \in \langle 3 \rangle$. But now $1 \in \langle 3 \rangle$ so by Theorem 5.12 $\langle 3 \rangle = \mathbb{Q}$.

We saw in Theorem 2.13 that every subgroup of $\mathbb{Z}$ (with usual addition) is a cyclic group. A parallel theorem in rings is next.

Theorem 5.18

In the ring $(\mathbb{Z}, +, \cdot)$, every ideal is a principal ideal.

Proof Let S be an ideal of $\mathbb{Z}$. First, we will consider the trivial ideals. If $S = \{0\}$ then clearly $S = \langle 0 \rangle$. If $S = \mathbb{Z}$ then $S = \langle 1 \rangle$ since by Theorem 5.12 $\langle 1 \rangle = \mathbb{Z}$. Thus in each of these cases our ideal is principal.

Now suppose we have a nontrivial ideal S of $\mathbb{Z}$. As S is nontrivial there is some nonzero integer $n \in S$. An ideal is closed under negatives, so we know that $-n$ is also in S. Either $n > 0$ or $-n > 0$, so there must exist some positive element of $\mathbb{Z}$ in S. Choose m to be the least positive element of $\mathbb{Z}$ that is in S, and we will show that $S = \langle m \rangle$.

Every element of $\langle m \rangle$ is of the form ym where y is an integer. Since S absorbs products and m is in S then clearly $\langle m \rangle \subseteq S$. To complete the proof, suppose we have $x \in S$ and we will show that $x \in \langle m \rangle$. By Theorem 0.18 there are integers q and r with $x = mq + r$ and $0 \leq r < m$. Since $m \in S$ then $mq \in S$ as S absorbs products. Thus $x - mq \in S$ as S is closed under subtraction, but $r = x - mq$ so $r \in S$. If $r > 0$ then there is a positive integer smaller than m in S, but m was chosen as the least positive element in S, so we must have $r = 0$, and thus $x = mq \in \langle m \rangle$. Hence $S \subseteq \langle m \rangle$ and therefore $S = \langle m \rangle$. □

5.3 Quotient Rings

In order to create a quotient ring, there must be an operation of multiplication on our cosets, along with the addition recalled before Example 5.7.

Definition 5.19 Let A be a ring and let S be a subring of A. For $a, b \in A$ we define $(a + S) * (b + S) = (ab) + S$ where ab uses the multiplication in the ring A.

In order to use this as our multiplication, we must know it is an *operation* (Definition 1.1) on ${}^{A}/_{S}$. Just as in quotient groups, this is more difficult than it seems. We must be able to verify that for $a+S, b+S, c+S, d+S \in {}^{A}/_{S}$ if $a+S=b+S$ and $c+S=d+S$ then $(a+S)*(c+S)=(b+S)*(d+S)$.

Example 5.20 Continuing with Example 5.7, we have $S = \{0,4,8\}$, a subring of $(\mathbb{Z}_{12}, +_{12}, \cdot_{12})$. The definition of the possible multiplication above gives the Cayley table:

$*$	$0+S$	$1+S$	$2+S$	$3+S$
$0+S$	$0+S$	$0+S$	$0+S$	$0+S$
$1+S$	$0+S$	$1+S$	$2+S$	$3+S$
$2+S$	$0+S$	$2+S$	$0+S$	$2+S$
$3+S$	$0+S$	$3+S$	$2+S$	$1+S$

Notice that $4+S=0+S$ and $3+S=11+S$. Also $(4+S)*(3+S)=0+S$ and $(0+S)*(11+S)=0+S$. This looks very promising.

Example 5.21 Consider the ring $(\mathbb{Q}, +, \cdot)$. We know that $\mathbb{Z}$ is subring of $\mathbb{Q}$ as well. Consider now some of the cosets of the form $a+\mathbb{Z}$. It should be clear that $\frac{1}{2}+\mathbb{Z}=\frac{3}{2}+\mathbb{Z}$ and $\frac{1}{4}+\mathbb{Z}=\frac{9}{4}+\mathbb{Z}$.

$$\left(\tfrac{1}{2}+\mathbb{Z}\right)*\left(\tfrac{9}{4}+\mathbb{Z}\right)=\tfrac{9}{8}+\mathbb{Z}$$
$$\left(\tfrac{3}{2}+\mathbb{Z}\right)*\left(\tfrac{1}{4}+\mathbb{Z}\right)=\tfrac{3}{8}+\mathbb{Z}$$

But $\frac{9}{8}-\frac{3}{8}=\frac{6}{8}$ is not in $\mathbb{Z}$, so $\frac{9}{8}+\mathbb{Z}\neq\frac{3}{8}+\mathbb{Z}$. This is a problem since the same cosets gave different answers under this "multiplication."

Thus there must be a difference in the examples above that caused the coset multiplication to fail as an operation in one and not the other. From Theorem 5.16 you can verify that in $\mathbb{Z}_{12}$ the set $S=\{0,4,8\}$ is the principal ideal $\langle 4\rangle$. Also in Example 5.11 we saw that while $\mathbb{Z}$ is a subring of $\mathbb{Q}$ it is not an ideal of $\mathbb{Q}$.

Theorem 5.22

Let A be a ring and S a subring of A. If S is an ideal of A then the multiplication $*$ on cosets in Definition 5.19 is an operation on ${}^{A}/_{S}$.

Proof Suppose A is a ring and S is an ideal of A. By Definition 5.19 for all $a+S, b+S \in {}^{A}/_{S}$, it is clear that an answer exists for $(a+S)*(b+S)=(ab)+S$ and that the answer is another coset. The problem is whether a *unique* answer exists for each pair. This was exactly the question normal subgroups solved for us in Theorem 3.24 for the addition of cosets. Let $a + S = c + S$ and $b + S = d + S$ where $a, b, c, d \in A$. We need to show that $(a+S)*(b+S) = (c+S)*(d+S)$, i.e., $ab+S = cd+S$. Since these are cosets, we can use Theorem 5.5 to help. We will prove that $ab - cd \in S$.

Consider the element $(a-c)b + c(b-d) \in A$. We know that $a - c \in S$ and $b - d \in S$ by Theorem 5.5. Since S is an ideal it absorbs products from A, so we know $(a-c)b \in S$ and $c(b-d) \in S$ as well. Thus $(a-c)b + c(b-d) \in S$ as S is closed under addition. The calculation below shows us $ab - cd \in S$ as needed, and $*$ is an operation on ${}^{A}/_{S}$.

$$(a-c)b + c(b-d) = ab - cb + cb - cd = ab - cd$$

□

We can now say ${}^{A}/_{S}$ is a ring when S is an ideal of A.

Theorem 5.23

Let A be a ring and S an ideal of A. With the addition and multiplication of cosets previously defined, ${}^{A}/_{S}$ is a ring, called the ***quotient ring***. (In words A *mod* S.)

Proof Assume A is a ring and S is an ideal of A. We know from Theorem 3.24 that the set ${}^{A}/_{S}$ is a group under coset addition. Let $a+S, b+S \in {}^{A}/_{S}$. Then $(a+S)+(b+S) = (a+_A b)+S$ and $(b+S)+(a+S) = (b+_A a)+S$. In a ring, the addition is commutative so $a+_A b = b+_A a$. Thus $(a+S)+(b+S) = (a+_A b)+S = (b+_A a)+S = (b+S)+(a+S)$ and we have an abelian group under addition. Theorem 5.22 showed us that coset multiplication is an operation on ${}^{A}/_{S}$, so we only need to show that:

(i) $*$ is associative.
(ii) The distributive laws hold.

(i) Let $a+S, b+S, c+S \in {}^{A}/_{S}$. We need to show that

$$[(a+S)*(b+S)]*(c+S) = (a+S)*[(b+S)*(c+S)].$$

By definition of the coset operations we have the following equations:

$$\begin{aligned} [(a+S)*(b+S)]*(c+S) &= (ab+S)*(c+S) = (ab)c+S \\ (a+S)*[(b+S)*(c+S)] &= (a+S)*(bc+S) = a(bc)+S. \end{aligned}$$

Since multiplication on A is associative we know $a(bc) = (ab)c$, so $a(bc)+S = (ab)c+S$. Thus $*$ is associative.

(ii) Let $a+S, b+S, c+S \in {}^{A}/_{S}$. We need to show the following two equalities:

$$[(a+S)+(b+S)]*(c+S) = [(a+S)*(c+S)] + [(b+S)*(c+S)]$$
$$(a+S)*[(b+S)+(c+S)] = [(a+S)*(b+S)] + [(a+S)*(c+S)].$$

By definition:

$$[(a+S)+(b+S)]*(c+S) = ((a+b)+S)*(c+S) = ((a+_A b)c)+S$$
$$(a+S)*(c+S)+(b+S)*(c+S) = (ac+S)+(bc+S) = (ac+_A bc)+S.$$

Since the distributive laws hold in A, we know $(a+_A b)c = ac+_A bc$, so $(a+_A b)c + S = (ac+_A bc)+S$. Thus, the first equality we needed to show is complete. The proof of $(a+S)*[(b+S)+(c+S)] = (a+S)*(b+S)+(a+S)*(c+S)$ is an exercise at the end of the chapter.

Thus ${}^{A}/_{S}$ is a ring under these operations. □

Example 5.24 Consider the ring $(\mathbb{Z}, +, \cdot)$. If we choose an element such as $a=-5$, we know the set $\langle -5\rangle$ is an ideal. Thus we can create the ring ${}^{\mathbb{Z}}/_{\langle -5\rangle}$. Consider the cosets

$$0+\langle -5\rangle \qquad 1+\langle -5\rangle \qquad 2+\langle -5\rangle \qquad 3+\langle -5\rangle \qquad 4+\langle -5\rangle.$$

First, we will show that these are distinct cosets in ${}^{\mathbb{Z}}/_{\langle -5\rangle}$. If $i,j \in \{0,1,2,3,4\}$ we will show that $i \neq j$ guarantees us $i+\langle -5\rangle \neq j+\langle -5\rangle$ and thus they are distinct. Suppose instead that $i+\langle -5\rangle = j+\langle -5\rangle$, then by Theorem 5.5 we have $i-j \in \langle -5\rangle$, and thus $i-j$ is a multiple of -5. The table below shows the answers to $i-j$ for $i,j \in \{0,1,2,3,4\}$.

$-$	0	1	2	3	4
0	0	−1	−2	−3	−4
1	1	0	−1	−2	−3
2	2	1	0	−1	−2
3	3	2	1	0	−1
4	4	3	2	1	0

Notice that the none of $-1,-2,-3,-4,1,2,3,$ or 4 can be written as an integer multiple of -5, thus $i-j \in \langle -5\rangle$ implies $i=j$. We now know that these cosets are distinct, but are there others in the ring ${}^{\mathbb{Z}}/_{\langle -5\rangle}$?

Suppose we have a coset $x+\langle -5\rangle$ in the ring ${}^{\mathbb{Z}}/_{\langle -5\rangle}$. As $x \in \mathbb{Z}$, by Theorem 0.18, $x = 5n+y$ where n, y are integers and $0 \leq y < 5$. But since $5n = -5(-n)$ is in $\langle -5\rangle$, then $x-y \in \langle -5\rangle$. Thus $x+\langle -5\rangle = y+\langle -5\rangle$, but y is either $0,1,2,3,$ or 4 so $x+\langle -5\rangle$ is one of the five cosets we found earlier.

$$\mathbb{Z}/_{\langle -5\rangle} = \{0+\langle -5\rangle\,, 1+\langle -5\rangle\,, 2+\langle -5\rangle\,, 3+\langle -5\rangle\,, 4+\langle -5\rangle\}$$

The Cayley tables for this ring are shown below (as you can verify).

$+$	$0+\langle -5\rangle$	$1+\langle -5\rangle$	$2+\langle -5\rangle$	$3+\langle -5\rangle$	$4+\langle -5\rangle$
$0+\langle -5\rangle$	$0+\langle -5\rangle$	$1+\langle -5\rangle$	$2+\langle -5\rangle$	$3+\langle -5\rangle$	$4+\langle -5\rangle$
$1+\langle -5\rangle$	$1+\langle -5\rangle$	$2+\langle -5\rangle$	$3+\langle -5\rangle$	$4+\langle -5\rangle$	$0+\langle -5\rangle$
$2+\langle -5\rangle$	$2+\langle -5\rangle$	$3+\langle -5\rangle$	$4+\langle -5\rangle$	$0+\langle -5\rangle$	$1+\langle -5\rangle$
$3+\langle -5\rangle$	$3+\langle -5\rangle$	$4+\langle -5\rangle$	$0+\langle -5\rangle$	$1+\langle -5\rangle$	$2+\langle -5\rangle$
$4+\langle -5\rangle$	$4+\langle -5\rangle$	$0+\langle -5\rangle$	$1+\langle -5\rangle$	$2+\langle -5\rangle$	$3+\langle -5\rangle$

$*$	$0+\langle -5\rangle$	$1+\langle -5\rangle$	$2+\langle -5\rangle$	$3+\langle -5\rangle$	$4+\langle -5\rangle$
$0+\langle -5\rangle$	$0+\langle -5\rangle$	$0+\langle -5\rangle$	$0+\langle -5\rangle$	$0+\langle -5\rangle$	$0+\langle -5\rangle$
$1+\langle -5\rangle$	$0+\langle -5\rangle$	$1+\langle -5\rangle$	$2+\langle -5\rangle$	$3+\langle -5\rangle$	$4+\langle -5\rangle$
$2+\langle -5\rangle$	$0+\langle -5\rangle$	$2+\langle -5\rangle$	$4+\langle -5\rangle$	$1+\langle -5\rangle$	$3+\langle -5\rangle$
$3+\langle -5\rangle$	$0+\langle -5\rangle$	$3+\langle -5\rangle$	$1+\langle -5\rangle$	$4+\langle -5\rangle$	$2+\langle -5\rangle$
$4+\langle -5\rangle$	$0+\langle -5\rangle$	$4+\langle -5\rangle$	$3+\langle -5\rangle$	$2+\langle -5\rangle$	$1+\langle -5\rangle$

You may notice that if we hide the "$+\langle -5\rangle$" in these tables, it would look exactly like the ring $(\mathbb{Z}_5, +_5, \cdot_5)$. It is precisely for that reason we use the word "mod" when dealing with quotient groups and rings. In fact, the rings $\mathbb{Z}_n$ could have been defined using equivalence classes $[0], [1], \ldots, [n-1]$, which are really our cosets, instead of the numbers $0, 1, \ldots, n-1$.

Properties of quotient rings will be important in later chapters so we will look at a few next.

Theorem 5.25

Suppose that A is a nontrivial ring and S is an ideal of A.

(i) If A is commutative then $A/_S$ is also commutative.
(ii) If A contains a unity 1_A then $1_A + S$ is the unity of $A/_S$.
(iii) $S = A$ if and only if $A/_S = \{0_A + S\}$.

Proof Suppose that A is a ring and S is an ideal of A.
(i) Left as an exercise at the end of the chapter.

(ii) Left as an exercise at the end of the chapter.

(iii) $(\rightarrow)$ Suppose $S = A$. By definition we know that $0_A+S \in A/_S$ so we have $\{0_A+S\} \subseteq A/_S$ and need only prove $\{0_A + S\} \supseteq A/_S$. Let $a + S \in A/_S$. Since we know $a \in S$ then

$a + S = 0_A + S$ by Theorem 5.5 and so $a + S \in \{0_A + S\}$. Thus $\{0_A + S\} \supseteq {}^{A}/_{S}$ and we have ${}^{A}/_{S} = \{0_A + S\}$.

($\leftarrow$) Now suppose that ${}^{A}/_{S} = \{0_A + S\}$. By definition we know $S \subseteq A$ so we need to prove $A \subseteq S$. Let $a \in A$ then $a + S \in {}^{A}/_{S}$. But ${}^{A}/_{S} = \{0_A + S\}$ so we have $a + S = 0_A + S$, thus by Theorem 5.5 $a - 0_A \in S$. Hence $a \in S$ and $A \subseteq S$ showing $S = A$. ☐

Example 5.26 Consider $A = M_2(\mathbb{Z})$ with its usual operations, a ring we know is not commutative. Choose S to be the whole ring $M_2(\mathbb{Z})$ which is an ideal of A. Then ${}^{M_2(\mathbb{Z})}/_{M_2(\mathbb{Z})}$ has only one element by (iii) of Theorem 5.25. However, the trivial ring is commutative. So it is possible to have ${}^{A}/_{S}$ commutative even if A was not.

Also consider the ring E of even integers (which has no unity) and $S = \{6n : n \in \mathbb{Z}\}$. It is straightforward to show S is an ideal of E. Then ${}^{E}/_{S} = \{0 + S, 2 + S, 4 + S\}$ has unity $4 + S$ as seen in the Cayley table below.

$*$	$0+S$	$2+S$	$4+S$
$0+S$	$0+S$	$0+S$	$0+S$
$2+S$	$0+S$	$4+S$	$2+S$
$4+S$	$0+S$	$2+S$	$4+S$

Again, using the ring E of even integers, which has no zero divisors, let $S = \{4n : n \in \mathbb{Z}\}$. You should show that S is an ideal of E. ${}^{E}/_{S} = \{0 + S, 2 + S\}$ and $(2 + S) * (2 + S) = 0 + S$ so the quotient ring does have zero divisors!

Other properties can hold in the quotient ring that did not hold in the original ring as well, as seen in the next example.

Example 5.27 Consider the ring $(\mathbb{Z}_8, +_8, \cdot_8)$ and $S = \{0, 2, 4, 6\}$. It is easy to see that S is a subring of $\mathbb{Z}_8$ using the Cayley tables.

$+_8$	0	2	4	6
0	0	2	4	6
2	2	4	6	0
4	4	6	0	2
6	6	0	2	4

$\cdot_8$	0	2	4	6
0	0	0	0	0
2	0	4	0	4
4	0	0	0	0
6	0	4	0	4

S also absorbs products from $\mathbb{Z}_8$, as shown in the table below. Thus S is an ideal of $\mathbb{Z}_8$.

$\cdot_8$	0	1	2	3	4	5	6	7
0	0	0	0	0	0	0	0	0
2	0	2	4	6	0	2	4	6
4	0	4	0	4	0	4	0	4
6	0	6	4	2	0	6	4	2

The only distinct cosets of S are $0+S$ and $1+S$ so ${}^{\mathbb{Z}_8}/_S = \{0+S, 1+S\}$. The quotient ring ${}^{\mathbb{Z}_8}/_S$ has some interesting properties that fail in $\mathbb{Z}_8$, which can all be seen in the Cayley tables below.

$+$	$0+S$	$1+S$
$0+S$	$0+S$	$1+S$
$1+S$	$1+S$	$0+S$

$*$	$0+S$	$1+S$
$0+S$	$0+S$	$0+S$
$1+S$	$0+S$	$1+S$

(i) Every element of ${}^{\mathbb{Z}_8}/_S$ is its own negative, i.e., $(a+S)+(a+S) = 0+S$.
(ii) For each element $a+S$ in ${}^{\mathbb{Z}_8}/_S$, $(a+S)^2 = a+S$.
(iii) The ring ${}^{\mathbb{Z}_8}/_S$ is an integral domain.
(iv) The ring ${}^{\mathbb{Z}_8}/_S$ is a field.

Analogous to Theorem 3.9 we have the following, with identically the same proof.

Theorem 5.28

Let A be a <u>finite</u> ring and S an ideal of A. Then $\left| {}^{A}/_S \right| = \frac{|A|}{|S|}$.

Notice in Example 5.27 we have $|\mathbb{Z}_8| = 8$, $|S| = 4$ and $\left| {}^{\mathbb{Z}_8}/_S \right| = 2$, which illustrates the theorem above.

5.4 The Fundamental Homomorphism Theorem

Recall from Definition 4.36 that a ring homomorphism $f : A \to K$ is a function where $f(a+b) = f(a) + f(b)$ and $f(ab) = f(a)f(b)$. Also Chapter 4 ended with the following definition.

Definition 4.42 Let A and K denote rings and $f : A \to K$ a ring homomorphism.

(i) The kernel of f is $ker(f) = \{x \in A : f(x) = 0_K\}$.
(ii) The image of f is $f(A) = \{y \in K : y = f(x) \text{ for some } x \in A\}$.

The proof of the next theorem is identical (when translated into additive notation) to the proof of Theorem 3.32. The operation of multiplication plays no part.

Theorem 5.29

Let A and K denote rings and $f : A \to K$ a ring homomorphism.

(i) f is one to one if and only if $ker(f) = \{0_A\}$.
(ii) f is onto if and only if $f(A) = K$.

Example 5.30 In Example 4.38 we saw that the function $g : \mathbb{Z}_6 \to \mathbb{Z}_{12}$ by $g(x) = (4x)(\text{mod } 12)$ is a ring homomorphism. Since $g(0) = 0, g(1) = 4, g(2) = 8, g(3) = 0, g(4) = 4, g(5) = 8$ we have $ker(g) = \{0, 3\}$ and $g(\mathbb{Z}_6) = \{0, 4, 8\}$. Notice that under the operations of $\mathbb{Z}_6$, we can create the Cayley tables for $ker(g)$.

$+_6$	0	3
0	0	3
3	3	0

$\cdot_6$	0	3
0	0	0
3	0	3

We know from Theorem 3.33 $ker(g)$ is a group under addition. The table above shows it is also closed under multiplication, and thus is a subring of $\mathbb{Z}_6$. To see that $ker(g)$ is an ideal notice that $1(0) = 0$, $1(3) = 3$, $2(0) = 0$, $2(3) = 0$, $4(0) = 0$, $4(3) = 0$, $5(0) = 0$, $5(3) = 3$. Thus $ker(g)$ absorbs products from $\mathbb{Z}_6$.

The example above motivates the next theorem, which is similar to Theorem 3.33.

Theorem 5.31

Let A and K denote rings and suppose $f : A \to K$ is a ring homomorphism.

(i) $ker(f)$ is an ideal of A.
(ii) $f(A)$ is a subring of K.

Proof Assume that A and K denote rings and $f : A \to K$ is a ring homomorphism.

(i) We know from Theorem 3.33 that $ker(f)$ is an additive subgroup of $(A, +)$. Thus we only need to show it absorbs products from A to have an ideal of A, which is as an exercise at the end of the chapter.

(ii) Again, from Theorem 3.33 we know that $f(A)$ is an additive subgroup of K. We only need to show it is closed under multiplication. Suppose $c, d \in f(A)$. Thus there exist elements $x, y \in A$ with $f(x) = c$ and $f(y) = d$. Now $f(xy) = f(x)f(y) = cd$, as f is a ring homomorphism. Thus $cd \in f(A)$. Similarly, $dc \in f(A)$, so $f(A)$ is closed under multiplication and is a subring of K. $\square$

Example 5.32 Using the rings $(\mathbb{Z}, +, \cdot)$ and $(\mathbb{Q}, +, \cdot)$, define $f : \mathbb{Z} \to \mathbb{Q}$ by $f(x) = x$. This is easily a ring homomorphism, and $f(\mathbb{Z}) = \mathbb{Z}$. We know that $\mathbb{Z}$ is a subring of $\mathbb{Q}$ but is <u>not</u> an ideal of $\mathbb{Q}$ from Example 5.11. Thus in Theorem 5.31 we cannot say that $f(A)$ is always an ideal of K.

Example 5.33 With rings $(\mathbb{Z}_8, +_8, \cdot_8)$ and $(\mathbb{Z}_6, +_6, \cdot_6)$, consider the function $f : \mathbb{Z}_8 \to \mathbb{Z}_6$ defined by $f(x) = (3x)(\text{mod}6)$. We see that $f(0) = 0, f(1) = 3, f(2) = 0, f(3) = 3, f(4) = 0, f(5) = 3, f(6) = 0$, and $f(7) = 3$. To verify that f is a ring homomorphism we use the Cayley tables below.

$+_8$	0	1	2	3	4	5	6	7
0	0	1	2	3	4	5	6	7
1	1	2	3	4	5	6	7	0
2	2	3	4	5	6	7	0	1
3	3	4	5	6	7	0	1	2
4	4	5	6	7	0	1	2	3
5	5	6	7	0	1	2	3	4
6	6	7	0	1	2	3	4	5
7	7	0	1	2	3	4	5	6

$\rightarrow$

$+_6$	0	3	0	3	0	3	0	3
0	0	3	0	3	0	3	0	3
3	3	0	3	0	3	0	3	0
0	0	3	0	3	0	3	0	3
3	3	0	3	0	3	0	3	0
0	0	3	0	3	0	3	0	3
3	3	0	3	0	3	0	3	0
0	0	3	0	3	0	3	0	3
3	3	0	3	0	3	0	3	0

$\cdot_8$	0	1	2	3	4	5	6	7
0	0	0	0	0	0	0	0	0
1	0	1	2	3	4	5	6	7
2	0	2	4	6	0	2	4	6
3	0	3	6	1	4	7	2	5
4	0	4	0	4	0	4	0	4
5	0	5	2	7	4	1	6	3
6	0	6	4	2	0	6	4	2
7	0	7	6	5	4	3	2	1

$\rightarrow$

$\cdot_6$	0	3	0	3	0	3	0	3
0	0	0	0	0	0	0	0	0
3	0	3	0	3	0	3	0	3
0	0	0	0	0	0	0	0	0
3	0	3	0	3	0	3	0	3
0	0	0	0	0	0	0	0	0
3	0	3	0	3	0	3	0	3
0	0	0	0	0	0	0	0	0
3	0	3	0	3	0	3	0	3

Since both of the final tables are correct in $\mathbb{Z}_6$, f is a ring homomorphism. Notice that $ker(f) = \{0, 2, 4, 6\}$ and $f(\mathbb{Z}_8) = \{0, 3\}$. Using the ideal $ker(f)$ notice that we have the cosets $0 + ker(f)$ and $1 + ker(f)$, with Cayley tables for ${}^{\mathbb{Z}_8}/_{ker(f)}$ as shown.

$+$	$0 + ker(f)$	$1 + ker(f)$
$0 + ker(f)$	$0 + ker(f)$	$1 + ker(f)$
$1 + ker(f)$	$1 + ker(f)$	$0 + ker(f)$

$*$	$0 + \text{ker}(f)$	$1 + \text{ker}(f)$
$0 + ker(f)$	$0 + ker(f)$	$0 + ker(f)$
$1 + ker(f)$	$0 + ker(f)$	$1 + ker(f)$

Compare these to the Cayley tables for the ring $f(\mathbb{Z}_8)$:

$+_6$	0	3
0	0	3
3	3	0

$\cdot_6$	0	3
0	0	0
3	0	3

By matching $0 + ker(f)$ to 0 and $1 + ker(f)$ to 3, the rings are identical.

This leads us to the Fundamental (Ring) Homomorphism Theorem, analogous to Theorem 3.34.

Theorem 5.34

The Fundamental Ring Homomorphism Theorem (FHT) Let A and K be rings and $f : A \to K$ a ring homomorphism. Then the image of f, $f(A)$ is isomorphic to the quotient ring ${}^{A}/_{ker(f)}$, i.e., $f(A) \cong {}^{A}/_{ker(f)}$.

Proof Suppose A and K are rings and $f : A \to K$ is a ring homomorphism. From Theorem 5.31 we know $ker(f)$ is an ideal of A, and thus ${}^{A}/_{ker(f)}$ is a ring by Theorem 5.23. Define the map φ (exactly as was done in Theorem 3.34).

$$\varphi : {}^{A}/_{ker(f)} \to f(A) \text{ with } \varphi(a + ker(f)) = f(a)$$

The steps needed to prove φ is well-defined, one to one, and onto and are identical to those in Theorem 3.34 when translated into additive notation, so we will not repeat them here.

Finally, consider $a + ker(f), b + ker(f) \in {}^{A}/_{ker(f)}$. Using coset rules and that f is a homomorphism we calculate the following.

$$\begin{aligned}
\varphi((a + ker(f)) + (b + ker(f))) &= \varphi((a + b) + ker(f)) \\
&= f(a + b) \\
&= f(a) + f(b) \\
&= \varphi(a + ker(f)) + \varphi(b + ker(f))
\end{aligned}$$

$$\begin{aligned}
\varphi((a + ker(f))(b + ker(f))) &= \varphi((ab) + ker(f)) \\
&= f(ab) \\
&= f(a)f(b) \\
&= \varphi(a + ker(f))\varphi(b + ker(f))
\end{aligned}$$

Thus φ is a homomorphism, so is an isomorphism, and $f(A) \cong {}^{A}/_{ker(f)}$. □

Example 5.35 Consider the rings $(M_2(\mathbb{Z}), +, \cdot)$ and $(\mathbb{Z}, +, \cdot)$, and define

$$f : \mathbb{Z} \to M_2(\mathbb{Z}) \text{ by } f(a) = \begin{bmatrix} a & 0 \\ 0 & a \end{bmatrix}.$$

In Example 4.39, we saw that this function is a homomorphism. For an integer a to be in $ker(f)$, we need $f(a)$ to equal the zero matrix. This can only occur when $a = 0$. Thus $ker(f) = \{0\}$. Finally, we can see that:

$$f(\mathbb{Z}) = \left\{ \begin{bmatrix} a & 0 \\ 0 & a \end{bmatrix} : a \in \mathbb{Z} \right\}$$

is the set of all diagonal matrices in $M_2(\mathbb{Z})$. Thus by Theorem 5.34,

$$\mathbb{Z}/_{\{0\}} \cong \left\{ \begin{bmatrix} a & 0 \\ 0 & a \end{bmatrix} : a \in \mathbb{Z} \right\}.$$

Just as was done in Chapter 3, we can also use the FHT to discover what the homomorphic images of a ring can be. It completely depends on the ideals of our ring.

Example 5.36 Consider the ring $(\mathbb{Z}_5, +_5, \cdot_5)$. What can the image of $\mathbb{Z}_5$ be under a homomorphism? Suppose we have a homomorphism $f : \mathbb{Z}_5 \to K$ for some ring K. In order for this to be a homomorphism, Theorem 5.31 tells us that $ker(f)$ is an ideal of $\mathbb{Z}_5$. However, the only ideals of $\mathbb{Z}_5$ must have cardinality that divides 5 by Theorem 3.9, so the only possible sizes are 1 and 5. The only ring of size 1 is the trivial ring, and $\mathbb{Z}_5$ has cardinality 5. Hence the only ideals are $\{0\}$ and $\mathbb{Z}_5$. Thus the only homomorphic images are $\mathbb{Z}_5/_{\{0\}}$ or $\mathbb{Z}_5/_{\mathbb{Z}_5}$, i.e., only $\mathbb{Z}_5$ and $\{0\}$.

We will see more uses of Theorem 5.34 in later chapters.

Exercises for Chapter 5

Sections 5.1 and 5.2 Cosets and Ideals

1. Verify that the set $S = \{\emptyset, \{1\}, \{2\}, \{1, 2\}\}$ in Example 5.3 is a subring of $\wp(C)$ when $C = \{1, 2, 3\}$. The operations on $\wp(C)$ are defined by $U \oplus V = (U - V) \cup (V - U)$ and $U \otimes V = U \cap V$.
2. Determine if the set $S = \{\emptyset, \{1\}, \{2\}, \{1, 2\}\}$ in Example 5.3 is an ideal of $\wp(C)$ when $C = \{1, 2, 3\}$. Either prove that S is an ideal or find a counterexample showing S is not an ideal. The operations on $\wp(C)$ are defined by $U \oplus V = (U - V) \cup (V - U)$ and

$U \otimes V = U \cap V$.

3. Verify that the set $S = \{(x, 2y) : x, y \in \mathbb{Z}\}$ is a subring of $\mathbb{Z} \times \mathbb{Z}$ with usual addition and multiplication in each coordinate. Find the distinct cosets of S.
4. Determine if $S = \{(x, 2y) : x, y \in \mathbb{Z}\}$ is an ideal of $\mathbb{Z} \times \mathbb{Z}$ with usual addition and multiplication in each coordinate. Either prove that S is an ideal or find a counterexample showing S is not an ideal.
5. Verify that $S = \{(0, 0), (1, 4), (0, 4), (1, 0)\}$ is an ideal of $\mathbb{Z}_2 \times \mathbb{Z}_8$ (usual operations $+_2$, $\cdot_2$, or $+_8$, $\cdot_8$ in appropriate coordinates). Is it a principal ideal? Explain your answer.
6. Verify that $S = \{(0, 0), (2, 0), (0, 2), (2, 2)\}$ is an ideal of $\mathbb{Z}_4 \times \mathbb{Z}_4$ ($+_4$, $\cdot_4$ in each coordinate). Is it a principal ideal? Explain your answer.
7. Find the elements of the principal ideal $\langle 4 \rangle$ in the ring $(\mathbb{Z}_{16}, +_{16}, \cdot_{16})$. Also find the distinct cosets of $\langle 4 \rangle$.
8. Find the elements of the principal ideal $\langle 2 \rangle$ in the ring $(\mathbb{Z}_{12}, +_{12}, \cdot_{12})$. Also find the distinct cosets of $\langle 2 \rangle$.
9. Find the elements of the principal ideal $\langle 6 \rangle$ in the ring $(\mathbb{Z}_{10}, +_{10}, \cdot_{10})$. Also find the distinct cosets of $\langle 6 \rangle$.
10. Find the elements of the principal ideal $\langle 8 \rangle$ in the ring $(\mathbb{Z}_{16}, +_{16}, \cdot_{16})$. Also find the distinct cosets of $\langle 8 \rangle$.
11. Find the elements of the principal ideal $\langle 3 \rangle$ in the ring $(\mathbb{Z}_{12}, +_{12}, \cdot_{12})$. Also find the distinct cosets of $\langle 3 \rangle$.
12. Find the elements of the principal ideal $\langle (1, 0) \rangle$ in the ring $\mathbb{Z}_3 \times \mathbb{Z}_3$ with $+_3$ and $\cdot_3$ in each coordinate. Also find the distinct cosets of $\langle (1, 0) \rangle$.
13. Find the elements of the principal ideal $\langle (1, 3) \rangle$ in the ring $\mathbb{Z}_2 \times \mathbb{Z}_9$ with $+_2$ and $\cdot_2$, or $+_9$ and $\cdot_9$ in appropriate coordinates. Also find the distinct cosets of $\langle (1, 3) \rangle$.
14. Find the elements of the principal ideal $\langle (2, 3) \rangle$ in the ring $\mathbb{Z}_4 \times \mathbb{Z}_6$ with $+_4$ and $\cdot_4$, or $+_6$ and $\cdot_6$ in appropriate coordinates. Also find the distinct cosets of $\langle (2, 3) \rangle$.
15. Determine if the set $S = \left\{ \begin{bmatrix} a & b \\ b & a \end{bmatrix} : a, b \in \mathbb{Z} \right\}$ is an ideal of $M_2(\mathbb{Z})$ (under usual matrix operations) and if so find the distinct cosets of S.
16. Determine if the set $S = \left\{ \begin{bmatrix} 0 & 0 \\ 0 & 0 \end{bmatrix}, \begin{bmatrix} 1 & 1 \\ 1 & 1 \end{bmatrix}, \begin{bmatrix} 1 & 1 \\ 0 & 0 \end{bmatrix}, \begin{bmatrix} 0 & 0 \\ 1 & 1 \end{bmatrix} \right\}$ is a subring of the ring $M_2(\mathbb{Z}_2)$. This ring uses standard matrix operations, but each entry is computed using $+_2$ and $\cdot_2$. Is it also an ideal?
17. Determine if the set $S = \left\{ \frac{a}{3} : a \in \mathbb{Z} \right\}$ is a subring of $(\mathbb{Q}, +, \cdot)$ and if so find the left cosets of S.
18. Prove: If A is a nontrivial ring with unity and S is an ideal of A with $1_A \in S$ then $S = A$.
19. Suppose A is a nontrivial ring with unity and S is an ideal of A. Prove if $a \in A$ is a unit with $a \in S$ then $S = A$.
20. Prove that the intersection of ideals of a ring A is also an ideal of A.

21. Determine if $S = \left\{ \begin{bmatrix} a & 0 \\ 0 & a \end{bmatrix} : a \in \mathbb{Z} \right\}$ is an ideal of $M_2(\mathbb{Z})$ with usual addition and multiplication of matrices.
22. Determine if $S = \left\{ \begin{bmatrix} 0 & 0 \\ a & 0 \end{bmatrix} : a \in \mathbb{Z} \right\}$ is an ideal of $M_2(\mathbb{Z})$ with usual addition and multiplication of matrices. Prove it is true or find a nontrivial counterexample.
23. Determine if $S = \left\{ \begin{bmatrix} a & b \\ 0 & 0 \end{bmatrix} : a, b \in \mathbb{Z} \right\}$ is an ideal of $M_2(\mathbb{Z})$ with usual addition and multiplication of matrices. Prove it is true or find a nontrivial counterexample.

Section 5.3 Quotient Rings

24. Consider the ring $(\mathbb{Z}_9, +_9, \cdot_9)$ and ideal $S = \{0, 3, 6\}$. Create the Cayley tables for the quotient ring ${}^{\mathbb{Z}_9}/_S$ and identify any zero divisors or units.
25. Consider the ring $(\mathbb{Z}_{14}, +_{14}, \cdot_{14})$ and ideal $S = \{0, 7\}$. Create the Cayley tables for the quotient ring ${}^{\mathbb{Z}_{14}}/_S$ and identify any zero divisors or units.
26. Consider the ring $(\mathbb{Z}_{18}, +_{18}, \cdot_{18})$ and ideal $S = \langle 6 \rangle$. Create the Cayley tables for the quotient ring ${}^{\mathbb{Z}_{18}}/_S$ and identify any zero divisors or units.
27. Consider the ring $(\mathbb{Z}_{12}, +_{12}, \cdot_{12})$ and ideal $S = \langle 4 \rangle$. Create the Cayley tables for the quotient ring ${}^{\mathbb{Z}_{12}}/_S$ and identify any zero divisors or units.
28. Consider the ring $(\mathbb{Z}_8, +_8, \cdot_8)$ and ideal $S = \langle 2 \rangle$. Create the Cayley tables for the quotient ring ${}^{\mathbb{Z}_8}/_S$ and identify any zero divisors or units.
29. Consider the ring $(\mathbb{Z}_{20}, +_{20}, \cdot_{20})$ and ideal $S = \langle 8 \rangle$. Create the Cayley tables for the quotient ring ${}^{\mathbb{Z}_{20}}/_S$ and identify any zero divisors or units.
30. Consider the ring $(\mathbb{Z}_{13}, +_{13}, \cdot_{13})$ and ideal $S = \langle 4 \rangle$. Create the Cayley tables for the quotient ring ${}^{\mathbb{Z}_{13}}/_S$ and identify any zero divisors or units.
31. Consider the ring $(\mathbb{Z}_{16}, +_{16}, \cdot_{16})$ and ideal $S = \langle 8 \rangle$. Create the Cayley tables for the quotient ring ${}^{\mathbb{Z}_{16}}/_S$ and identify any zero divisors or units.
32. Consider the ring $(\wp(C), \oplus, \otimes)$ with $C = \{1, 2, 3\}$ and ideal $\mathrm{S} = \{\emptyset, \{1\}, \{2\}, \{1, 2\}\}$. Create the Cayley tables for the quotient ring ${}^{\wp(C)}/_S$ and identify any zero divisors or units. The operations on $\wp(C)$ are defined by $U \oplus V = (U - V) \cup (V - U)$ and $U \otimes V = U \cap V$.
33. Consider the ring $\mathbb{Z}_4 \times \mathbb{Z}_4$ with $+_4$ and $\cdot_4$ in each coordinate and $S = \{(0, 0), (2, 0), (0, 2), (2, 2)\}$. Create the Cayley tables for the quotient ring ${}^{\mathbb{Z}_4 \times \mathbb{Z}_4}/_S$ and identify any zero divisors or units.
34. Consider the ring $\mathbb{Z}_4 \times \mathbb{Z}_6$ with $+_4$, $+_6$, $\cdot_4$ and $\cdot_6$ in appropriate coordinates and $S = \{(0, 0), (2, 0), (0, 3), (2, 3)\}$. Create the Cayley tables for the quotient ring ${}^{\mathbb{Z}_4 \times \mathbb{Z}_4}/_S$ and identify any zero divisors or units.
35. Consider the ring $\mathbb{Z} \times \mathbb{Z}$ with usual addition and multiplication in each coordinate and $S = \{(x, 2y) : x, y \in \mathbb{Z}\}$. Create the Cayley tables for the quotient ring ${}^{\mathbb{Z} \times \mathbb{Z}}/_S$ and identify any zero divisors or units.

36. Complete Theorem 5.23 by proving $(a+S)*[(b+S)+(c+S)] = (a+S)*(b+S)+(a+S)*(c+S)$.
37. Prove (i) of Theorem 5.25.
38. Prove (ii) of Theorem 5.25.
39. Suppose A is a nontrivial commutative ring with unity and S is an ideal of A so that $S \neq \{0_A\}$, $S \neq A$. Prove or find a counterexample to the statement: If $a \in A$ is a unit in A then $a+S$ is a unit in ${}^{A}/_{S}$.
40. Suppose A is a nontrivial commutative ring with unity and S is an ideal of A so that $S \neq \{0_A\}$, $S \neq A$. Prove or find a counterexample to the statement: If $a \in A$ is a zero divisor in A then $a+S$ is a zero divisor in ${}^{A}/_{S}$.
41. Suppose A is a nontrivial commutative ring with unity and S is an ideal of A so that $S \neq \{0_A\}$, $S \neq A$. Prove or find a counterexample to the statement: If $a+S$ is a unit in ${}^{A}/_{S}$ then a is a unit in A.
42. Suppose A is a nontrivial commutative ring with unity and S is an ideal of A so that $S \neq \{0_A\}$, $S \neq A$. Prove or find a counterexample to the statement: If $a+S$ is a zero divisor in ${}^{A}/_{S}$ then a is a zero divisor in A.
43. Suppose A is a nontrivial commutative ring with unity and S is an ideal of A so that $S \neq \{0_A\}$, $S \neq A$. Prove or find a counterexample to the statement: If A is an integral domain then ${}^{A}/S$ is also an integral domain.

Section 5.4 The Fundamental Homomorphism Theorem

44. Complete the proof of (i) of Theorem 5.31 by showing that $ker(f)$ absorbs products from A.
45. Suppose A and K are rings with $f : A \to K$, a homomorphism. Prove: If $a, b \in A$ with $f(a) = f(b)$ then there is an element $z \in ker(f)$ for which $a - b = z$
46. Suppose A and K are rings with $f : A \to K$ a homomorphism. Prove: For any $x \in a + ker(f)$ we have $f(x) = f(a)$. This will help when trying to create homomorphisms in later exercises.
47. Let A be a ring. Use the FHT to prove: ${}^{A}/_{\{0_A\}} \cong A$ and ${}^{A}/_{A} \cong \{0_A\}$.
48. Prove: The only ideals of $(\mathbb{Q}, +, \cdot)$ are $\{0\}$ and $\mathbb{Q}$.
49. Prove: The only ideals of a field K are $\{0_K\}$ and K.
50. Use the kernel to help prove: If A and K are nontrivial rings and $f : A \to K$ is an onto ring homomorphism then $f(1_A) \neq 0_K$.
51. Use the kernel to help prove: If A and K are nontrivial rings and $f : A \to K$ is a one to one ring homomorphism then $f(1_A) \neq 0_K$.
52. Prove: If A and K are fields and $f : A \to K$ is a ring isomorphism then $f(y^{-1}) = (f(y))^{-1}$ for any nonzero $y \in A$.
53. Prove: If $f : \mathbb{Q} \to \mathbb{Q}$ is an isomorphism then $f(x) = x$ for all $x \in \mathbb{Q}$. Use the standard operations on $\mathbb{Q}$.
54. Use the FHT to prove that ${}^{\mathbb{Z}_8}/_{\langle 2 \rangle} \cong \mathbb{Z}_2$. Be sure to define the homomorphism from $\mathbb{Z}_8$

to $\mathbb{Z}_2$ you are using and verify all needed properties. Use $+_8$ and $\cdot_8$, or $+_2$ and $\cdot_2$ for the appropriate rings.

55. Use the FHT to prove that ${}^{\mathbb{Z}_4\times\mathbb{Z}_4}/_{S} \cong \mathbb{Z}_2 \times \mathbb{Z}_2$ where $S = \{(0,0),(2,0),(0,2),(2,2)\}$. Be sure to define the homomorphism from $\mathbb{Z}_4 \times \mathbb{Z}_4$ to $\mathbb{Z}_2 \times \mathbb{Z}_2$ you are using and verify all needed properties. Use the standard operations on each ring.
56. Use the FHT to prove that ${}^{\mathbb{Z}_{16}}/_{\langle 8\rangle} \cong \mathbb{Z}_8$. Be sure to define the homomorphism from $\mathbb{Z}_{16}$ to $\mathbb{Z}_8$ you are using and verify all needed properties. Use the standard operations on each ring.
57. Use the FHT to prove that ${}^{\mathbb{Z}_4\times\mathbb{Z}_6}/_{S} \cong \mathbb{Z}_2 \times \mathbb{Z}_3$ where $S = \langle(2,3)\rangle$. Be sure to define the homomorphism from $\mathbb{Z}_4 \times \mathbb{Z}_6$ to $\mathbb{Z}_2 \times \mathbb{Z}_3$ you are using and verify all needed properties! Use the standard operations on each ring.
58. Use the FHT to prove that ${}^{\wp(C)}/_{S} \cong \mathbb{Z}_2$ for $C = \{1,2,3\}$ and $S = \{\emptyset,\{1\},\{2\},\{1,2\}\}$. Be sure to define the homomorphism from $\wp(C)$ to $\mathbb{Z}_2$ you are using and verify all needed properties. Use $\oplus$ and $\otimes$ for $\wp(C)$ as defined in Example 4.6, along with $+_2$ and $\cdot_2$ on $\mathbb{Z}_2$.
59. Find all of the ideals of $\mathbb{Z}_2 \times \mathbb{Z}_4$ and use them to find all possible homomorphic images of $\mathbb{Z}_2 \times \mathbb{Z}_4$. Use the standard operations on each ring.
60. Find all of the ideals of $(\mathbb{Z}_{10}, +_{10}, \cdot_{10})$ and use them to find all possible homomorphic images of $\mathbb{Z}_{10}$.

Projects for Chapter 5

Project 5.1

Let A be a commutative ring with unity and S an ideal of A.

1. Circle all of the logical flaws in the following "proof," and explain why any circled step is incorrect. Remember to look at the logic of each step to see if it follows from the previous steps. A mistake does not make each step following it incorrect.

   Prove: If S contains a unit of A then $S = A$.

   "Proof": Suppose that S is an ideal and there is a unit $a \in S$. We must show $S = A$. We already know that $A \subseteq S$ by definition of an ideal, so we only need to show that $A \subseteq S$. Let $x \in A$. Since S is an ideal then S is closed under multiplication so $x \in S$. Thus $xa \in S$. But since a is a unit we have $a^{-1} \in S$ and so $xaa^{-1} \in S$. Hence $x \in S$ and so $A \subseteq S$. Thus $S = A$ as needed.

2. Write a correct proof for the statement above.
3. If S does not contain a unit can we still have $A = S$? Explain!

Project 5.2

Remember that the group $(\mathbb{Z},+)$ is a very important example, namely an infinite cyclic group. The ring $(\mathbb{Z},+,\cdot)$ is also very important; an integral domain that is not a field. In the group $(\mathbb{Z},+)$ every subgroup of $\mathbb{Z}$ is cyclic, while in the ring $(\mathbb{Z},+,\cdot)$ we find that every ideal is principal! In the group $(\mathbb{Z},+)$, the cyclic subgroup generated by a is $\{a^n : n \in \mathbb{Z}\}$, but since our operation is addition the notation a^n really means $n \bullet a$. Thus we can rewrite the cyclic group generated by a as $\{n \bullet a : n \in \mathbb{Z}\}$.

1. In the ring $(\mathbb{Z},+,\cdot)$, show that the principal ideal generated by a is the same as $T = \{n \bullet a : n \in \mathbb{Z}\}$.

   Thus these two sets, the cyclic group generated by a and the principal ideal generated by a, are the same for the integers. Unfortunately, the group $\mathbb{Z}$ and the ring $\mathbb{Z}$ do not have the same properties. As groups we must have $\mathbb{Z} \cong \langle 2 \rangle$ since both are infinite cyclic groups. We will see that the principal ideal $\langle 2 \rangle$ is not isomorphic to the ring $(\mathbb{Z},+,\cdot)$ in the next part.

2. Suppose we had a ring isomorphism $f : \mathbb{Z} \to \langle 2 \rangle$.
   (i) Explain why we must have $f(1) \neq 0$.
   (ii) Show that if $f(1) = k$ then $k^2 = k$.
   (iii) Use the elements of $\langle 2 \rangle$ to explain why no such ring isomorphism can exist.
3. Similarly as groups, the cyclic subgroups $\langle 2 \rangle$ and $\langle 3 \rangle$ are both infinite cyclic groups and thus isomorphic. Now suppose that we had a ring isomorphism $f : \langle 2 \rangle \to \langle 3 \rangle$. Find a condition about $f(2)$ which must be true using homomorphism properties, but which is impossible in $\langle 3 \rangle$.

   There are other differences between cyclic subgroups and principal ideals of $\mathbb{Z}$ that we will see later in the text.

Project 5.3

Let A and B be nontrivial commutative rings in each of the following.

1. Prove: If $f : A \to B$ is a homomorphism then $ker(f)$ is an ideal of A.
2. Show that the converse of #1 is false using A as $(\mathbb{Z}_4, +_4, \cdot_4)$ and B as $(\mathbb{Z}_3, +_3, \cdot_3)$. Be sure the function you define is nontrivial and that you verify any claims you make.
3. Recall from Theorem 5.31: If $f : A \to B$ is a homomorphism then $f(A)$ is a subring of B. Is the converse also true? Either prove that it is or find a counterexample showing a function where $f(A)$ is a subring of B but f is not a homomorphism. Be sure to verify any claims.

Project 5.4

Recall that not every subring of a ring must also be an ideal, but as before $(\mathbb{Z}, +, \cdot)$ is special!

1. Fill in the blanks to complete the proof of the statement below.
   Prove: Every subring of $\mathbb{Z}$ is also an ideal of $\mathbb{Z}$.

   Consider a subring S of $\mathbb{Z}$. Then $(S, +)$ is a ________ of $(\mathbb{Z}, +)$ and so it is a cyclic subgroup, generated by some element a. To see if S is an ideal we must determine if S __________. Let $x \in S$ and $b \in$ ________. Then $bx = b \bullet x$ under usual multiplication in $\mathbb{Z}$. Now as $x =$ ________ for some $m \in \mathbb{Z}$ then $bx = b\bullet(m \bullet a)$. By Theorem ________ we have $bx = (nm) \bullet a$ so ________. Also $xb \in S$ since multiplication is __________ in $\mathbb{Z}$. Thus S absorbs products and is an________ of $\mathbb{Z}$.

   In Theorem 5.18 we saw that every ideal of $\mathbb{Z}$ is a principal ideal. Is this also true for $(\mathbb{Z}_n, +_n, \cdot_n)$ when $n > 1$?
2. Fill in the blanks to complete the proof of the statement below.
   Prove: For $n > 1$, every ideal of $\mathbb{Z}_n$ is principal.

   Let $n > 1$ and S an ideal of $\mathbb{Z}_n$. We want to show that S is ________. If $S = \{0\}$ then $S = \langle$________$\rangle$ and is principal, so we will assume for the rest of this proof that $S \neq \{0\}$. Since $S \subseteq \{0, 1, \ldots, n-1\}$ choose m to be the least positive element of S. The goal is to show that $S = \langle m \rangle$.

   Since S is an __________, we know that $xm \in S$ for every $x \in$ ________. Thus as we saw in the previous problem $x \bullet m \in S$ and so $\langle m \rangle$________S. Now suppose $y \in S$ and $y \neq 0$. By the ________, there exist $q, r \in \mathbb{Z}$ with $y = mq + r$ and $0 \leq r < m$. But as S is an ________ of $\mathbb{Z}_n$ then $y - mq \in$ ________, or $r \in$ ________. However, m is the least positive element of S, which tells us that $r =$ ________. Thus $y = mq$ so $y \in \langle m \rangle$. Thus $S \subseteq \langle m \rangle$ and so $S = \langle m \rangle$. Thus S is a __________.

   This brings up the question of whether the same facts will hold in rings of the form $\mathbb{Z}_n \times \mathbb{Z}_m$ for $m, n > 1$. (This ring uses $+_n$, $\cdot_n$, $+_m$, or $\cdot_m$ in appropriate coordinates.)
3. Find $n, m > 1$ and a subring S of $\mathbb{Z}_n \times \mathbb{Z}_m$ which is not an ideal. Be sure to show S is a subring, and find a counterexample showing it does not absorb products.

Project 5.5

Suppose A is a commutative ring with unity with subrings S and T.

1. Suppose $S \subseteq T \subseteq A$ and S is an ideal of T. Can we also say that S is an ideal of A? Either prove that it is true or find a counterexample with a specific ring A.
2. Suppose $S \subseteq T \subseteq A$ and S is an ideal of A. Can we also say that S is an ideal of T? Either prove that it is true or find a counterexample with a specific ring A.

3. Suppose $S \subseteq T \subseteq A$ and T is an ideal of A. Can we also say that S is an ideal of T? Either prove that it is true or find a counterexample with a specific ring A.
4. Suppose $S \subseteq T \subseteq A$ and S is an ideal of A. Can we also say that T is an ideal of A? Either prove that it is true or find a counterexample with a specific ring A.
5. Suppose $S \subseteq T \subseteq A$, T is an ideal of A and S is an ideal of T. Prove that ${}^{T}/_{S}$ is a subring of ${}^{A}/_{S}$.
6. Suppose $S \subseteq T \subseteq A$, T is an ideal of A and S is an ideal of T. Is it also true that ${}^{A}/_{S} \subseteq {}^{A}/_{T}$?

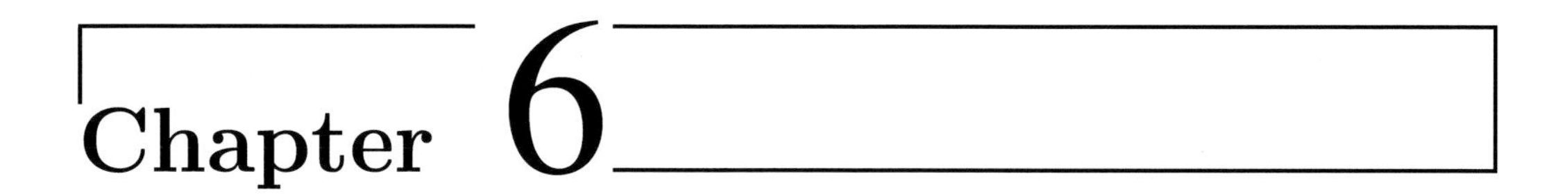

Chapter 6

Domains

Although there are many interesting theorems and properties of rings (see [3], for example), the focus of the rest of this book is on domains and fields. We will now assume ***every ring contains a unity*** unless otherwise stated. This does not, however, assume that every *subring* also contains the unity, otherwise by Theorem 5.12 the only ideals of a ring A would be $\{0_A\}$ and A. This would exclude one of our most familiar examples, the ring of integers and the ideal of even integers.

6.1 Characteristic of a Ring

In Chapter 1 we saw the following definition:

> **Definition 1.24** Suppose we have a group G and an element $a \in G$. If there exists a positive integer n with $a^n = e_G$, then the smallest positive n for which $a^n = e_G$ is called the ***order of a***, denoted $ord(a)$. If no positive integer n has $a^n = e_G$ then we say that a has infinite order.

When we translated the definitions and theorems of groups into the notation of rings we saw that group "exponentiation" a^n changed to an additive form $n \bullet a$ using Definition 4.23. Thus in a ring A, $ord(a)$ is the least positive integer n for which $n \bullet a = 0_A$. The order of elements has a special role in rings as seen in the next definition.

Definition 6.1 Let A be a nontrivial ring with unity 1_A. If there exists a nonzero integer n for which $n \bullet 1_A = 0_A$, then the ***characteristic*** of A, $char(A)$, is the smallest positive integer n for which $n \bullet 1_A = 0_A$. If no such nonzero integer exists then we say $char(A) = 0$.

The definition above is given for nontrivial rings. In the case of the trivial ring, $\{0_A\}$ where we have $0_A = 1_A$, we say that $char(A) = 1$. This is the *only* ring that has characteristic 1.

Example 6.2 Consider the ring of integers $\mathbb{Z}$, with its usual operations. In Example 2.11, we found that under addition $\mathbb{Z}$ is an infinite cyclic group generated by 1. Thus, there is no nonzero integer with $n \bullet 1 = 0$, and so $char(\mathbb{Z}) = 0$. It should be obvious that this is also the case for the rational and real numbers.

Example 6.3 Consider the ring $(\mathbb{Z}_n, +_n, \cdot_n)$ for $n > 1$. It is relatively easy to see why $char(\mathbb{Z}_n) = n$. By definition of $+_n$, we know $n \bullet 1 = 0$, thus we only need to know that n is the smallest such positive integer. Suppose $0 < m < n$. Then $m \bullet 1 = 1 +_n 1 +_n \cdots +_n 1 = m(\text{mod} n)$. Since $0 < m < n$, we know n does not evenly divide m so $m(\text{mod} n) \neq 0$, and $m \bullet 1 \neq 0$. Thus n is the least positive integer with $n \bullet 1 = 0$ and $char(\mathbb{Z}_n) = n$.

Example 6.4 Consider the ring $\mathbb{Z}_3 \times \mathbb{Z}_5$ with $+_3$ and $\cdot_3$ or $+_5$ and $\cdot_5$ in appropriate coordinates. The unity of this ring is $(1, 1)$, and since the addition is by coordinates it is easy to see that for any $n \in \mathbb{Z}$, $n \bullet (1, 1) = (n \bullet 1, n \bullet 1)$. Since in both rings we use the number 1 as unity, it is important to keep track of which ring we are working in, $\mathbb{Z}_3$ or $\mathbb{Z}_5$. Thus we will use $1_{\mathbb{Z}_3}$ and $1_{\mathbb{Z}_5}$ to denote the unities when appropriate.

In order to have $n \bullet (1_{\mathbb{Z}_3}, 1_{\mathbb{Z}_5}) = (0, 0)$ then $n \bullet 1_{\mathbb{Z}_3} = 0$ and $n \bullet 1_{\mathbb{Z}_5} = 0$. By Example 6.3 we know $char(\mathbb{Z}_3) = 3$ and $char(\mathbb{Z}_5) = 5$, so $3 \bullet 1_{\mathbb{Z}_3} = 0$ and $5 \bullet 1_{\mathbb{Z}_5} = 0$. By Theorem 1.26 using that $ord(1_{\mathbb{Z}_3}) = 3$, we can only have $n \bullet 1_{\mathbb{Z}_3} = 0$ if n is a multiple of 3. Similarly, in $\mathbb{Z}_5$, $n \bullet 1_{\mathbb{Z}_5} = 0$ only when n is a multiple of 5. The smallest positive integer which is a multiple of both 3 and 5 is 15, so 15 is the least positive integer with $15 \bullet (1_{\mathbb{Z}_3}, 1_{\mathbb{Z}_5}) = (0, 0)$. Hence $char(\mathbb{Z}_3 \times \mathbb{Z}_5) = 15$.

In Example 3.50, we saw that $\mathbb{Z}_3 \times \mathbb{Z}_5$ is isomorphic to $\mathbb{Z}_{15}$. By Example 6.3 above $char(\mathbb{Z}_{15}) = 15$ which is consistent with $char(\mathbb{Z}_3 \times \mathbb{Z}_5) = 15$.

Example 6.5 We saw in Example 4.6 that, for a nonempty set C, $\wp(C)$ is a ring under the operations $U \oplus V = (U - V) \cup (V - U)$ and $U \otimes V = U \cap V$. An exercise of Chapter 4 also tells us that C is the unity of $\wp(C)$, so $char(\wp(C)) = ord(C)$.

Notice that $C \oplus C = (C - C) \cup (C - C)$. But $C - C = \emptyset$ so we have $C \oplus C = \emptyset$. In $\wp(C)$ we have $0_{\wp(C)} = \emptyset$, so $2 \bullet C = 0_{\wp(C)}$. Thus $char(\wp(C)) = 2$.

The characteristic of a ring has many properties we will use in later chapters, some are given in the next few theorems.

Theorem 6.6

Suppose A is a nontrivial ring with unity.

(i) If $char(A) = n$ for some $n \in \mathbb{Z}$ with $n > 0$, then for every element $a \in A$, $n \bullet a = 0_A$.
(ii) If $char(A) = 0$ and $a \neq 0_A$ is not a zero divisor then $n \bullet a \neq 0_A$ for any positive $n \in \mathbb{Z}$.

Proof Assume A is a nontrivial ring, so $0_A \neq 1_A$ by Theorem 4.22.

(i) Left as an exercise at the end of the chapter.

(ii) Suppose that $char(A) = 0$, and there is an element $a \in A$ so that $a \neq 0_A$ and a is not a zero divisor. Let n be a positive integer and suppose that $n \bullet a = 0_A$. By Theorem 4.28, we have $n \bullet a = (n \bullet 1_A)a$. Thus $(n \bullet 1_A)a = 0_A$. But since $char(A) = 0$ then $n \bullet 1_A \neq 0_A$, and we have two nonzero elements of A (a and $n \bullet 1_A$) and whose product is 0_A. This contradicts that a is not a zero divisor. Hence $n \bullet a \neq 0_A$. □

Example 6.7 Consider the ring $A = M_2(\mathbb{Z})$ with usual matrix addition and multiplication. We know the zero and unity of the ring as shown below.

$$0_A = \begin{bmatrix} 0 & 0 \\ 0 & 0 \end{bmatrix} \qquad 1_A = \begin{bmatrix} 1 & 0 \\ 0 & 1 \end{bmatrix}$$

For each positive integer n, we can calculate $n \bullet 1_A$ as shown, which cannot be 0_A for any positive $n \in \mathbb{Z}$.

$$n \bullet 1_A = \begin{bmatrix} n & 0 \\ 0 & n \end{bmatrix}$$

Thus $char(M_2(\mathbb{Z})) = 0$. The element a below is a zero divisor, but we still have $n \bullet a \neq 0_A$ for every positive n.

$$a = \begin{bmatrix} 0 & 0 \\ 0 & 1 \end{bmatrix} \qquad n \bullet a = \begin{bmatrix} 0 & 0 \\ 0 & n \end{bmatrix}$$

Example 6.8 Consider the ring $B = \mathbb{Z} \times \mathbb{Z}_3$ with usual addition and multiplication in the first coordinate and $+_3$ and $\cdot_3$ in the second. We have $1_B = (1, 1)$ and $0_B = (0, 0)$. For each positive integer n, $n \bullet 1_B = (n \bullet 1, n \bullet 1)$, but $char(\mathbb{Z}) = 0$ so $n \bullet 1_B \neq (0, 0)$. Thus $n \bullet 1_B \neq 0_B$ and $char(B) = 0$. This time the element $a = (0, 1)$ is a zero divisor since $(1, 0)(0, 1) = (0, 0)$. However, $3 \bullet a = (3 \bullet 0, 3 \bullet 1)$ and in $\mathbb{Z}_3$, $3 \bullet 1 = 0$, thus $3 \bullet a = 0_B$. Thus this zero divisor, in a ring of characteristic 0, does have a positive integer n with $n \bullet a = 0$.

The previous examples show that in Theorem 6.6 knowing $char(A) = 0$ is not enough to know whether there will be a positive n with $n \bullet a = 0_A$ when a is a zero divisor.

For each $n > 1$ we saw that $char(\mathbb{Z}_n) = n$, so these finite rings cannot have characteristic 0. The next theorem shows that this is not a coincidence; it also holds for any other finite ring.

Theorem 6.9

Suppose A is a ring. If A is *finite*, then $char(A) \neq 0$.

Proof Suppose A is a finite ring, say A contains exactly n elements, with $n \geq 1$. Thus $(A, +_A)$ is a finite group, so by Theorem 3.10 we know the order of 1_A evenly divides n. Thus there exists some positive integer k with $k \bullet 1_A = 0_A$, and by definition $char(A) \neq 0$. □

A natural question to ask after this theorem is whether an infinite ring must have characteristic 0. The next example shows that an infinite ring can have finite characteristic, and in Chapter 7 we will define large a collection of infinite rings with nonzero characteristic. Thus the converse of Theorem 6.9 above is false!

Example 6.10 Using the set $\mathbb{Z}$ as C in Example 6.5 we have a ring $\wp(\mathbb{Z})$. We saw that for any nonempty C, $char(\wp(C)) = 2$, and thus $char(\wp(\mathbb{Z})) = 2$. However, $\wp(\mathbb{Z})$ is an infinite ring since it contains $\{n\}$ for every $n \in \mathbb{Z}$.

6.2 Domains

Many of the ideas introduced in this book have been motivated by properties of the integers. One special property that we have occasionally used is that $\mathbb{Z}$ does not have zero divisors. Recall from Definition 4.14 that a nontrivial ring with unity which has no zero divisors is called a ***domain***. If we also have commutativity we call it an ***integral domain***.

It should be clear that many familiar rings, the integers, rational numbers, and real numbers are integral domains. However, $\mathbb{Z}_8$ is not a domain (but is commutative) with its usual operations since 2, 4, and 6 are zero divisors. Also in Example 6.7 we found a zero divisor in $M_2(\mathbb{Z})$ so it is not a domain (nor is it commutative).

Example 6.11 Consider the ring $\mathbb{Q} \times \mathbb{Z}$ where both $\mathbb{Q}$ and $\mathbb{Z}$ have their usual addition and multiplication. Although both $\mathbb{Q}$ and $\mathbb{Z}$ are domains individually, the product $\mathbb{Q} \times \mathbb{Z}$ is not a domain. Notice that $\left(\frac{1}{3},\ 0\right) \cdot_{\mathbb{Q}\times\mathbb{Z}} (0,\ -2) = (0, 0)$. This is similar to $\mathbb{Z} \times \mathbb{Z}_3$ in Example 6.7. Will there ever be a way for the direct product of two domains to form a domain?

Recall from Theorem 4.17 that the ring $(\mathbb{Z}_n, +_n, \cdot_n)$ is an integral domain exactly when n is prime; in other words, when $char(\mathbb{Z}_n)$ is prime. The same will be true for any domain whose characteristic is nonzero, as seen in the theorem below.

Theorem 6.12

Let A be domain. Then $char(A)$ must either be 0 or a prime number.

Proof Suppose A is domain, and thus we know A is a nontrivial ring with unity 1_A. Assume that $char(A) \neq 0$, and we must show that $char(A)$ is a prime number.

Since $char(A) \neq 0$, there is a positive integer n with $char(A) = n$. As A is nontrivial $char(A) \neq 1$, and so $n > 1$. If n is not prime then we can write $n = mq$ where $m, q \in \mathbb{Z}$ and $1 < q, m < n$. Now by Theorem 4.28:

$$0_A = n \bullet 1_A = (mq) \bullet 1_A = m \bullet (q \bullet 1_A) = (m \bullet 1_A)(q \bullet 1_A).$$

But $char(A) = n$, and $1 < q, m < n$, so we must have $m \bullet 1_A \neq 0_A$ and $q \bullet 1_A \neq 0_A$. This gives us zero divisors $m \bullet 1_A$ and $q \bullet 1_A$ which contradicts that A is a domain. Thus n must instead be prime and $char(A)$ is either 0 or a prime number. □

This theorem reinforces why we found that $\mathbb{Z}_8$ is not a domain under $+_8$ and $\cdot_8$. Does the converse of Theorem 6.12 hold, i.e., does having characteristic 0 or a prime guarantee a ring is a domain? We saw that the answer is no in Example 6.7.

We can say even more about *finite* integral domains, motivated by what we learned in Chapter 2 about the group of units in $\mathbb{Z}_n$, $U(n)$, under $\cdot_n$. One of the key pieces of information was how to find which elements of $\mathbb{Z}_n$ were units, and we used the ϕ function to help. When p is prime we calculate $\phi(p)$ as follows:

$$\phi(p) = p \left(\prod_{\substack{n \text{ prime and} \\ n \text{ divides } p}} \left(1 - \tfrac{1}{n}\right) \right) = p\left(1 - \tfrac{1}{p}\right) = p\left(\tfrac{p-1}{p}\right) = p - 1.$$

This tells us there are $p-1$ units in $\mathbb{Z}_p$, but $|\mathbb{Z}_p| = p$, so every element other than 0 (that makes $p-1$ of them) is a unit. If every nonzero element is a unit we call the ring a *field.* Thus, whenever $(\mathbb{Z}_p, +_p, \cdot_p)$ is an integral domain it is also a field. We generalize this result in the next theorem.

Theorem 6.13

Every *finite* integral domain is a field.

Proof Suppose that A is a finite integral domain, with exactly n elements, $|A| = n$. Since an integral domain must be nontrivial we know that $n \geq 2$. Let $a \in A$ be a nonzero element. We need to show that a is a unit in A. If $a = 1_A$ then clearly $aa = 1_A$ and so a is a unit in A.

Suppose now that we have an element $a \in A$ with $a \neq 0_A$ and $a \neq 1_A$. Thus A has $n \geq 3$ elements. If there are exactly three elements, i.e., $A = \{0_A, 1_A, a\}$ then since $a^2 \neq 0_A$, then either $a^2 = 1_A$ or $a^2 = a$. However, if $a^2 = a$ then $a^2 - a = 0_A$ so $a(a - 1_A) = 0_A$. Thus either $a = 0_A$ or $a = 1_A$ since we have no zero divisors. Thus $a^2 = 1_A$ and a is a unit.

Now suppose $n \geq 4$. We can list the elements of A distinctly as shown, and then multiply each element in the list by a.

$$0_A, 1_A, a, b_4, b_5, \ldots, b_n$$

$$0_A, a, aa, ab_4, ab_5, \ldots, ab_n$$

The key to this proof is to see why our second list cannot contain repeated elements and so one of them is equal to 1_A. Thus we will assume first that the elements in the second list are not all distinct. A is an integral domain so there are no zero divisors, and as none of $1_A, a, b_4, b_5, \ldots, b_n$ are equal to 0_A, none of $a, aa, ab_4, ab_5, \ldots, ab_n$ can be 0_A, so two of the following must be equal.

$$a, aa, ab_4, ab_5, \ldots, ab_n$$

If for some $4 \leq i \leq n$ we have $ab_i = a$ we subtract ab_i from both sides and use the distributive law to find:

$$0_A = a - ab_i = a(1_A - b_i).$$

As A is an integral domain and $a \neq 0_A$ then $1_A - b_i = 0_A$ so we have $1_A = b_i$, contradicting that our original elements were distinct. If instead $ab_i = ab_j$ for some $4 \leq i \neq j \leq n$, then we find a similar equation.

$$0_A = ab_j - ab_i = a(b_j - b_i)$$

But again we must conclude that $b_j = b_i$ which is impossible since they were distinct.

Thus the list $0_A, a, aa, ab_4, ab_5, \ldots, ab_n$ contains exactly n distinct elements of A, i.e., this list is <u>all</u> of A. However, A has a unity 1_A so one of $0_A, a, aa, ab_4, ab_5, \ldots, ab_n$ is equal to 1_A. As $0_A \neq 1_A$ and $a \neq 1_A$, then one of $aa, ab_4, ab_5, \ldots, ab_n$ is equal to 1_A and thus one of

$a, b_4, b_5, \ldots, b_n$ is the inverse of a. Therefore each nonzero element of A is a unit and A is a field. □

The next example considers a ring homomorphism to see if having one ring as an integral domain will also force the other to be an integral domain as well.

Example 6.14 Suppose that A and K are rings with unity, and $f : A \to K$ is a ring homomorphism.

If A is an integral domain must K also be an integral domain? Clearly the function $f : A \to K$ defined by $f(x) = 0_K$ for every $x \in A$ is a ring homomorphism and K does not need to be an integral domain. What about a nonzero homomorphism? This time Theorem 4.41 shows us it will still fail since there is a nonzero ring homomorphism from $\mathbb{Z}$ to $\mathbb{Z}_4$.

Consider instead the opposite question. If we know that K is an integral domain must A also be one as well? Again for $f : A \to K$ with $f(x) = 0_K$ for every $x \in A$ it will not matter if A is an integral domain. Finally, we consider a nonzero homomorphism where K is an integral domain, using the rings $(\mathbb{Z}_6, +_6, \cdot_6)$ and $(\mathbb{Z}_3, +_3, \cdot_3)$.

Define $f : \mathbb{Z}_6 \to \mathbb{Z}_3$ by $f(x) = x(\text{mod}3)$. Then $f(0) = 0, f(1) = 1, f(2) = 2, f(3) = 0, f(4) = 1, f(5) = 2$.

$+_6$	0	1	2	3	4	5		$+_3$	0	1	2	0	1	2
0	0	1	2	3	4	5		0	0	1	2	0	1	2
1	1	2	3	4	5	0		1	1	2	0	1	2	0
2	2	3	4	5	0	1	$\to$	2	2	0	1	2	0	1
3	3	4	5	0	1	2		0	0	1	2	0	1	2
4	4	5	0	1	2	3		1	1	2	0	1	2	0
5	5	0	1	2	3	4		2	2	0	1	2	0	1

$\cdot_6$	0	1	2	3	4	5		$\cdot_3$	0	1	2	0	1	2
0	0	0	0	0	0	0		0	0	0	0	0	0	0
1	0	1	2	3	4	5		1	0	1	2	0	1	2
2	0	2	4	0	2	4	$\to$	2	0	2	1	0	2	1
3	0	3	0	3	0	3		0	0	0	0	0	0	0
4	0	4	2	0	4	2		1	0	1	2	0	1	2
5	0	5	4	3	2	1		2	0	2	1	0	2	1

The tables above show that f is a ring homomorphism, and thus it is still not required that A be an integral domain.

Theorem 6.15

Let A and K be commutative rings with unity and $f : A \to K$ a *nonzero* ring homomorphism.

(i) If K is an integral domain then $f(1_A) = 1_K$.
(ii) If f is one to one and K is an integral domain, then A is an integral domain.

Proof Assume A and K are commutative rings with unity and $f : A \to K$ is a nonzero ring homomorphism.

(i) Suppose K is an integral domain. If $f(1_A) = 0_K$ then for every $a \in A$ the following must hold:

$$f(a) = f(a1_A) = f(a)f(1_A) = f(a)0_K = 0_K.$$

This would tell us that f is the zero homomorphism, a contradiction. Thus $f(1_A) = b$ where $b \neq 0_K$. But now $f(1_A) = f(1_A1_A) = f(1_A)f(1_A)$, so we have $b = b^2$. Subtracting and using the distributive law we see $b(1_K - b) = 0_K$. But as K is a domain and $b \neq 0_K$ we must have $1_K - b = 0_K$ or $b = 1_K$ and $f(1_A) = 1_K$.

(ii) Left as an exercise at the end of the chapter.

□

Example 6.16 Consider the ring $(\mathbb{Z}_6, +_6, \cdot_6)$ and define the function $f : \mathbb{Z}_6 \to \mathbb{Z}_6$ by $f(x) = x^6$. We know that $f(0) = 0, f(1) = 1, f(2) = 4, f(3) = 3$, $f(4) = 4$, and $f(5) = 1$. Now $f(1 + 2) = f(3) = 3$, and $f(1) + f(2) = 1 + 4 = 5$, so $f(1 + 2) \neq f(1) + f(2)$, and f is not a homomorphism.

In contrast, consider the ring $\mathbb{Z}_5$ and $f : \mathbb{Z}_5 \to \mathbb{Z}_5$ by $f(x) = x^5$. This time we have $f(0) = 0, f(1) = 1, f(2) = 2, f(3) = 3, f(4) = 4$, and $f(5) = 5$ which is the identity function and clearly a ring homomorphism.

The difference in the two rings above is that one is an integral domain and the other is not!

Theorem 6.17

Suppose A is a finite integral domain with $char(A) = p$. Then the function f with $f : A \to A$ defined by $f(x) = x^p$ is an isomorphism.

Proof Suppose A is a finite integral domain with $char(A) = p$. Since the trivial ring is not an integral domain then $p > 1$. By Theorem 6.9 and Theorem 6.12, we know that p must be prime.

First, we will see why f is a ring homomorphism. Let $a, b \in A$ then $f(ab) = (ab)^p = a^p b^p = f(a)f(b)$ since multiplication is commutative in an integral domain. Also $f(a+b) = (a+b)^p$ so we use Theorem 4.30 to expand this expression.

$$(a+b)^p = a^p + \binom{p}{1} \bullet a^{p-1}b + \ldots + \binom{p}{p-1} \bullet ab^{p-1} + b^p$$

But as $char(A) = p$, and when $j < p$ each $\binom{p}{j}$ is a multiple of p (see [7]) then $\binom{p}{j} \bullet a^{p-j}b^j = 0_A$ for $j = 1, 2, \ldots, p-1$. Thus $f(a+b) = (a+b)^p = a^p + b^p = f(a) + f(b)$ and f is a ring homomorphism.

Using Theorem 5.29, we need to show that $ker(f) = \{0_A\}$. Suppose we have $a \in ker(f)$, then $a^p = 0_A$. But if $a \neq 0_A$ then $a(a^{p-1}) = 0_A$. As A has no zero divisors we must have $a^{p-1} = 0_A$. Repeating this process, eventually we find $aa = 0_A$. But A has no zero divisors which gives us a contradiction. Hence $a = 0_A$, so $ker(f) = \{0_A\}$ and f is one to one.

Finally, since A is a finite integral domain and $f : A \to A$ is one to one, it must be onto by Theorem 0.15. Thus $f(A) = A$ and f is an isomorphism. □

There will be many homomorphisms throughout the book. Now we turn our attention to ideals and integral domains.

6.3 Prime and Maximal Ideals

Ideals were introduced in Definition 5.8, and Principal Ideals in Definition 5.15. Two more important types of ideals are defined next.

Definition 6.18 Let A be a commutative ring with unity and let S be a *proper* ideal of A, i.e., $S \neq A$.

(i) The ideal S is called a ***prime ideal*** of A if whenever there are $a, b \in A$ with $ab \in S$, then either $a \in S$ or $b \in S$.
(ii) The ideal S is called a ***maximal ideal*** of A if there does not exist an ideal T with $S \subset T \subset A$. (Note that these are strict subsets.)

Example 6.19 Consider the ring $(\mathbb{Z}_{12}, +_{12}, \cdot_{12})$. We will see that the ideal $S = \langle 3 \rangle = \{0, 3, 6, 9\}$ is a prime ideal of $\mathbb{Z}_{12}$. Suppose we have two elements $a, b \in \mathbb{Z}_{12}$ with $ab \in S$; then we need to show that at least one of a or b is in S. Thus we need to see if there is a way to create an element of S by multiplying two elements of $\mathbb{Z}_{12}$ that are not in S. The elements of $\mathbb{Z}_{12}$ that are not is S are 1, 2, 4, 5, 7, 8, 10, 11, and we multiply every pair of them in the Cayley table below (using $\cdot_{12}$, of course).

$\cdot_{12}$	1	2	4	5	7	8	10	11
1	1	2	4	5	7	8	10	11
2	2	4	8	10	2	4	8	10
4	4	8	4	8	4	8	4	8
5	5	10	8	1	11	4	2	7
7	7	2	4	11	1	8	10	5
8	8	4	8	4	8	4	8	4
10	10	8	4	2	10	8	4	2
11	11	10	8	7	5	4	2	1

As none of these answers are in S, we know $ab \in S$ does in fact imply that at least one of a or b is in S. Thus S is a prime ideal of $\mathbb{Z}_{12}$.

Example 6.20 Using the same ring and ideal as above, is S a maximal ideal? Suppose there exists an ideal T with $S \subset T \subset \mathbb{Z}_{12}$. Thus there is an element of T that is not in S, so either 1, 2, 4, 5, 7, 8, 10, or 11 is in T. If $1 \in T$ we know that $T = \mathbb{Z}_{12}$, by Theorem 5.12. Thus 1 cannot be an element in T.

- Since $3 \in S$, then $3 \in T$. Now if $10 \in T$ then $3 + 10 = 1$ is in T, and so $10 \notin T$.
- By $6 \in S$ we have $6 \in T$, but $6 + 7 = 1$ and $1 \notin T$ so $7 \notin T$.
- As $9 \in S$ we have $9 \in T$, but $9 + 4 = 1$ so $4 \notin T$.
- Notice that $2 + 2 = 4$ and $4 \notin T$ so $2 \notin T$ as well.
- Since $8 + 8 = 4$ and $5 + 5 = 10$, but $4, 10 \notin T$ then $8 \notin T$ and $5 \notin T$.
- Finally $11 + 11 = 2$ but we just showed that $2 \notin T$, so $11 \notin T$.

Thus, no ideal T with $S \subset T \subset \mathbb{Z}_{12}$ can exist and S is a maximal ideal as well.

There is a close relationship between prime ideals and maximal ideals. It may seem from the previous example that they are the same, but a simple way to see that this fails is to consider the ring of integers $(\mathbb{Z}, +, \cdot)$. In $\mathbb{Z}$, $\{0\}$ is a prime ideal (since $\mathbb{Z}$ is a domain), but the principal ideal $\langle 2 \rangle$ has $\{0\} \subset \langle 2 \rangle \subset \mathbb{Z}$. Thus $\{0\}$ is not maximal.

Example 6.21 Notice that $\langle 2 \rangle$ is also a prime ideal of $(\mathbb{Z}, +, \cdot)$. The set $\langle 2 \rangle$ consists of all of the even integers, so to have a product xy in $\langle 2 \rangle$, the product must be even. The product of two odd integers can never be even, so one of x or y must be even and in $\langle 2 \rangle$.

Now $\langle 2 \rangle$ is also a maximal ideal of $\mathbb{Z}$, since if we had an ideal T with $\langle 2 \rangle \subset T \subset \mathbb{Z}$ then T must contain an odd number m. But 2 and m must be relatively prime so by Theorem 0.22 there are integers x, y with $2x + my = 1$. Since $2, m \in T$ and T absorbs products from $\mathbb{Z}$ we have $2x, my \in T$. Thus by T closed under addition $1 = 2x + my \in T$. However, $1 \in T$ tells us that $T = \mathbb{Z}$ by Theorem 5.12 so no such T can exist. Hence $\langle 2 \rangle$ is a maximal ideal.

This leads to the question of whether maximal ideals are always prime ideals. To help answer this question we use quotient rings.

Theorem 6.22

Suppose A is a nontrivial commutative ring with unity, and S is an ideal of A.

(i) S is a prime ideal of A if and only if $^{A}/_{S}$ is an integral domain.
(ii) S is a maximal ideal of A if and only if $^{A}/_{S}$ is a field.

Proof Suppose A is a nontrivial commutative ring with unity and S is an ideal of A.

(i) We prove one implication here and leave the other as an exercise at the end of the chapter.

($\rightarrow$) Suppose that S is a prime ideal of A. We know that $^{A}/_{S}$ is a commutative ring with unity by Theorem 5.25. Also S is a proper ideal of A by Definition 6.18 so by Theorem 5.25 we know $^{A}/_{S}$ is not a trivial ring. We only need to prove that $^{A}/_{S}$ has no zero divisors.

Assume that $^{A}/_{S}$ has a zero divisor $a + S$. Thus $a + S \neq 0_A + S$, and there exists $b + S \neq 0_A + S$ with $(a + S) * (b + S) = 0_A + S$. By Definition 5.19 $ab + S = 0_A + S$ so by Theorem 5.5, $ab \in S$. But S is a prime ideal which means that either $a \in S$ or $b \in S$, and thus either $a + S = 0_A + S$ or $b + S = 0_A + S$. This contradiction tells us that $^{A}/_{S}$ cannot have zero divisors and is an integral domain.

($\leftarrow$) Left as an exercise at the end of the chapter.

(ii) Again, we need two proofs, and one is an exercise at the end of the chapter.

($\rightarrow$) Suppose S is a maximal ideal of A. We know that $^{A}/_{S}$ is a commutative ring with unity by Theorem 5.25. Also S is a proper ideal of A by Definition 6.18 so by Theorem 5.25 we know $^{A}/_{S}$ is not a trivial ring. We only need to show that every nonzero element of $^{A}/_{S}$ is a unit.

Let $a + S \in {}^{A}/_{S}$ with $a + S \neq 0_A + S$, so we know $a \notin S$. Define the following set, which we will show is an ideal of A:

$$T = \{x + ay : x \in S, y \in A\}.$$

Since S is an ideal, $0_A \in S$. Also $1_A \in A$, so $0_A + a1_A = a \in T$, or $T \neq \emptyset$. Let $u + av, w + az \in T$ with $u, w \in S$ and $v, z \in A$. Also assume $c \in A$.

$$(u+av)-(w+az)=(u-w)+(av-az)=(u-w)+a(v-z)$$

$$(u+av)c=uc+(av)c=uc+a(vc)$$

But $u+w, uc \in S$ by S an ideal, and $v-z, vc \in A$, so $(u+av)-(w+az) \in T$ and $(u+av)c \in T$. Thus T is closed under subtraction and absorbs products, so T is an ideal of A.

Also for each $x \in S$ using $y = 0_A$ we have $x + a0_A = x \in T$ and thus $S \subseteq T \subseteq A$. But since $a \in T$ and $a \notin S$ then $S \subset T \subseteq A$. As we assumed S is maximal, we cannot have $T \subset A$, and so $T = A$.

However, we now have $1_A \in T$ so there exist $u \in S$ and $y \in A$ with $1_A = u + av$, or $1_A + S = (u+av)+S = (u+S)+(av+S)$. By Theorem 5.5 we know that $u+S = 0_A + S$, so $1_A + S = (av+S) = (a+S)(v+S)$ and $a+S$ is a unit in ${}^{A}/_{S}$.

Therefore every nonzero element of ${}^{A}/_{S}$ is a unit, and we have shown that ${}^{A}/_{S}$ is a field.

($\leftarrow$) This implication is left as an exercise at the end of the chapter, but is not as difficult to prove!

□

A direct consequence of Theorem 6.22 settles the question we asked before the previous theorem. The proof is left as an exercise at the end of the chapter.

Theorem 6.23

Let A be a nontrivial commutative ring with unity. Every maximal ideal of A is also a prime ideal of A.

Example 6.24 Consider the ring $(\mathbb{Z}, +, \cdot)$ of integers and p a prime integer. We know by Theorem 4.41 that $f : \mathbb{Z} \to \mathbb{Z}_p$ defined by $f(x) = x(\text{mod} p)$ is a ring homomorphism, and is clearly onto. By the Fundamental Ring Homomorphism Theorem (Theorem 5.34) we know ${}^{\mathbb{Z}}/_{ker(f)} \cong f(\mathbb{Z})$. But by the definition of f, $ker(f) = \langle p \rangle$, and $f(\mathbb{Z}) = \mathbb{Z}_p$. Thus we have ${}^{\mathbb{Z}}/_{\langle p \rangle} \cong \mathbb{Z}_p$. By Theorems 4.17 and 6.13 $\mathbb{Z}_p$ is a field since p is prime, so ${}^{\mathbb{Z}}/_{\langle p \rangle}$ is a field. By Theorem 6.22 we then see that $\langle p \rangle$ is a maximal ideal of $\mathbb{Z}$. Thus for any prime p, $\langle p \rangle$ is a maximal (and prime) ideal of $\mathbb{Z}$.

For the rest of this chapter, we will consider what makes the ring of integers unique. We have seen already that $\mathbb{Z}$ is an integral domain that is not a field and $char(\mathbb{Z}) = 0$, but there is more that makes it special since in Chapter 7 we will see a large collection of such integral domains that are not fields.

6.4 Ordered Integral Domains (∗ optional)

One special property of the integers is that it has a natural *ordering.* We will look at how to define an ordering on any integral domain, and what consequences follow.

Definition 6.25 Suppose that A is an integral domain. An ***ordering*** on A is a relation $<$ which satisfies the following properties:

(i) For all $a, b \in A$, exactly one of $a = b, a < b$, or $b < a$ holds.
(ii) For all $a, b, c \in A$ if $a < b$ and $b < c$, then $a < c$.
(iii) For all $a, b, c \in A$ if $a < b$, then $a + c < b + c$.
(iv) For all $a, b, c \in A$ if $a < b$ and $0_A < c$, then $ac < bc$.

(We may also use $<_A$ to clarify that this is the ordering associated with A if another ordering is also being discussed.) When such an ordering exists on an integral domain A we say that A is an ***ordered integral domain***.

These properties are very familiar in the integers, rational numbers, and real numbers! Can we do the same for other integral domains?

Example 6.26 Consider the integral domain $\mathbb{Z}_5$ with its usual operations. Suppose we had an ordering in which $2 < 1$. Then we can repeatedly use (iii) in Definition 6.25 with the element $c = 1$.

$$2 + 1 < 1 + 1 \text{ or } 3 < 2$$
$$3 + 1 < 2 + 1 \text{ or } 4 < 3$$
$$4 + 1 < 3 + 1 \text{ or } 0 < 4$$

Since $0 < 4$ and $2 < 1$, we can use (iv) of Definition 6.25 to find $2(4) < 1(4)$ and thus $3 < 4$. But now $4 < 3$ and $3 < 4$, which is impossible by (i) of Definition 6.25 so an ordering on $\mathbb{Z}_5$ cannot have $2 < 1$.

Thus if an ordering exists on $\mathbb{Z}_5$ we must have $1 < 2$. Using (iii) of Definition 6.25 with $c = 1$ we have $2 < 3$, so by (ii) we see that $1 < 3$. Again, using (iii) with $c = 4$ this time we have $1 + 4 < 3 + 4$, giving us $0 < 2$. Using (ii) we now have $0 < 3$ so we can use (iv) to find $3(1) < 3(2)$ or $3 < 1$. But $1 < 3$ and $3 < 1$ contradict (i) of Definition 6.25 so we cannot have $1 < 2$ either.

Thus no ordering can exist on $\mathbb{Z}_5$.

We would like to know when an integral domain *can* be ordered by our definition, but need a few more theorems to get us there. The next theorem helps us understand how negatives will interact with an ordering.

Theorem 6.27

Let A be an ordered integral domain.

(i) For all $a \in A$, if $0_A < a$, then $-a < 0_A$.
(ii) For all $a, b \in A$, if $a < b$, then $-b < -a$.
(iii) For all $a, b, c \in A$, if $a < b$ and $c < 0_A$, then $bc < ac$.

Proof Let A be an ordered integral domain.

(i) This part is left as an exercise at the end of the chapter.

(ii) Suppose we have $a, b, \in A$ and $a < b$. Thus using (iii) of Definition 6.25 as well as associativity and commutativity of addition we find the following inequalities:

$$a + (-a - b) < b + (-a - b)$$
$$(a - a) + (-b) < (b - b) + (-a).$$

Thus $-b < -a$ as needed.

(iii) Assume now that we have $a, b, c \in A$, with $a < b$ and $c < 0_A$. Using (ii) of Theorem 6.27 we know that $0_A < -c$. Thus using (i) of Definition 6.25 $a(-c) < b(-c)$. By Theorem 4.20 $a(-c) = -ac$ and $b(-c) = -bc$. Thus we have $-ac < -bc$ so using part (ii) of Theorem 6.27, again $bc < ac$. □

We could also show that $(\mathbb{Z}_5, +_5, \cdot_5)$ cannot be ordered using the previous theorem. If we assume that $2 < 3$ then (ii) tells us that $(-2) < (-3)$ so in $\mathbb{Z}_5$ this means $3 < 2$, a contradiction. Similarly, if we assume $3 < 2$ then by (ii) we have $-3 < -2$ or $2 < 3$, again a contradiction. Hence neither $2 < 3$ nor $3 < 2$ is possible!

One critical part to understanding an ordering is the relationship between 1_A and 0_A. The question of whether to choose $1_A < 0_A$ or $0_A < 1_A$ in an ordering is completely settled by the next theorem.

Theorem 6.28

Let A be an ordered integral domain.

(i) For all $a \in A$ with $a \neq 0_A$, $0_A < a^2$.
(ii) $0_A < 1_A$.

Proof Notice that (ii) follows immediately from (i) since $1_A = (1_A)^2$, thus we only need to prove (i).

Suppose A is an ordered integral domain and $a \in A$ with $a \neq 0_A$. By (i) of Definition 6.25 either $0_A < a$ or $a < 0_A$. Thus we have two possibilities to consider. Suppose that $0_A < a$ first. Then using a as the element in (iv) of Definition 6.25 we get $0_A a < a^2$ and so $0_A < a^2$. You should complete the proof by showing that if $a < 0_A$ then $0_A < a^2$. □

This gives us more information to help us rule out orderings for integral domains.

Example 6.29 Can there be an ordering on the integral domain $(\mathbb{Z}_7, +_7, \cdot_7)$? If there is an ordering we must have $0 < 1$. Also as $4 = 2^2$ and $2 = 4^2$ then $0 < 2$ and $0 < 4$ by Theorem 6.28. Thus by adding 2 to both sides of $0 < 4$ we get $2 < 4$ so by (ii) of Definition 6.25 we have $0 < 6$. Now simply adding 6 to both sides of $0 < 1$ we see that $6 < 0$, a contradiction. Thus it is impossible to create an ordering on $\mathbb{Z}_7$.

We have now seen two different finite integral domains that cannot be ordered. Is that a coincidence? The problem is that the operations on these rings "wrap around" to 0 when they run out of elements, i.e., they have $n \bullet 1_A = 0_A$ for $n > 0$.

Theorem 6.30

Let A be an ordered integral domain. Then A must have characteristic 0.

Proof Let A be an ordered integral domain. We know by Theorem 6.12 that $char(A)$ is either 0 or a prime number. We need to show that $char(A) = 0$ so suppose instead that $char(A) = p$ for some prime number p.

Using Theorem 6.28 we know we must have $0_A < 1_A$. Now by adding 1_A as in Definition 6.25 we see that $1_A < (1_A + 1_A)$ or $1_A < 2 \bullet 1_A$. Hence by Definition 6.25 (ii) $0_A < 2 \bullet 1_A$. We again add 1_A to both sides and using Definition 6.25 (ii) get $0_A < 3 \bullet 1_A$. Repeating these steps we get to $0_A < (p-1) \bullet 1_A$. However, adding 1_A once more to both sides gives $1_A < p \bullet 1_A$, and $p \bullet 1_A = 0_A$ so we have $1_A < 0_A$ contradicting (i) of Definition 6.25. Thus $char(A) = 0$. □

The previous theorem guarantees that no finite integral domain can have an ordering, since the characteristic for a finite domain cannot be 0 by Theorem 6.9. In the previous proof we used elements of the form $n \bullet 1_A$. It should seem natural to ask if $n < m$ in the integers will always tell us that $n \bullet 1_A <_A m \bullet 1_A$. Notice that $<_A$ is used for the ordering on A since the ordering $<$ on $\mathbb{Z}$ is part of the statement as well.

Theorem 6.31

Let A be an ordered integral domain.

(i) For any integer n, $n \bullet 1_A <_A (n+1) \bullet 1_A$.
(ii) For any integer n and positive integer k, $n \bullet 1_A <_A (n+k) \bullet 1_A$.
(iii) For any integers n and m, if $n < m$ then $n \bullet 1_A <_A m \bullet 1_A$.

Proof Let A be an ordered integral domain.

(i) If $n = 0$ then $n \bullet 1_A = 0_A$, and since $n + 1 = 1$, we know $(n+1) \bullet 1_A = 1_A$. By Theorem 6.28 $0_A < 1_A$ so $n \bullet 1_A <_A (n+1) \bullet 1_A$. Now suppose we have a positive integer n, then by Theorem 4.28 we get the following:

$$(n+1) \bullet 1_A = (n \bullet 1_A) + 1 \bullet 1_A.$$

As $1 \bullet 1_A = 1_A$ and $0_A <_A 1_A$, we use Definition 6.25 (iii) to find the next inequality.

$$n \bullet 1_A + 0_A <_A (n \bullet 1_A) + 1 \bullet 1_A$$

Thus for any positive integer n we have $n \bullet 1_A <_A (n+1) \bullet 1_A$.

Finally suppose we have a negative integer n. Then $-n > 0$, and so $(-n) \bullet 1_A <_A (-n+1) \bullet 1_A$ using the first part of this proof. This inequalities implies the next by Theorem 4.28.

$$(-n) \bullet 1_A <_A (-n) \bullet 1_A + 1 \bullet 1_A$$

Now we add $-(1 \bullet 1_A)$ to both sides to find the next inequalities.

$$(-n) \bullet 1_A + (-1) \bullet 1_A <_A (-n) \bullet 1_A$$
$$(-n-1) \bullet 1_A <_A (-n) \bullet 1_A$$

Finally using (ii) of Theorem 6.27, $n \bullet 1_A <_A (n+1) \bullet 1_A$ as we needed to show.

(ii) Now suppose that n is an integer and k is a positive integer. We will use induction on k to prove this part. Consider the statement to prove:

$$P(k) \text{ is } \text{“}n \bullet 1_A <_A (n+k) \bullet 1_A.\text{”}$$

We know from (i) of this theorem that P(1) is true since $n \bullet 1_A <_A (n+1) \bullet 1_A$. Now assume for some arbitrary positive integer m that $P(m)$ is true and we need to show that $P(m+1)$ is true. Thus we assume that:

$$n \bullet 1_A <_A (n+m) \bullet 1_A.$$

Using the first part of this theorem, and then by (ii) of Definition 6.25 we get the next two inequalities.

$$(n+m) \bullet 1_A <_A (n+m+1) \bullet 1_A$$
$$n \bullet 1_A <_A (n+m+1) \bullet 1_A$$

Hence $P(m+1)$ is true and thus $P(k)$ holds for any positive integer k.

(iii) This part is left as an exercise at the end of the chapter. □

For an ordered integral domain A we know that $\{n \bullet 1_A : n \in \mathbb{Z}\} \subseteq A$. Theorem 6.31 tells us that all of these elements are different since for any two integers $m \neq n$ we now see that $m \bullet 1_A \neq n \bullet 1_A$. Thus we have a "copy" of the integers sitting inside of any ordered integral domain. We have not yet ruled out that more elements can exist; the final step to understanding the unique properties of the integers is the "empty space" between 0 and 1.

If A is an ordered integral domain we define $A^+ = \{a \in A : 0_A <_A a\}$ and call the elements of this set the *positive* elements of A. By Theorem 6.27 if $0_A <_A a$ then $-a <_A 0_A$, so we cannot have both a and $-a$ in A^+.

Definition 6.32 Let A be an ordered integral domain. A is called an ***Integral System*** if every nonempty subset of A^+ contains a least element with respect to the ordering $<_A$.

Notice that $(\mathbb{Z}, +, \cdot)$ is clearly an integral system, but $(\mathbb{Q}, +, \cdot)$ is not. In $\mathbb{Q}$ the set $\{x \in \mathbb{Q} : 0 < x < 1\}$ contains $\frac{1}{2}$, but does not have a least element. The theorem that will guarantee an integral system must be isomorphic to $\mathbb{Z}$ is next.

Theorem 6.33

Let A be an integral system.

(i) The set $\{x \in A : 0_A <_A x <_A 1_A\}$ is empty.
(ii) For every integer n, the set $\{x \in A : n \bullet 1_A <_A x <_A (n+1) \bullet 1_A\}$ is empty.

Proof Suppose A is an integral system.

(i) Let $B = \{x \in A : 0_A <_A x <_A 1_A\}$, and assume that B is not empty.

Since each element $x \in B$ satisfies $0_A <_A x$, then $B \subseteq A^+$. Thus by Definition 6.32, B contains a least element, call it u. By Theorem 6.28 $0_A <_A u^2$. Also by $u < 1_A$, and $0_A <_A u$ Definition 6.25 gives us $u^2 <_A u$. Thus by (ii) of Definition 6.25, $0_A <_A u^2 <_A 1_A$ and so $u^2 \in B$. But now $u^2 \in B$ and $u^2 <_A u$, contradicting that u is the least element of B. Thus B must be empty as we wanted to prove.

(ii) Fix an integer n, and let $B = \{x \in A : n \bullet 1_A <_A x <_A (n+1) \bullet 1_A\}$. If $n = 0$, then B is the same as the set in (i) and thus is empty, so we will assume that $n \neq 0$.

Suppose that the set B is nonempty, say $u \in B$. By definition of B we then have the following:

$$n \bullet 1_A <_A u <_A (n+1) \bullet 1_A.$$

Subtracting $n \bullet 1_A$ in this inequality using Definition 6.25 we have:

$$0_A <_A u - n \bullet 1_A <_A 1_A$$

But this implies that the set defined in (i) is nonempty which is impossible. Hence the set B must be empty completing the proof. □

This now gives us a clear indication that the only elements of an integral system will be those of the form $n \bullet 1_A$ and thus the integral system looks exactly like $\mathbb{Z}$.

Theorem 6.34

Let A be an integral system. Then A is isomorphic to the ring $\mathbb{Z}$ under usual addition and multiplication.

Proof Suppose A is an integral system.

$$\text{Define } f : \mathbb{Z} \to A \text{ by } f(n) = n \bullet 1_A$$

Theorem 4.28 tells us that $(n+m)\bullet 1_A = n\bullet 1_A +_A m\bullet 1_A$ and so $f(n+m) = f(n) +_A f(m)$. By definition of the function we have:

$$f(nm) = (nm) \bullet 1_A$$
$$f(n)f(m) = (n \bullet 1_A)(m \bullet 1_A).$$

So by Theorem 4.28 $f(nm) = f(n)f(m)$. Thus f is a ring homomorphism.

Since A is an integral system, $char(A) = 0$, and thus $n \bullet 1_A$ will only equal 0_A when $n = 0$. This means that $ker(f) = \{0\}$ so that f is one to one. Finally, we need to know that

f is onto, i.e., every element of A is of the form $n \bullet 1_A$.

Consider the set B defined by $B = \{y \in A : \ y \neq n \bullet 1_A \text{ for all } n \in \mathbb{Z}\}$. Our goal is to show that this set must be empty, so assume instead that B is nonempty. Notice that if we have $y \in B$ then also $-y \in B$ (you should verify this). Thus as the set B is nonempty, it must contain an element y with $y > 0_A$. Hence the set $C = B \cap A^+$ is nonempty. By the definition of an integral system there must be a least element of C, call it u.

We know that $u > 0_A$, and $u \neq 1_A$ so by Theorem 6.33 $u > 1_A$. Thus $u - 1_A >_A 0_A$ and $u - 1_A \in A^+$. However, since $0_A <_A 1_A$, we can conclude that:

$$0_A + u - 1_A <_A 1_A + u - 1_A$$

$$u - 1_A <_A u.$$

But u is the least element of C so we must have $u - 1_A \notin C$. Thus as $u - 1_A \in A^+$ we know $u - 1_A$ is not in B so we have $u - 1_A = n \bullet 1_A$ for some $n \in \mathbb{Z}$. Thus $u = (n+1) \bullet 1_A$ and $u \notin B$. But since $u \in C$ we must have $u \in B$, a contradiction. Therefore the set B is empty, and every element of A is of the form $n \bullet 1_A$. Thus $f(Z) = A$ and f is onto, so we have a ring isomorphism and $\mathbb{Z} \cong A$. □

We have now discovered *exactly* what is needed to identify the ring of integers, it is an integral system! The rational numbers and real numbers are not integral systems; instead, they are fields which will be the focus of most of the rest of this text.

Exercises for Chapter 6

Section 6.1 Characteristic of a Ring

1. Determine the characteristic of the ring $\mathbb{Z}_2 \times \mathbb{Z}_4$ with $+_2$ and $\cdot_2$ or $+_4$ and $\cdot_4$ in appropriate coordinates.
2. Determine the characteristic of the ring $M_2(\mathbb{Z}_3)$ with usual matrix operations and $+_3$ and $\cdot_3$ operations in each entry.
3. Determine the characteristic of the ring $\mathbb{Z}_3 \times \mathbb{Z}_9$ with $+_3$ and $\cdot_3$ or $+_9$ and $\cdot_9$ in appropriate coordinates.
4. Determine the characteristic of the ring $END(\mathbb{Z}_3)$ defined in Project 4.3. Remember that the elements are functions.
5. Determine the characteristic of the ring $\Im(\mathbb{R})$ defined in Example 4.10.
6. Determine the characteristic of the ring ${}^{\mathbb{Z}_{12}}/_{\langle 4 \rangle}$ with usual coset operations.
7. Determine the characteristic of the ring ${}^{\mathbb{Z}_4 \times \mathbb{Z}_4}/_{\langle (2,0) \rangle}$ with usual coset operations.
8. Prove (i) of Theorem 6.6. Don't forget about Theorem 4.28.
9. Determine if the following statement is true or false, then either prove it or find a

counterexample: If A is a commutative ring with unity and $char(A) = n$ for $n \neq 0$, then for every nonzero element $a \in A$ $ord(a) = n$.

10. Determine if the following statement is true or false, then either prove it or find a counterexample: If A is a commutative ring with unity and $char(A) = 0$, then every nonzero element $a \in A$ has infinite order.
11. Suppose A is a nontrivial ring and $n \in \mathbb{Z}$. Determine if the following statement is true or false, then either prove it or find a counterexample: If $n \in \mathbb{Z}^+$ and $n \bullet a = 0_A$ for every $a \in A$, then $char(A) = n$.
12. Determine if the following statement is true or false, then either prove it or find a counterexample: If A is a commutative ring with unity and S is an ideal of A, then $char\left({}^{A}/_{S}\right) = char(A)$.
13. Determine if the following statement is true or false, then either prove it or find a counterexample: If A is a commutative ring with unity and S is an ideal of A with $A \neq S$ and $char(A) = 0$, then $char\left({}^{A}/_{S}\right) = 0$.
14. Determine if the following statement is true or false, then either prove it or find a counterexample: If A is a commutative ring with unity with $char(A) \neq 0$ and S is an ideal of A with $A \neq S$, then $char\left({}^{A}/_{S}\right)$ evenly divides $char(A)$.
15. Give an example of commutative rings A, B with unity for which $char(A \times B) = char(A) \cdot char(B)$.
16. Give an example of commutative rings A, B with unity for which $char(A \times B) \neq char(A) \cdot char(B)$.
17. Let A and B be commutative rings with unity where $char(A) = n$ and $char(B) = m$ ($m \neq 0$ and $n \neq 0$). Determine if the following statement is true or false, then either prove it or find a counterexample: If $d = gcd(m, n)$, then $char(A \times B) = d$.
18. Let A and B be commutative rings with unity where $char(A) = n$ and $char(B) = m$ ($m \neq 0$ and $n \neq 0$). Determine if the following statement is true or false, then either prove it or find a counterexample: If $k \in \mathbb{Z}^+$ and n, m both divide k, then $char(A \times B)$ divides k.
19. Let A and B be commutative rings with unity where $char(A) = n$ and $char(B) = m$ ($m \neq 0$ and $n \neq 0$). Determine if the following statement is true or false, then either prove it or find a counterexample: If m evenly divides n, then $char(A \times B) = m$.
20. Let A and B be commutative rings with unity where $char(A) = n$ and $char(B) = m$ ($m \neq 0$ and $n \neq 0$). Determine if the following statement is true or false, then either prove it or find a counterexample: If m evenly divides n, then $char(A \times B) = n$.

Section 6.2 Domains

21. Prove: If A and K are integral domains, then $A \times K$ is not a domain.
22. Prove: If A is an integral domain with $a \in A$, $a \neq 0$, and $n \bullet a = 0_A$ for some positive integer n, then $char(A)$ must divide n.
23. Show that the statement in the previous exercise fails if A is only required to be a

commutative ring with unity, but may have zero divisors.

24. Suppose A is an integral domain and $a \in A$ with $a \neq 0_A$. If $24 \bullet a = 0_A$, what are the possible choices for $char(A)$? Explain.
25. Suppose A is an integral domain and $a \in A$ with $a \neq 0_A$. If $50 \bullet a = 0_A$, what are the possible choices for $char(A)$? Explain.
26. Suppose A is an integral domain and $a \in A$ with $a \neq 0_A$. If $66 \bullet a = 0_A$, what are the possible choices for $char(A)$? Explain.
27. Prove: If A is an integral domain and there is $a \in A$, $a \neq 0_A$, with $9 \bullet a = 0_A$, then $char(A) = 3$.
28. Show that the statement in the previous exercise fails if A is only required to be a commutative ring with unity, but may have zero divisors.
29. Prove: If A is an integral domain and there is $a \in A$, $a \neq 0_A$, with $p^2 \bullet a = 0_A$ for some prime p, then $char(A) = p$.
30. Show that the previous exercise fails if we do not require p to be prime.
31. Suppose A is an integral domain and $a \in A$ with $a \neq 0_A$. Prove: If $10 \bullet a = 0_A$ and $15 \bullet a = 0_A$, then $char(A) = 5$.
32. Suppose A is an integral domain and $a \in A$ with $a \neq 0_A$. Prove: If $16 \bullet a = 0_A$, then $char(A) = 2$.
33. Suppose A is an integral domain and $a \in A$ with $a \neq 0_A$. If $14 \bullet a = 0_A$ and $8 \bullet a \neq 0_A$ what must $char(A)$ be equal to? Carefully explain your answer.
34. Suppose A is an integral domain and $a \in A$ with $a \neq 0_A$. If $30 \bullet a = 0_A$ and $6 \bullet a \neq 0_A$ what must $char(A)$ be equal to? Carefully explain your answer.
35. Suppose A is an integral domain and $a \in A$ with $a \neq 0_A$. If $210 \bullet a = 0_A$ but $6 \bullet a \neq 0_A$ and $10 \bullet a \neq 0_A$ what must $char(A)$ be equal to? Carefully explain your answer.
36. Prove (ii) of Theorem 6.15.
37. Carefully explain why for each prime $p \in \mathbb{Z}$ and $0 < j < p$ the number $\binom{p}{j}$, as defined in Theorem 4.30, is divisible by p.

Section 6.3 Prime and Maximal Ideals

38. Prove directly from the definition of a prime ideal: If A is an integral domain then $\{0_A\}$ is a prime ideal.
39. Prove directly from the definition of a maximal ideal: If A is a field then $\{0_A\}$ is a maximal ideal.
40. Prove directly from the definition of a prime ideal: If p is a prime in $\mathbb{Z}$ then $\langle p \rangle$ is a prime ideal.
41. Determine if $S = \langle 3 \rangle$ is a prime ideal of $(\mathbb{Z}_{18}, +_{18}, \cdot_{18})$. Either prove that it is or find elements $a, b \in S$ with $ab \in S$ but $a \notin S$ and $b \notin S$. Do not use Theorem 6.22.
42. Determine if $S = \langle 6 \rangle$ is a prime ideal of $(\mathbb{Z}_{18}, +_{18}, \cdot_{18})$. Either prove that it is or find elements $a, b \in S$ with $ab \in S$ but $a \notin S$ and $b \notin S$. Do not use Theorem 6.22.

43. Determine if $S = \langle 4 \rangle$ is a prime ideal of $(\mathbb{Z}_{12}, +_{12}, \cdot_{12})$. Either prove that it is or find elements $a, b \in S$ with $ab \in S$ but $a \notin S$ and $b \notin S$. Do not use Theorem 6.22.
44. Determine if $S = \langle (2, 2) \rangle$ is a prime ideal of $\mathbb{Z}_4 \times \mathbb{Z}_4$ using $+_4$ and $\cdot_4$ in both coordinates. Either prove that it is or find elements $a, b \in S$ with $ab \in S$ but $a \notin S$ and $b \notin S$. Do not use Theorem 6.22.
45. Determine if $S = \langle (1, 3) \rangle$ is a prime ideal of $\mathbb{Z}_3 \times \mathbb{Z}_6$ using $+_3$ and $\cdot_3$ or $+_6$ and $\cdot_6$ in appropriate coordinates. Either prove that it is or find elements $a, b \in S$ with $ab \in S$ but $a \notin S$ and $b \notin S$. Do not use Theorem 6.22.
46. Determine if $S = \langle 2 \rangle$ is a maximal ideal of $(\mathbb{Z}_8, +_8, \cdot_8)$. Either prove that it is or find an ideal T with $S \subset T \subset \mathbb{Z}_8$. Do not use Theorem 6.22.
47. Determine if $S = \langle 12 \rangle$ is a maximal ideal of $(\mathbb{Z}_{20}, +_{20}, \cdot_{20})$. Either prove that it is or find an ideal T with $S \subset T \subset \mathbb{Z}_{20}$. Do not use Theorem 6.22.
48. Determine if $S = \langle 6 \rangle$ is a maximal ideal of $(\mathbb{Z}_{30}, +_{30}, \cdot_{30})$. Either prove that it is or find an ideal T with $S \subset T \subset \mathbb{Z}_{12}$. Do not use Theorem 6.22.
49. Determine if $S = \langle 2 \rangle$ is a maximal ideal of $(\mathbb{Z}_{20}, +_{20}, \cdot_{20})$. Either prove that it is or find an ideal T with $S \subset T \subset \mathbb{Z}_{20}$. Do not use Theorem 6.22.
50. Determine if $S = \langle (1, 0) \rangle$ is a maximal ideal of $\mathbb{Z}_2 \times \mathbb{Z}_4$ using $+_2$ and $\cdot_2$ or $+_4$ and $\cdot_4$ in appropriate coordinates. Either prove that it is or find an ideal T with $S \subset T \subset \mathbb{Z}_2 \times \mathbb{Z}_4$. Do not use Theorem 6.22.
51. Determine if $S = \langle (1, 2) \rangle$ is a maximal ideal of $\mathbb{Z}_2 \times \mathbb{Z}_6$ using $+_2$ and $\cdot_2$ or $+_6$ and $\cdot_6$ in appropriate coordinates. Either prove that it is or find an ideal T with $S \subset T \subset \mathbb{Z}_2 \times \mathbb{Z}_6$. Do not use Theorem 6.22.
52. Complete the proof of (i) in Theorem 6.22.
53. Use Theorem 6.22 to determine if the ideal $S = \langle 4 \rangle$ is a prime ideal of $(\mathbb{Z}_{24}, +_{24}, \cdot_{24})$.
54. Use Theorem 6.22 to determine if the ideal $S = \langle (0, 1) \rangle$ is a prime ideal of $\mathbb{Z}_3 \times \mathbb{Z}_8$ using $+_3$ and $\cdot_3$ or $+_8$ and $\cdot_8$ in appropriate coordinates.
55. Complete the proof of (ii) in Theorem 6.22. Hint: Suppose S is not maximal, so there is an ideal T with $S \subset T \subset A$, then show there is a nonzero element of ${}^{A}/_{S}$ which is not a unit.
56. Use Theorem 6.22 to determine if the ideal $S = \langle (1, 2) \rangle$ is a maximal ideal of $\mathbb{Z}_4 \times \mathbb{Z}_4$ using $+_4$ and $\cdot_4$ in each coordinate.
57. Use Theorem 6.22 to determine if $S = \langle 5 \rangle$ is a maximal ideal of $(\mathbb{Z}_{25}, +_{25}, \cdot_{25})$.
58. Use Theorem 6.22 to determine if $S = \langle 8 \rangle$ is a maximal ideal of $(\mathbb{Z}_{12}, +_{12}, \cdot_{12})$.
59. Prove Theorem 6.23.

Section 6.4 Ordered Integral Domains and Integral Systems

60. Prove (i) of Theorem 6.27.
61. Suppose A is an ordered integral domain and $a, b, c, d \in A$. Determine if the following statement is true or false, then either prove it or find a counterexample: If $a < b$ and $c < d$ then $a + c < b + d$.

62. Suppose A is an ordered integral domain and $a, b, c \in A$. Determine if the following statement is true or false, then either prove it or find a counterexample: If $a < b$ and $a < c$ then $a < b + c$.
63. Suppose A is an ordered integral domain and $a, b \in A$. Determine if the following statement is true or false, then either prove it or find a counterexample: If $a < 0_A$ and $b < 0_A$ then $ab > 0_A$.
64. Suppose A is an ordered integral domain and $a, b, c \in A$. Determine if the following statement is true or false, then either prove it or find a counterexample: If $a + b < c$ then $a < c$ or $b < c$.
65. Suppose A is an ordered integral domain and $a, b, c, d \in A$. Determine if the following statement is true or false, then either prove it or find a counterexample: If $a+b < c+d$ then $a < c$ or $b < d$.
66. Suppose A is an ordered integral domain and $a, b, c \in A$. Determine if the following statement is true or false, then either prove it or find a counterexample: If $ab < cd$ then $a < c$ or $a < d$.
67. To complete the proof of Theorem 6.28, prove: If A is an ordered integral domain and $a < 0_A$ then $0_A < a^2$.
68. Prove (iii) of Theorem 6.31.
69. Let A be an integral system and $y \in A$. Prove: If $y \neq n \bullet 1_A$ for any $n \in \mathbb{Z}$, then $-y \neq n \bullet 1_A$ for any $n \in \mathbb{Z}$.

Projects for Chapter 6

Project 6.1

Characteristic of a Ring. Recall that for a nontrivial commutative ring with unity the characteristic is defined below.

$$char(A) = \begin{cases} n & n \text{ is the least positive integer with } n \bullet 1_A = 0_A \\ 0 & \text{no positive integer exists with } n \bullet 1_A = 0_A \end{cases}$$

1. For the ring $(\mathbb{Z}_9, +_9, \cdot_9)$, show that $char(\mathbb{Z}_9) = 9$. Be sure to show that 9 is least.
2. Find the order of each nonzero element of $(\mathbb{Z}_9, +_9, \cdot_9)$. Are they all the same?
3. For the ring $B = \mathbb{Z}_3 \times \mathbb{Z}_3$ with $+_3$ and $\cdot_3$ in both coordinates, show that $char(B) = 3$. Be sure to show that 3 is least.
4. Find the order of each nonzero element of $B = \mathbb{Z}_3 \times \mathbb{Z}_3$. Are they all the same?
5. Consider now the set $C = \{0_C, 1_C, a, b, c, d, e, f, g\}$ with the new operations defined by the tables at the end of this project. You may assume that $(C, \oplus, \otimes)$ is a commutative ring with unity. Find $char(C)$.

6. Find the order of each nonzero element of the ring C. Are they all the same?
7. Notice that in A the characteristic is $9 = |\mathbb{Z}_9|$, but not all of the elements had the same order. How does this differ from the other two rings of this project?
8. Finally, using characteristic and orders of nonzero elements the two rings B and C seem the same! What important difference between them guarantees these are not really the same ring?

$\oplus$	0_C	1_C	a	b	c	d	e	f	g
0_C	0_C	1_C	a	b	c	d	e	f	g
1_C	1_C	e	b	d	f	a	0_C	g	c
a	a	b	f	g	e	c	d	0_C	1_C
b	b	d	g	c	0_C	f	a	1_C	e
c	c	f	e	0_C	b	1_C	g	d	a
d	d	a	c	f	1_C	g	b	e	0_C
e	e	0_C	d	a	g	b	1_C	c	f
f	f	g	0_C	1_C	d	e	c	a	b
g	g	c	1_C	e	a	0_C	f	b	d

$\otimes$	0_C	1_C	a	b	c	d	e	f	g
0_C	0_C	0_C	0_C	0_C	0_C	0_C	0_C	0_C	0_C
1_C	0_C	1_C	a	b	c	d	e	f	g
a	0_C	a	b	g	d	1_C	f	c	e
b	0_C	b	g	e	1_C	a	c	d	f
c	0_C	c	d	1_C	e	f	b	g	a
d	0_C	d	1_C	a	f	c	g	e	b
e	0_C	e	f	c	b	g	1_C	a	d
f	0_C	f	c	d	g	e	a	b	1_C
g	0_C	g	e	f	a	b	d	1_C	c

Project 6.2

Let A denote an **integral domain** for this project.

1. Prove: For $a \in A$ with $a \neq 0_A$ we have $ord(1_A) = ord(a)$.
2. Prove: If $char(A) = 3$, $a \in A$, and $5 \bullet a = 0_A$ then $a = 0_A$.
3. Prove: If $a \in A$, $a \neq 0_A$ and $256 \bullet a = 0_A$ then $char(A) = 2$.
4. Prove: If $a, b \in A$ with $a \neq b$ and $125 \bullet a = 125 \bullet b$ then $char(A) = 5$.
5. Prove: If $char(A) = 2$ then for any $a, b \in A$ we have $(a + b)^2 = a^2 + b^2$.

Project 6.3

Suppose that A is a *finite* **integral domain** for this entire project.

1. Explain why $char(A) = p$ for some prime p.
2. Look back at Theorem 3.10 and explain why: If $char(A) = p$ then p evenly divides $|A|$.
3. Use the previous parts of this project to prove: If $|A| = p$ and p is prime then $char(A) = p$.
4. Fill in the blanks to complete the proof of the statement below.

   Prove: If $|A| = p$ and p is prime then $A \cong \mathbb{Z}_p$.

   Assume that $|A| = p$ and p is prime. We need to show that ________. Define the function $f : \mathbb{Z}_p \to A$ by $f(m) = m \bullet 1_A$ for all $m \in \mathbb{Z}_p$. We need to show that f is a ________________. To show f is one to one suppose we have $m, n \in \mathbb{Z}_p$ with ________________. Using the definition of f we get $m \bullet 1_A = n \bullet 1_A$. We need to show that $m = n$ so assume instead that ________. Thus either $m < n$ or ________ as elements of $\mathbb{Z}$. Suppose (WLOG) that $m < n$. Using $m \bullet 1_A = n \bullet 1_A$ subtracting gives us ________________ . Theorem 4.28 tells us that $m \bullet 1_A - n \bullet 1_A =$ __________ $\bullet$ 1_A. But by the previous part of this project $char(A) = p$, and $0 < m - n < p$ so ________________ $\bullet\, 1_A \neq 0_A$, a contradiction. Thus f is one to one. Since $|A| = p$ and $|\mathbb{Z}_p| = p$, by Theorem ________ f must also be onto.

   Finally we need to show that f is a ring homomorphism. Let $m, n \in \mathbb{Z}_p$, then $f(m +_p n) =$ ________________. Also $f(m) +_A f(n) =$ ________________ $= (m +_{\mathbb{Z}} n) \bullet 1_A$. If $m +_p n = r$, then $m +_{\mathbb{Z}} n = pq +_{\mathbb{Z}} r$ for some integer q. Thus $(m +_{\mathbb{Z}} n) \bullet 1_A =$ ________ $\bullet\, 1_A = (pq) \bullet 1_A +_A$ ________________. But since $char(A) = p$ then $pq \bullet 1_A = 0_A$ and so $(m +_{\mathbb{Z}} n) \bullet 1_A =$ ________________. Thus $f(m) +_A f(n) = r \bullet 1_A = (m +_p n) \bullet 1_A = f(m +_p n)$. Similarly $f(m \cdot_p n) =$ ________________ and $f(m) \cdot_A f(n) =$ ________________ $= (m \cdot_{\mathbb{Z}} n) \bullet 1_A$. If $m \cdot_p n = s$ then $m \cdot_{\mathbb{Z}} n = px +_{\mathbb{Z}} s$ for some integer x. Thus $(m \cdot_{\mathbb{Z}} n) \bullet 1_A =$ ________________. Again since $char(A) = p$ we know $(px) \bullet 1_A = 0_A$ so $(m \cdot_{\mathbb{Z}} n) \bullet 1_A =$ ________________. Thus we have $f(m) \cdot_A f(n) = r \bullet 1_A = (m \cdot_p n) \bullet 1_A = f(m \cdot_p n)$, and f is a ________________ .

   Thus f is an ________________ and $A \cong \mathbb{Z}_p$.

Now we can say that every finite integral domain of prime order is really one of our familiar rings $\mathbb{Z}_p$.

Project 6.4

In this project we will look at some maximal and prime ideals in a commutative ring with unity.

1. Consider the ring $\mathbb{Z}_3 \times \mathbb{Z}_4$ using $+_3$ and $\cdot_3$ or $+_4$ and $\cdot_4$ in appropriate coordinates and ideal $S = \{(0,0), (0,1), (0,2), (0,3)\}$. Prove that S is a maximal ideal of $\mathbb{Z}_3 \times \mathbb{Z}_4$ by

explaining why there does not exist an ideal T with $S \subset T \subset \mathbb{Z}_3 \times \mathbb{Z}_4$.

2. Prove that the ideal $S = \{(0,0),(0,1),(0,2),(0,3)\}$ is also a prime ideal of $\mathbb{Z}_3 \times \mathbb{Z}_4$.
3. Find a nontrivial ideal K of $\mathbb{Z}_3 \times \mathbb{Z}_4$ which is not prime. Show that K is an ideal and find elements $a, b \in \mathbb{Z}_3 \times \mathbb{Z}_4$ with $ab \in K$ but $a \notin K$ and $b \notin K$.
4. Is K a maximal ideal? Either prove that no ideal T has $K \subset T \subset \mathbb{Z}_3 \times \mathbb{Z}_4$ or actually find the ideal T.

Project 6.5

This project will concentrate on Theorems 6.22 and 6.23.

1. Consider $A = (\mathbb{Z}_{15}, +_{15}, \cdot_{15})$. For the ideal $S = \langle 3 \rangle$, show the two Cayley tables for the ring ${}^{A}/_{S}$. Is ${}^{A}/_{S}$ an integral domain?
2. Why do we know $S = \langle 3 \rangle$ is a prime ideal of $\mathbb{Z}_{15}$? Is it also a maximal ideal?
3. The converse of Theorem 6.22 was shown to be false before Example 6.21. We only need to modify it slightly however. Prove: If A is a commutative ring with unity, S is a prime ideal of A, and ${}^{A}/_{S}$ is finite, then S is a maximal ideal of A.
4. Explain why in a finite commutative ring with unity, every prime ideal is also maximal.
5. $\{0\}$ was an example of a prime ideal that is not maximal in $(\mathbb{Z}, +, \cdot)$. Why doesn't $\{0_A\}$ contradict the previous statement in a finite commutative ring with unity?

Project 6.6

For this project A will always denote an *Ordered Integral Domain.*

1. Suppose that $a, b \in A$ with $a > 1_A$ and $b > 1_A$. Fill in the blanks to complete the proof of the statement below.

   Prove: $ab + 1_A > a + b$

   Since $a > 1_A$ and $b > 1_A$, then we know $(a - 1_A) >$ __________ and $(b - 1_A) >$ __________ using (ii) of Definition 6.25. Now using ________________ of Theorem 6.25 we can say that $(a - 1_A)(b - 1_A) > 0_A(b - 1_A)$ so ________________. Using the Distributive Law we now have $ab - a1_A - b1_A + (-1_A)(-1_A) > 0_A$. Thus __________. Adding ________________ to both sides, (ii) of Theorem 6.25 tells us that $ab - a - b + 1_A + (a + b) > 0_A + (a + b)$. As addition in A is commutative and associative we get ________________ as needed.

2. Suppose $a \in A$ an $a > 0_A$. Prove by PMI that for any positive integer n, $n \bullet a > 0_A$.
3. Suppose $a \in A$ an $a < 0_A$. Explain why $n \bullet a < 0_A$ for every positive integer n. (Use the previous question to help.)
4. Suppose $a \in A$ an $a > 0_A$. Prove by PMI that for any positive integer n, $a^n > 0_A$.
5. Is it also true that for $a < 0_A$ and n a positive integer we can say $a^n < 0_A$? Explain!

Chapter 7

Polynomial Rings

Since we have discussed domains in great detail, the next logical step in algebraic structure is to learn more about fields such as the rational numbers and real numbers. One critical part of the study of fields will involve the solutions to polynomial equations, so in this chapter we carefully introduce the notation for polynomials and roots of polynomials. In order to do this we must consider x to be a ***fixed symbol*** used only to define polynomials. We will not use it to stand for an arbitrary element of a set, or an unknown to solve for, from now on.

Every ring in this chapter is assumed to be a ***commutative ring with unity***, unless otherwise indicated.

7.1 Polynomials over a Ring

The definition of a polynomial should be familiar, but usually with real number coefficients. In the summation form it is assumed that $x^0 = 1_A$.

Definition 7.1 Let A be a commutative ring with unity. For each non-negative integer n and elements $a_0, a_1, \ldots, a_n \in A$ we can define a ***polynomial over A***, $a(x)$, by:

$$a(x) = a_0 + a_1 x + \cdots + a_n x^n \qquad \text{or} \qquad a(x) = \sum_{i=0}^{n} a_i x^i.$$

The set of all polynomials over a ring A is denoted $A[x]$.

Note that in this definition the symbol + does not denote addition in any specific ring we currently have, so we treat it as just a symbol at the moment. Students are familiar with polynomials such as $1 + x + 3x^2$ or $\frac{2}{3} + \frac{1}{5}x$ from $\mathbb{Z}[x]$ and $\mathbb{Q}[x]$. However, the definition can be used for any ring of coefficients, as we will see in the next examples.

Example 7.2 Consider the ring of functions $\Im(\mathbb{R})$ in Example 4.10. The functions f_0, f_1, and f_2 in $\Im(\mathbb{R})$, which are defined by $f_0(y) = \sin(y)$, $f_1(y) = 2y^2$, and $f_2(y) = e^y$, define a polynomial in $\Im(\mathbb{R})[x]$ as $a(x) = f_0 + f_1x + f_2x^2$. Notice that x is not used in the definition of each function since x is a reserved symbol for polynomials.

Example 7.3 Consider the ring $\wp(C) = \{U : U \subseteq C\}$ defined in Example 4.6, where $C = \{1, 2, 3\}$. The sets $\{1\}, \{1, 2\}$, and $\{2, 3\}$ are elements of $\wp(C)$ so we can create the polynomial $b(x) = \{2, 3\} + \{1\}x + \{1, 2\}x^2 + \{1\}x^3$ in $\wp(C)[x]$.

There are several frequently used words when working with polynomials; the first few are defined next.

Definition 7.4 Suppose A is a commutative ring with unity and $a(x) \in A[x]$ with $a(x) = a_0 + a_1x + \cdots + a_nx^n$ for some nonnegative integer n.

(i) The elements $a_0, a_1, \ldots, a_n \in A$ are the ***coefficients*** of $a(x)$.
(ii) For each $0 \leq i \leq n$, a_ix^i is called a ***term*** of $a(x)$.
(iii) The largest nonnegative integer n with $a_n \neq 0_A$ (if one exists) is the ***degree*** of $a(x)$, denoted $deg(a(x)) = n$. So for $k > n$ we know $a_k = 0_A$.
(iv) If all coefficients of $a(x)$ are 0_A we say the degree of $a(x)$ is $-\infty$.
(v) For $n \geq 0$ if $deg(a(x)) = n$ then a_n is called the ***leading coefficient*** of $a(x)$.

In Example 7.2, we saw $a(x) \in \Im(\mathbb{R})[x]$ defined as $a(x) = f_0 + f_1x + f_2x^2$ Thus $a(x)$ has three terms, $deg(a(x)) = 2$, and the leading coefficient is the function $f_2(y) = \sin(y)$. Similarly, in Example 7.3, with $b(x) \in \wp(C)[x]$ defined by $b(x) = \{2, 3\} + \{1\}x + \{1, 2\}x^2 + \{1\}x^3$, we see that $deg(b(x)) = 3$ and the leading coefficient of $b(x)$ is $\{1\}$.

Some conventions will make this notation easier to use:

- The term 1_Ax^n can be abbreviated as x^n in a polynomial $a(x)$.
- If a term in a polynomial has coefficient 0_A we need not include the term when writing the polynomial.

- If a polynomial $a(x)$ has degree $n \geq 0$ then for any $k > n$ we can assume the term $0_A x^k$ is also part of the definition of $a(x)$ when needed. Thus in $\mathbb{Z}[x]$ we can write $1 + 2x$ as $1 + 2x + 0x^2$ if needed.

These conventions will make polynomials easier to read and write. For example in $\mathbb{Z}[x]$ it is easier to write $a(x) = 2 + x^3 + x^6$ than $a(x) = 2 + 0x + 0x^2 + 1x^3 + 0x^4 + 0x^5 + 1x^6$.

Constant polynomials (polynomials with degree 0) can have confusing notation since in $a(x) = a_0$ the symbol "x" never appears, even though it is a polynomial. We will frequently write $a(x) = a_0 + 0_A x$ when we want to make it clearer. Note that we do not consider the polynomial $0(x)$, with all coefficients equal to 0_A, to be a constant polynomial since it does not have degree 0.

Understanding how to know two polynomials are equal is important when proving statements about polynomials and when we discuss roots of polynomials.

Definition 7.5 Let A be a commutative ring with unity. For polynomials $a(x), b(x) \in A[x]$ we say $a(x) = b(x)$ if and only if they have the same degree and if the degree is equal to $n \geq 0$ then for every $i \leq n$, $a_i = b_i$.

To create a ring of polynomials we need operations of addition and multiplication between polynomials.

Definition 7.6 Let A be a commutative ring with unity and let $a(x), b(x) \in A[x]$ as shown below.

$$a(x) = \sum_{i=0}^{n} a_i x^i \qquad b(x) = \sum_{i=0}^{m} b_i x^i$$

We define the new polynomial $c(x) = a(x) + b(x)$ as follows where $k = max\{n, m\}$.

$$c(x) = \sum_{i=0}^{k} c_i x^i \quad \text{with} \quad c_i = a_i +_A b_i$$

Remember, if $i > n$ or $i > m$ we assume $a_i = 0_A$ or $b_i = 0_A$, respectively.

Example 7.7 Consider the ring $(\mathbb{Z}_3, +_3, \cdot_3)$. In $\mathbb{Z}_3[x]$ we can define $a(x) = 1 + 2x + x^2 + x^3$ and $b(x) = 2 + 2x + 2x^2$. According to our definition of addition we find $a(x) + b(x)$ as follows.

$$a(x)+b(x)=(1+_3 2)+(2+_3 2)x+(1+_3 2)x^2+(1+_3 0)x^3=x+x^3$$

If our ring A is unusual, for example, if its elements are not numbers, we need to be careful with notation when adding polynomials in $A[x]$. For example, $x+x$ is often simplified to $2x$, but 2 may not be in A and thus would not make sense in $A[x]$. Since 1_A is the unity of A we actually mean $(2 \bullet 1_A)x = 1_A x + 1_A x = x + x$. For ease of notation we will usually shorten $(n \bullet 1_A)x$ to nx if we do not have a specific ring A we are using.

Notice that with this addition of polynomials we can now consider the symbol $+$ in the definition of a polynomial to be the same as our addition. To see why, choose $a(x) = 2+3x+4x^2+x^3$ from $\mathbb{Z}[x]$. We have polynomials $a_0(x)=2$, $a_1(x)=3x$, $a_2(x)=4x^2$, and $a_3(x)=x^3$ in $\mathbb{Z}[x]$, and under our new addition $a(x)=a_0(x)+a_1(x)+a_2(x)+a_3(x)$ (we will prove associativity soon).

Polynomial multiplication is a bit more complicated as we see next.

Definition 7.8 Let A be a commutative ring with unity and polynomials $a(x), b(x) \in A[x]$ as shown below.

$$a(x)=\sum_{i=0}^{n} a_i x^i \qquad b(x)=\sum_{i=0}^{m} b_i x^i$$

Define the new polynomial $d(x)=a(x)b(x)$ as follows.

$$d(x)=\sum_{i=0}^{n+m} d_i x^i \quad \text{where} \quad d_i=\sum_{j+t=i} a_j \cdot_A b_t$$

Note: $0 \le j \le n$ and $0 \le t \le m$.

Example 7.9 Consider the ring $(\mathbb{Z}_8, +_8, \cdot_8)$ and polynomials in $\mathbb{Z}_8[x]$.

$$a(x)=3+2x+4x^2 \text{ and } b(x)=0+6x+5x^2+3x^3$$

Thus $c(x)=a(x)+b(x)$ and $d(x)=a(x)b(x)$ as follows.

$$c(x)=(3+_8 0)+(2+_8 6)x+(4+_8 5)x^2+(0+_8 3)x^3=3+0x+x^2+3x^3$$

$$\begin{array}{ll} d_0=3\cdot_8 0=0 & d_1=3\cdot_8 6+_8 2\cdot_8 0=2 \\ d_2=3\cdot_8 5+_8 2\cdot_8 6+_8 4\cdot_8 0=3 & d_3=3\cdot_8 3+_8 2\cdot_8 5+_8 4\cdot_8 6=3 \\ d_4=2\cdot_8 3+_8 4\cdot_8 5=2 & d_5=4\cdot_8 3=4 \end{array}$$

So $d(x) = 0 + 2x + 3x^2 + 3x^3 + 2x^4 + 4x^5$.

We will prove that addition and multiplication for polynomials are commutative and associative, and that the distributive laws hold. Assuming these for the moment we can more easily understand the process of multiplication, writing the steps in a more familiar way.

Example 7.10 Consider the polynomials $a(x) = 3 + 2x + 4x^2$ and $b(x) = 6x + 5x^2 + 3x^3$ in $\mathbb{Z}_8[x]$.

$$\begin{aligned}
&(3 + 2x + 4x^2)(0 + 6x + 5x^2 + 3x^3) \\
&= 3(0 + 6x + 5x^2 + 3x^3) + 2x(0 + 6x + 5x^2 + 3x^3) + 4x^2(0 + 6x + 5x^2 + 3x^3) \\
&= (0 + 2x + 7x^2 + x^3) + (0 + 4x^2 + 2x^3 + 6x^4) + (0 + 0x^3 + 4x^4 + 4x^5) \\
&= 0 + 2x + 3x^2 + 3x^3 + 2x^4 + 4x^5
\end{aligned}$$

Thus we get the same answer as found in Example 7.9. There will be many polynomials to multiply in exercises and throughout the text so be sure to practice this technique.

Using the definitions of addition and multiplication defined in $A[x]$, it is important to verify that they are operations on the set $A[x]$. Both rules are clearly well defined since addition and multiplication in A tell us the coefficients of the sum and product. Also, both rules give us new coefficients in A so the result is again a polynomial in $A[x]$. Thus they are operations on our set.

We are, of course, aiming to show that $A[x]$ is a ring but because of our notation, the most complicated properties to prove are *associativity* for the two different operations.

Theorem 7.11

Let A be a commutative ring with unity. The operations of polynomial addition and polynomial multiplication from Definitions 7.6 and 7.8 are associative in $A[x]$.

Proof Assume A is a commutative ring with unity. Consider three polynomials $a(x), b(x), c(x) \in A[x]$. Since we can add more terms with coefficients of 0_A to any of the polynomials, we can express all three using the same nonnegative integer n:

$$a(x) = \sum_{i=0}^{n} a_i x^i \qquad b(x) = \sum_{i=0}^{n} b_i x^i \qquad c(x) = \sum_{i=0}^{n} c_i x^i$$

$$(a(x) + b(x)) + c(x) = g(x) = \sum_{i=0}^{n} g_i x^i \text{ where } g_i = (a_i +_A b_i) +_A c_i.$$

However, $+_A$ is associative in A, so $g_i = a_i +_A (b_i +_A c_i)$ as well. Thus associativity of polynomial addition follows easily as seen below.

$$g(x^i) = \sum_{i=0}^{n}([a_i +_A b_i] +_A c_i) = \sum_{i=0}^{n}(a_i +_A [b_i +_A c_i])x^i = a(x) + (b(x) + c(x))$$

Now consider $h(x) = [a(x)b(x)]c(x)$, and let $q(x) = a(x)b(x)$. Thus $h(x) = q(x)c(x)$. For each $0 \le j \le 2n$, $q_j = \sum_{k+s=j}(a_k \cdot_A b_s)$ where $0 \le k, s \le n$ and so for each $i \le 3n$ we find $h_i \in A$ as below, where $0 \le k, t, s \le n$ and $0 \le j \le 2n$.

$$h_i = \sum_{j+t=i} q_j \cdot_A c_t = \sum_{j+t=i}\left(\sum_{k+s=j}(a_k \cdot_A b_s)\right) \cdot_A c_t$$

Using the distributive law, commutativity, and associativity in the ring A we can rewrite these finite sums $(0 \le r \le 2n)$.

$$h_i = \sum_{k+s+t=i}(a_k \cdot_A b_s) \cdot_A c_t = \sum_{k+s+t=i} a_k \cdot_A (b_s \cdot_A c_t) = \sum_{k+r=i} a_k \cdot_A \left(\sum_{s+t=r}(b_s \cdot_A c_t)\right)$$

Thus $h(x) = a(x)[b(x)c(x)]$ and polynomial multiplication is associative. □

Example 7.12 The notation in the theorem above is confusing, so we will consider a concrete example in $\mathbb{Z}_4[x]$ to illustrate associativity of multiplication. Let $a(x) = 2 + 3x, b(x) = 0 + x + 2x^2$, and $c(x) = 1 + x$. Remember the coefficients are in the ring $(\mathbb{Z}_4, +_4, \cdot_4)$.

$$\begin{aligned} b(x)c(x) &= (0 + x + 2x^2)(1 + x) \\ &= 0 + 0x + x + x^2 + 2x^2 + 2x^3 \\ &= 0 + x + 3x^2 + 2x^3 \end{aligned}$$

$$\begin{aligned} a(x)\,[b(x)c(x)] &= (2 + 3x)(0 + x + 3x^2 + 2x^3) \\ &= 0 + 2x + 2x^2 + 0x^3 + 0x + 3x^2 + x^3 + 2x^4 \\ &= 0 + 2x + x^2 + x^3 + 2x^4 \end{aligned}$$

$$\begin{aligned} a(x)b(x) &= (2 + 3x)(0 + x + 2x^2) \\ &= 0 + 2x + 0x^2 + 0 + 3x^2 + 2x^3 \\ &= 0 + 2x + 3x^2 + 2x^3 \end{aligned}$$

$$\begin{aligned} [a(x)b(x)]c(x) &= (0 + 2x + 3x^2 + 2x^3)(1 + x) \\ &= 0 + 2x + 3x^2 + 2x^3 + 0x + 2x^2 + 3x^3 + 2x^4 \\ &= 0 + 2x + x^2 + x^3 + 2x^4 \end{aligned}$$

Thus we see that $a(x)[b(x)c(x)] = [a(x)b(x)]c(x)$.

The next theorem is left as an exercise at the end of the chapter.

Theorem 7.13

Let A be a commutative ring with unity. In $A[x]$, polynomial addition and polynomial multiplication are both commutative.

We have one more theorem with this confusing notation, but it is the last complicated part of showing $A[x]$ is a ring.

Theorem 7.14

Let A be a commutative ring with unity. Then the distributive laws hold in $A[x]$.

Proof Let A be a commutative ring with unity. Consider the three polynomials in $A[x]$ below.

$$a(x) = \sum_{i=0}^{n} a_i x^i \qquad b(x) = \sum_{i=0}^{n} b_i x^i \qquad c(x) = \sum_{i=0}^{n} c_i x^i$$

Since by Theorem 7.13 polynomial multiplication is commutative we only need to show that $a(x)[b(x) + c(x)] = a(x)b(x) + a(x)c(x)$. For help with notation we will use $d(x) = b(x) + c(x)$ and $a(x)d(x) = p(x)$. Thus $d_t = b_t +_A c_t$ for each $0 \leq t \leq n$. Remember that the coefficients are all from A and the distributive law holds in A, so for each $0 \leq i \leq 2n$ we can calculate p_i as seen below.

$$p_i = \sum_{j+t=i} (a_j \cdot_A d_t) = \sum_{j+t=i} (a_j \cdot_A [b_t +_A c_t]) = \sum_{j+t=i} (a_j \cdot_A b_t) +_A (a_j \cdot_A c_t)$$

If we call $s(x) = a(x)b(x)$ and $u(x) = a(x)c(x)$ then for $0 \leq i \leq 2n$ we have:

$$s_i = \sum_{j+t=i} (a_j \cdot_A b_t) \qquad u_i = \sum_{j+t=i} (a_j \cdot_A c_t).$$

Clearly (as these sums are finite), $p_i = s_i +_A u_i$ for each i and so $p(x) = s(x) + u(x)$. Thus we have $a(x)[b(x) + c(x)] = a(x)b(x) + a(x)c(x)$ and the distributive laws hold. □

Finally, we can prove that $A[x]$ is a commutative ring with unity. The only pieces not complete are to prove that $0(x) = 0_A + 0_A x$ is the zero, the unity is $1(x) = 1_A + 0_A x$, and every element of $A[x]$ has a negative in $A[x]$. These steps are left as an exercise at the end of the chapter.

Theorem 7.15

Let A be a commutative ring with unity. Then the set $A[x]$ of polynomials over A is a commutative ring with unity.

We now have a large collection of polynomial rings we can consider, each of $\mathbb{Z}[x]$, $\mathbb{Q}[x]$, $\mathbb{R}[x]$, $\mathbb{Z}_n[x]$ for each n, as well as other unusual examples such as $\wp(C)[x]$. We must be careful not to make too many assumptions about polynomials, since just as rings have a variety of properties so can the polynomial rings. We will look at some of these properties in the next section.

7.2 Properties of Polynomial Rings

We saw in the previous section that using a commutative ring with unity A we can say that $A[x]$ is also a commutative ring with unity. Will $A[x]$ always have the same properties as A?

Example 7.16 The ring $(\mathbb{Z}, +, \cdot)$ is an integral domain. Does that mean $\mathbb{Z}[x]$ is an integral domain as well?

To verify $\mathbb{Z}[x]$ is an integral domain we must show it has no zero divisors. For a contradiction suppose we have two nonzero polynomials $a(x), b(x) \in \mathbb{Z}[x]$ with $a(x)b(x) = 0(x)$.

By $a(x) \neq 0(x)$, there is a coefficient which is nonzero. Suppose that n is the degree of $a(x)$, then $a_n \neq 0$ but for all $k > n, a_k = 0$. Similarly, $b(x) \neq 0(x)$ so assume m is the degree of $b(x)$ and $b_m \neq 0$ but for all $k > m, b_k = 0$. From $a(x)b(x) = 0(x)$ we know the polynomial $d(x) = a(x)b(x)$ has coefficient $d_{n+m} = 0$.

$$d_{n+m} = \sum_{j+t=n+m} a_j b_t$$

For each $j \geq 0$, $a_j b_t$ must have $j+t = n+m$, so if $j < n$ then $t > m$ which tells us $b_t = 0$ and $a_j b_t = 0$. Similarly, if $j > n$ then $a_j = 0$ and so $a_j b_t = 0$ again. Thus $0 = d_{n+m} = a_n b_m$. But $a_n b_m \neq 0$ since $\mathbb{Z}$ is an integral domain! This contradiction tells us $\mathbb{Z}[x]$ is an integral domain.

The example above motivates the next theorem, whose proof mimics the steps in the example exactly.

Theorem 7.17

If A is an integral domain then $A[x]$ is also an integral domain.

Proof Suppose that A is an integral domain. Thus $A[x]$ is a commutative ring with unity by Theorem 7.15 so we only need to show that $A[x]$ has no zero divisors.

Assume instead we have a zero divisor $a(x) \in A[x]$ so $a(x) \neq 0(x)$ and $deg(a(x)) = n \geq 0$. There must exist $b(x) \in A[x]$ so that $b(x) \neq 0(x)$, but $a(x)b(x) = 0(x)$. Again, by $b(x) \neq 0(x)$, we have $deg(b(x)) = m \geq 0$. Let $d(x) = a(x)b(x)$.

$$d_{n+m} = \sum_{j+t=n+m} a_j \cdot_A b_t$$

We know $a_n \neq 0_A$ but for all $k > n$, $a_k = 0_A$. Also, $b_m \neq 0_A$, but for all $k > m$, $b_k = 0_A$. Thus $a_j \cdot_A b_t = 0_A$ except when $j = n$ and $t = m$, so $d_{n+m} = a_n \cdot_A b_m$.

Since A is an integral domain then $d_{n+m} \neq 0_A$ and $d(x) \neq 0(x)$, a contradiction. Therefore $A[x]$ has no zero divisors and $A[x]$ is an integral domain. □

Example 7.18 If the original ring A is not an integral domain, strange things can happen in $A[x]$. Consider $(\mathbb{Z}_4, +_4, \cdot_4)$ and the polynomials below in $\mathbb{Z}_4[x]$:

$$a(x) = 2 + 2x + 2x^2 \quad b(x) = 2 + 2x \quad c(x) = 3 + 2x.$$

Calculating $(a(x))^2$ and $b(x)c(x)$ we find:

$$\begin{aligned} (a(x))^2 &= (2 + 2x + 2x^2)(2 + 2x + 2x^2) \\ &= 0 + 0x + 0x^2 + 0x^3 + 0x^4 \\ &= 0(x) \end{aligned}$$

$$\begin{aligned} b(x)c(x) &= (2 + 2x)(3 + 2x) \\ &= 2 + 0x + 2x + 0x^2 \\ &= 2 + 2x. \end{aligned}$$

Thus $(a(x))^2 = 0(x)$ and $a(x)$ is a zero divisor, while $b(x)c(x) = b(x)$. Also notice that $deg(b(x)) = 1$, $deg(c(x)) = 1$, and $deg(b(x)c(x)) = 1$.

The next logical question to ask is: *If A is a field will $A[x]$ also be a field?*

Example 7.19 Using the field $(\mathbb{Q}, +, \cdot)$, consider the polynomial $a(x) = x$ in $\mathbb{Q}[x]$. If $\mathbb{Q}[x]$ is a field then there must exist a polynomial $b(x) \in \mathbb{Q}[x]$ so that $a(x)b(x) = 1(x)$. The coefficients of the product $d(x) = a(x)b(x)$ must have:

$$d_i = \sum_{j+t=i} a_j b_t = \begin{cases} 1 & i = 0 \\ 0 & i > 0. \end{cases}$$

However, $d_0 = a_0 b_0$ and $a_0 = 0$, so we must have $d_0 = 0$ and $d(x) \neq 1(x)$. Thus for any $b(x) \in \mathbb{Q}[x]$, $a(x)b(x) \neq 1(x)$. Since the polynomial $a(x) = x$ is not a unit, $\mathbb{Q}[x]$ is not a field. However, a field is an integral domain, so as $\mathbb{Q}$ is a field then $\mathbb{Q}[x]$ is still an integral domain.

We have discussed the degree of a nonzero polynomial $a(x)$, i.e., the nonnegative integer n so that $a_n \neq 0_A$ (if one exists), but for all $m > n, a_m = 0_A$. Recall when we defined the product of two polynomials $a(x) = \sum_{i=0}^{n} a_i x^i$ and $b(x) = \sum_{i=0}^{m} b_i x^i$ in Definition 7.8, the summation for the product was from $i = 0$ to $i = n + m$. This suggests that the degree of the product should be the sum of the degrees, leading us to the next theorem.

Theorem 7.20

Let A be an *integral domain*, and nonzero $a(x), b(x) \in A[x]$. If $deg(a(x)) = n$ and $deg(b(x)) = m$, then $deg(a(x)b(x)) = n + m$.

Proof Suppose we have A, an integral domain, and $a(x), b(x) \in A[x]$ are nonzero polynomials with $deg(a(x)) = n$ and $deg(b(x)) = m$. Since $a(x)$ and $b(x)$ are nonzero we know that $n \geq 0$, $m \geq 0$, $a_n \neq 0_A$, and $b_m \neq 0_A$. By definition $d(x) = a(x)b(x)$ has

$$d(x) = \sum_{i=0}^{m+n} d_i x^i \text{ where } d_i = \sum_{j+t=i} a_j \cdot_A b_t.$$

As we have done before, $d_{n+m} = a_n \cdot_A b_m$ since any other term $a_j \cdot_A b_t$ with $j+t = n+m$ has either $j > n$ or $t > m$ and thus $a_j \cdot_A b_t = 0_A$. Since A is an integral domain we know $a_n \cdot_A b_m \neq 0_A$, but if $k > n + m$ and $j + t = k$ either $j > n$ or $t > m$ so $d_k = 0_A$. Thus $deg(a(x)b(x)) = n + m$. □

It is critical that we have an integral domain in Theorem 7.20, since in Example 7.18 we saw a polynomial of degree 2, which when squared results in the zero polynomial whose degree is not 4. In the definition of polynomial addition the summation of the result was from 0 to $k = max\{n, m\}$, or in other words, $max\{deg(a(x)), deg(b(x))\}$. Is it similarly true that $deg(a(x)+b(x)) = max\{deg(a(x)), deg(b(x))\}$ for nonzero polynomials over an integral domain? This question is an exercise at the end of the chapter.

Another property of a ring A that will be important for polynomials in $A[x]$ is the characteristic.

Example 7.21 Consider the ring $(\mathbb{Z}_4[x], +_4, \cdot_4)$. We know $\mathbb{Z}_4[x]$ is a commutative ring with unity, and the unity is the polynomial $1(x) = 1+0x$. To find $char(A[x])$ we calculate $n\bullet 1(x)$ as shown below.

$$\begin{array}{rclcl} 1\bullet 1(x) & = & 1+0x & & \\ 2\bullet 1(x) & = & (1+0x)+(1+0x) & = & 2+0x \\ 3\bullet 1(x) & = & (1+0x)+(2+0x) & = & 3+0x \\ 4\bullet 1(x) & = & (1+0x)+(3+0x) & = & 0+0x \end{array}$$

Thus we see that $char(\mathbb{Z}_4[x]) = 4 = char(\mathbb{Z}_4)$.

Will A and $A[x]$ always have the same characteristic? The key to understanding characteristic for polynomial rings is realizing that for any integer n and $a(x) \in A[x]$ we find $n \bullet a(x) = c(x)$ where $c_i = n \bullet a_i$. The proof of this fact is an exercise at the end of the chapter. The next theorem quickly follows once $n \bullet 1(x)$ is determined.

Theorem 7.22

If A is a commutative ring with unity then $char(A) = char(A[x])$.

Example 7.23 Consider the ring $\mathbb{Z}_5[x]$. We know that $(\mathbb{Z}_5, +_5, \cdot_5)$ is an integral domain since 5 is prime (Theorem 4.17) and $|\mathbb{Z}_5| = 5$. We know $\mathbb{Z}_5[x]$ is an integral domain by Theorem 7.17 and our discussion before this example tells us that $char(\mathbb{Z}_5[x]) = 5$.

By our definition there exist polynomials $x, x^2, x^3, \ldots, x^n, x^{n+1}, \ldots$ in $\mathbb{Z}_5[x]$ (each polynomial with leading coefficient 1). These are distinct polynomials, by Definition 7.5, since they all have different degrees. Thus $\mathbb{Z}_5[x]$ is an <u>infinite</u> integral domain with <u>finite</u> characteristic. We saw in Example 6.10, an infinite integral domain with nonzero characteristic, namely $\wp(\mathbb{Z})$, but now we see how to create infinite integral domains with any prime characteristic we choose!

Another important algebraic property we will use in polynomial rings is the ability to "divide," i.e., get a quotient and reminder. This parallels algebra that can be done in the integral domain $\mathbb{Z}$ (which is not a field) as described in the Division Algorithm (Theorem 0.18). There is a similar theorem, we will also call the Division Algorithm, for $K[x]$ when K is a *field.*

Theorem 7.24

The Division Algorithm Let K be a field and $a(x), b(x) \in K[x]$. If $b(x) \neq 0(x)$ then there exist unique polynomials $q(x), r(x) \in K[x]$, for which $a(x) = b(x)q(x) + r(x)$ and either $deg(r(x)) < deg(b(x))$ or $r(x) = 0(x)$.

Proof To make the proof easier to read we will not put $\cdot_K$ or $\cdot_K$ in our equations, so make sure you know what multiplication or addition is being discussed. Suppose that K is a field and $a(x), b(x) \in K[x]$, with $b(x) \neq 0(x)$. We will look at different cases, and in each case find the appropriate $q(x)$ and $r(x)$ in $K[x]$. The uniqueness of $q(x)$ and $r(x)$ is an exercise at the end of the chapter.

(i) $deg(a(x)) < deg(b(x))$
(ii) $deg(a(x)) = deg(b(x))$
(iii) $deg(a(x)) > deg(b(x))$

(i) This case is left as an exercise at the end of the chapter.

(ii) Assume that $deg(a(x)) = deg(b(x)) = n$. Thus $b_n \neq 0_K$ and since K is a field, we have $b_n^{-1} \in K$. Let $q(x) = (a_n b_n^{-1}) + 0_K x$, then $q(x) \in K[x]$ and $d(x) = q(x)b(x)$ has degree n.

$$d_n = q_0 b_n = (a_n b_n^{-1}) b_n = a_n$$

Now define $r(x) = a(x) - q(x)b(x)$, then clearly $a(x) = b(x)q(x) + r(x)$. We only need to show that either $deg(r(x)) < deg(b(x))$ or $r(x) = 0(x)$ to complete this part of the proof.

Assume $r(x) \neq 0(x)$ then $r_n = a_n - a_n = 0_K$, and for all $j > n$, $r_j = 0_K$ since both $a(x)$ and $d(x)$ have degree n. Thus $deg(r(x)) < n$ and $deg(r(x)) < deg(b(x))$ as needed.

(iii) Suppose that $deg(a(x)) > deg(b(x))$. Let $n = deg(a(x))$ and $m = deg(b(x))$. As $n > m$, then $n - m$ is a positive integer. As in case (ii), since $b_m \neq 0_K$ there is an element $b_m^{-1} \in K$.

$$\text{Define } q(x) \in K[x] \text{ by } q(x) = (a_n b_m^{-1}) x^{n-m}.$$

Then $deg(q(x)) = n - m$ and $q_{n-m} = a_n b_m^{-1}$, but for all other nonnegative integers t, $q_t = 0_K$. Now the polynomial $d(x) = q(x)b(x)$ has degree $(n - m) + m = n$.

$$d_n = q_{n-m} b_m = (a_n b_m^{-1}) b_m = a_n$$

Exactly as in (ii) we let $r(x) = a(x) - q(x)b(x)$, so that $a(x) = b(x)q(x) + r(x)$. We know from the same steps as (ii) that $deg(r(x)) \leq n - 1$, but if $deg(r(x)) > m$ we can now repeat this process on $r(x)$ and $b(x)$ to produce $c(x), s(x) \in K[x]$ with $r(x) = c(x)b(x) + s(x)$ and $deg(s(x)) < n - 1$ or $s(x) = 0(x)$.

$$a(x) = q(x)b(x) + c(x)b(x) + s(x) = [q(x) + c(x)]b(x) + s(x)$$

Thus we now have a remainder with an even lower degree than $r(x)$. Repeating this at most $n - m$ times we will eventually find remainder $0(x)$ or a remainder $r'(x)$ with $deg(r'(x)) < m$ as needed. □

Example 7.25 Suppose we have $a(x), b(x) \in \mathbb{Z}_7[x]$ where $a(x) = 3 + 2x + 3x^2 + 5x^3 + x^4$ and $b(x) = 6 + 4x$. By Theorem 7.24, we can find $q(x), r(x) \in \mathbb{Z}_7[x]$ with $a(x) = b(x)q(x) + r(x)$. To find $q(x)$ and $r(x)$ we can use "long division" in $\mathbb{Z}_7[x]$. Don't forget to use $+_7$ and $\cdot_7$ on the coefficients.

$$\begin{array}{r|l} & \underline{2 + 6x + 0x^2 + 2x^3} \\ 6 + 4x & 3 + 2x + 3x^2 + 5x^3 + x^4 \\ & \underline{\qquad\qquad\qquad 5x^3 + x^4} \\ & 3 + 2x + 3x^2 \\ & \underline{\qquad x + 3x^2} \\ & 3 + x \\ & \underline{5 + x} \\ & 4 \end{array}$$

Thus we have $q(x) = 2 + 6x + 2x^3$ and $r(x) = 4$ which give us $a(x) = (2 + 6x + 2x^3)(6 + 4x) + 4$. If we had used coefficients in $(\mathbb{Z}_8, +_8, \cdot_8)$ then for any polynomial $q(x) \in \mathbb{Z}_8[x]$ the coefficient of x^4 in $(6 + 4x)q(x)$ must be of the form $6q_4 + 4q_3$. But there is no solution to $6q_4 + 4q_3 = 1$ in $(\mathbb{Z}_8, +_8, \cdot_8)$ so we could not find $q(x)$ or $r(x)$ in $\mathbb{Z}_8[x]$. Hence we need coefficients from a field to apply the theorem.

Another very useful property that is shared by $\mathbb{Z}$ and $K[x]$, when K is a field, parallels Theorem 5.18.

Theorem 7.26

Let K be a field. Then every ideal of $K[x]$ is a principal ideal.

Proof Let K be a field, and suppose S is an ideal of $K[x]$. As in the proof of Theorem 5.18, if $S = \{0(x)\}$ or $S = K[x]$, then $S = \langle 0(x) \rangle$ or $S = \langle 1(x) \rangle$, respectively, and S is principal. Thus assume that S is not one of these trivial ideals and so there is polynomial $a(x) \in S$ with $a(x) \neq 0(x)$.

As $a(x) \neq 0(x)$ then $deg(a(x)) \geq 0$, so first assume $deg(a(x)) = 0$. Thus $a(x)$ is a nonzero constant polynomial, $a(x) = a_0$ with $a_0 \neq 0_K$. We know a_0 is a unit with inverse $a_0^{-1} \in K$, and so the polynomial $b(x) = (a_0)^{-1}$ is in $K[x]$. But S absorbs products from $K[x]$ so $a(x)b(x) \in S$, or $1(x) \in S$. This contradicts that S is not equal to $K[x]$ by Theorem 5.12 and so S cannot contain a polynomial of degree 0.

Now S contains a nonzero polynomial, $a(x)$, but cannot contain a polynomial of degree 0. Thus for every nonzero $a(x) \in S$, $deg(a(x))$ is a positive integer.

$$\text{Define } B = \{n \in \mathbb{Z} : deg(q(x)) = n \text{ for some nonzero } q(x) \in S\}.$$

Clearly, $B \subseteq \mathbb{Z}^+$, but $\mathbb{Z}$ is an integral system (Definition 6.32) which tells us that B has a least element, call it m. By definition of B there exists some polynomial $p(x) \in S$ with $deg(p(x)) = m$. We will now show that $S = \langle p(x) \rangle$.

As $p(x) \in S$ and S is an ideal it is clear that $\langle p(x) \rangle \subseteq S$, so we only need to show that $\langle p(x) \rangle \supseteq S$. Let $b(x) \in S$. If $b(x) = 0(x)$ then $b(x) = 0(x)p(x)$ and $b(x) \in \langle p(x) \rangle$, so assume instead that $b(x) \neq 0(x)$.

Now we know $deg(b(x)) \in B$, which tells us that either $deg(b(x)) = m$ or $deg(b(x)) > m$. Using Theorem 7.24 we find polynomials $q(x), r(x) \in K[x]$ with $b(x) = p(x)q(x) + r(x)$ and $0 < deg(r(x)) < deg(p(x))$ or $r(x) = 0(x)$. But $r(x) = b(x) - p(x)q(x)$ and $b(x), p(x) \in S$, so as S is an ideal $r(x) \in S$. If $r(x) \neq 0(x)$ we would have a contradiction since m is the least element of B. Thus we must have $r(x) = 0(x)$ and so $b(x) = p(x)q(x)$. Hence $b(x) \in \langle p(x) \rangle$, $S = \langle p(x) \rangle$, and S is a principal ideal. □

When we use constant polynomials, we often write $a_0 + 0_A x$ instead of just a_0 to remind us we have a polynomial, not just an element of A. The link between constant polynomials and elements of A is made precise in the next theorem.

Theorem 7.27

Let A be a commutative ring with unity. Then the function $f : A \to A[x]$ defined by $f(a) = a + 0_A x$ is an injective ring homomorphism.

Proof Let A be a commutative ring with unity, and define $f : A \to A[x]$ by $f(a) = a + 0_A x$. To see that f is a ring homomorphism, let $a, b \in A$.

$$f(a +_A b) = (a +_A b) + 0_A x$$
$$f(a) + f(b) = (a + 0_A x) + (b + 0_A x) = (a +_A b) + 0_A x$$

Thus $f(a +_A b) = f(a) + f(b)$. Verification of $f(ab) = f(a)f(b)$ using polynomial multiplication is an exercise at the end of the chapter. Thus the function f is a ring homomorphism.

To see that f is injective, suppose we have $c \in A$ with $c \in ker(f)$. Thus $f(c) = 0(x)$ since $0_{A[x]} = 0(x)$, and so $c + 0_A x = 0(x)$. This can only be true if $c = 0_A$ so $ker(f) = \{0_A\}$. Hence f is injective by Theorem 5.29. □

With the function defined in Theorem 7.27 together with Theorem 5.31, we know that $f(A)$ is a subring of $A[x]$. By $ker(f) = \{0_A\}$ we know $A \cong f(A)$ using Therem 5.34. So we can consider $A[x]$ to have an isomorphic *copy of* A inside of it. Hence we will often use the idea that A is a "subset" of $A[x]$ by identifying A with the constant polynomials.

The strong relationship between a ring A and the polynomial ring $A[x]$ we have discussed in this section will be important throughout the rest of the text. The next theorem gives us a way to extend any homomorphism between rings to the associated polynomial rings as well.

Theorem 7.28

Let A and K be commutative rings with unity, and suppose that $f : A \to K$ is a ring homomorphism. Then the function $\overline{f} : A[x] \to K[x]$ defined below is also a ring homomorphism.

$$\overline{f}(a_0 + a_1 x + \cdots + a_n x^x) = f(a_0) + f(a_1)x + \cdots + f(a_n)x^n$$

Proof Suppose A and K are commutative rings with unity, and $f : A \to K$ is a ring homomorphism. Clearly, $\overline{f} : A[x] \to K[x]$ is well defined since f is well defined, and so each coefficient of $\overline{f}(a(x))$ is unique. To see that $\overline{f}$ is a homomorphism, for $a(x), b(x) \in A[x]$ we need to show that $\overline{f}(a(x) + b(x)) = \overline{f}(a(x)) + \overline{f}(b(x))$ and $\overline{f}(a(x)b(x)) = \overline{f}(a(x))\overline{f}(b(x))$.

As an exercise at the end of the chapter you will show the first of these, so we will look at the second. Suppose $a(x), b(x) \in A[x]$ and $d(x) = a(x)b(x)$.

$$a(x) = \sum_{i=0}^{n} a_i x^i \qquad b(x) = \sum_{i=0}^{m} b_i x^i$$

$$d(x) = \sum_{i=0}^{n+m} d_i x^i \text{ where } d_i = \sum_{j+t=i} a_j b_t x^i$$

Thus using that f is a homomorphism we see that $\overline{f}(a(x)b(x)) = \overline{f}(a(x))\overline{f}(b(x))$ following the steps below.

$$\overline{f}(a(x)b(x)) = \overline{f}(d(x)) = \sum_{i=0}^{n+m} f(d_i)x^i$$

$$= \sum_{i=0}^{n+m} f\left(\sum_{j+t=i}(a_jb_t)\right)x^i = \sum_{i=0}^{n+m}\left(\sum_{j+t=i} f(a_jb_t)\right)x^i$$

$$= \sum_{i=0}^{n+m}\left(\sum_{j+t=i} f(a_j)f(b_t)\right)x^i = \overline{f}(a(x))\overline{f}(b(x))$$

Thus $\overline{f}$ is a homomorphism. □

Example 7.29 Consider the rings $(\mathbb{Z}, +, \cdot)$ and $(\mathbb{Z}_5, +_5, \cdot_5)$. We saw in Theorem 4.41 $f : \mathbb{Z} \to \mathbb{Z}_5$ defined by $f(x) = x(\bmod 5)$ is a homomorphism and so $\overline{f} : \mathbb{Z}[x] \to \mathbb{Z}_5[x]$, as defined in Theorem 7.28, will be a homomorphism. Consider the polynomials shown below.

$$a(x) = 0 + 5x \qquad b(x) = 0 + 10x + 5x^3 \qquad c(x) = 7 + 4x + 8x^2 + 10x^3$$

Notice that $\overline{f}(a(x)) = 0(x)$ and $\overline{f}(b(x)) = 0(x)$, thus the function $\overline{f}$ is not injective. Also $\overline{f}(c(x)) = 2 + 4x + x^2$ so $deg(c(x)) \neq deg\left(\overline{f}(c(x))\right)$.

In our previous example the function f was not injective and neither was $\overline{f}$. If f is an isomorphism, will the extension $\overline{f}$ also be an isomorphism? The next theorem is left as an exercise at the end of the chapter.

Theorem 7.30

Let A, K be commutative rings with unity, and suppose that $f : A \to K$ is an isomorphism. Then the extension $\overline{f} : A[x] \to K[x]$ is also an isomorphism.

Suppose f is an isomorphism as in Theorem 7.30, and $a(x) \in A[x]$. It is straightforward to show that $deg(a(x)) = deg(\overline{f}(a(x))$. This fact will be used repeatedly and is an exercise at the end of the chapter.

Our method of extending a homomorphism of rings to a homomorphism of polynomial rings will be used often. We will not repeat the definition of the extension each time but just put a bar above the name of the function to denote it.

7.3 Polynomial Functions and Roots

The notation defined in this chapter for polynomials over a ring looks very similar to the notation used for many functions. Thus it is important to make a distinction between polynomials and functions. Equality of two polynomials compared to equality between functions is the most important difference as seen in the example below.

Example 7.31 Consider the ring $(\mathbb{Z}_6, +_6, \cdot_6)$. The polynomial $a(x) = 3x + 3x^2$ in $\mathbb{Z}_6[x]$ has degree 2 with two nonzero coefficients while $0(x)$ has degree $-\infty$ with all of its coefficients equal to 0. As polynomials these two are very different. However, look closely at the functions defined below on $\mathbb{Z}_6$.

$$h : \mathbb{Z}_6 \to \mathbb{Z}_6 \text{ defined by } h(a) = 3a + 3a^2 \text{ for all } a \in \mathbb{Z}_6$$
$$g : \mathbb{Z}_6 \to \mathbb{Z}_6 \text{ defined by } g(a) = 0 \text{ for all } a \in \mathbb{Z}_6$$

We can calculate $h(a)$ and $g(a)$ for each $a \in \mathbb{Z}_6$.

$$\begin{array}{lclcl} h(0) & = & 0 & & \\ h(1) & = & 3(1)+3(1) & = & 0 \\ h(2) & = & 3(2)+3(4) & = & 0 \\ h(3) & = & 3(3)+3(3) & = & 0 \\ h(4) & = & 3(4)+3(4) & = & 0 \\ h(5) & = & 3(5)+3(1) & = & 0 \end{array} \qquad \begin{array}{lcl} g(0) & = & 0 \\ g(1) & = & 0 \\ g(2) & = & 0 \\ g(3) & = & 0 \\ g(4) & = & 0 \\ g(5) & = & 0 \end{array}$$

Thus the function h maps the elements of $\mathbb{Z}_6$ to the ***same elements*** as the function g. The functions h and g are equal, but the polynomials used to define them are drastically different.

Example 7.32 Even if the coefficients are from a field this difference in functions and polynomials can be seen. Consider instead $(\mathbb{Z}_7, +_7, \cdot_7)$ and polynomial $a(x) = x^7$ in $\mathbb{Z}_7[x]$. Define the function $f : \mathbb{Z}_7 \to \mathbb{Z}_7$ by $f(b) = b^7$.

$$\begin{array}{llll} f(0) = 0^7 = 0 & f(1) = 1^7 = 1 & f(2) = 2^7 = 2 & f(3) = 3^7 = 3 \\ f(4) = 4^7 = 4 & f(5) = 5^7 = 5 & f(6) = 6^7 = 6 & \end{array}$$

As a function, f is identical to the identity function which can be expressed as a polynomial by $e(x) = x$. But as polynomials $a(x) \neq e(x)$ since the degrees are different.

We will use polynomials to help us define functions, but as the examples above show us, we must be careful discussing equality in this situation.

Definition 7.33 Let A be a commutative ring with unity and $a(x) \in A[x]$ with $a(x) \neq 0(x)$. If $c \in A$ and $deg(a(x)) = n$, we define the element $a(c) \in A$ as follows:

$$a(c) = a_0 +_A (a_1 \cdot_A c) +_A (a_2 \cdot_A c^2) +_A \cdots +_A (a_n \cdot_A c^n).$$

If $a(x) = 0(x)$ we say $a(c) = 0_A$ for all $c \in A$.

Example 7.34 Consider $a(x) = 2 + 3x + x^2$ in $\mathbb{Z}_5[x]$ (with $+_5$ and $\cdot_5$ on the coefficients).

$$a(0) = 2 +_5 0 +_5 0 = 2 \quad a(1) = 2 +_5 3 +_5 1 = 1 \quad a(2) = 2 +_5 1 +_5 4 = 2$$
$$a(3) = 2 +_5 4 +_5 4 = 0 \quad a(4) = 2 +_5 2 +_5 1 = 0$$

Theorem 7.35

Let A be an integral domain. The substitution function $h_c : A[x] \to A$ defined by $h_c(a(x)) = a(c)$ is a ring homomorphism.

Proof In this proof, we will again stop writing $+_A$ and $\cdot_A$ for the operations in A to keep the notation readable.

Let A be a commutative ring with unity, and define $h_c : A[x] \to A$ by $h_c(a(x)) = a(c)$. Consider two polynomials $a(x), b(x) \in A[x]$.

$$a(x) = \sum_{i=0}^{n} a_i x^i \qquad b(x) = \sum_{i=0}^{m} b_i x^i$$

The proof of $h_c(a(x) + b(x)) = a(c) + b(c) = h_c(a(x)) + h_c(b(x))$ is left as an exercise at the end of the chapter. Let $g(x) = a(x)b(x)$. Then $h_c(a(x)b(x)) = h_c(g(x)) = g(c)$.

$$g(c) = \sum_{i=0}^{n+m} g_i c^i = \sum_{i=0}^{n+m} \left(\sum_{j+t=i} a_j b_t \right) c^i = \sum_{j+t=0}^{n+m} (a_j b_t c^{j+t})$$
$$= \sum_{j+t=0}^{n+m} (a_j c^j)(b_t c^t)$$

Using commutativity and associativity in A we can rearrange this and notice a pattern.

$$g(c) = a_0(b_0 + b_1c + \cdots + b_mc^m) + a_1c(b_0 + b_1c + \cdots + b_mc^m) \\ + \cdots + a_nc^n(b_0 + b_1c + \cdots + b_mc^m)$$

Finally, using commutativity and the distributive laws, we can factor out the expression for $b(c)$ and find $g(c) = a(c)b(c)$ Thus $h_c(a(x)b(x)) = h_c(a(x))h_c(b(x))$ and h_c is a ring homomorphism. □

We must be careful not to confuse defining a function using a polynomial with the homomorphism h_c.

Example 7.36 For the ring $(\mathbb{Z}, +, \cdot)$ consider the function $f : \mathbb{Z} \to \mathbb{Z}$ defined by $f(c) = a(c)$ using substitution with the polynomial $a(x) = 2 + 3x + x^2$. For the elements $2, 3, 5 \in \mathbb{Z}$ we have $f(2) = 2+6+4 = 12$, $f(3) = 2+9+9 = 20$, and $f(5) = 2+15+25 = 42$. Notice that $f(2+3) \neq f(2) + f(3)$. Thus, this function is not a ring homomorphism.

Suppose A is a commutative ring with unity, and $c \in A$. We know that h_c from Definition 7.3 is a ring homomorphism, thus $ker(h_c)$ is a *subring* of $A[x]$.

$$ker(h_c) = \{a(x) \in A[x] : h_c(a(x)) = 0_A\} = \{a(x) \in A[x] : a(c) = 0_A\}$$

Definition 7.37 Let A be a commutative ring with unity, $c \in A$, and $a(x) \in A[x]$ $a(x) \neq 0(x)$. We say that ***c is a root*** of the polynomial $a(x)$ exactly when $a(c) = 0_A$. We do *not* say any element of A is a root of $0(x)$ even though $0(c) = 0_A$ for each $c \in A$.

Example 7.38 Consider the ring $(\mathbb{Z}_6, +_6, \cdot_6)$ with polynomial $a(x) = 2 + 3x + x^2$ in $\mathbb{Z}_6[x]$. By calculating $a(c)$ for each $c \in \mathbb{Z}_6$, we can determine if any are roots.

$$a(0) = 2 \quad a(1) = 0 \quad a(2) = 0 \quad a(3) = 2 \quad a(4) = 0 \quad a(5) = 0$$

Thus, the elements 1, 2, 4, and 5 are roots of $a(x)$ in $\mathbb{Z}_6[x]$. This may seem odd to some who learned that a quadratic equation should only have two roots. The obvious answer is that $\mathbb{Z}_6$ is not an integral domain. The same polynomial, considered as an element of $\mathbb{Z}_5[x]$, has exactly two roots. We will see this again in Chapter 8.

Exercises for Chapter 7

Section 7.1 Polynomial Rings

1. Consider $A = \mathbb{Z} \times \mathbb{Z}$ with usual addition and multiplication in each coordinate. Create $a(x), b(x) \in A[x]$ with $deg(a(x)) = deg(b(x)) = 3$ and $a(x) \neq b(x)$. Compute $a(x) + b(x)$ and $a(x) - b(x)$. (Don't use $(0,0)$ as a coefficient in the polynomials.)
2. Consider $A = \mathbb{Z}_3 \times \mathbb{Z}_3$ with $+_3$ and $\cdot_3$ in each coordinate. Create $a(x), b(x) \in A[x]$ with $deg(a(x)) = deg(b(x)) = 3$ and $a(x) \neq b(x)$. Compute $a(x) + b(x)$ and $a(x) - b(x)$. (Don't use $(0,0)$ as a coefficient in the polynomials.)
3. Consider $A = {}^{\mathbb{Z}_6}/_S$ with $S = \{0, 3\}$ and the usual coset addition and multiplication. (Don't forget to use $+_6$ and $\cdot_6$ for elements of $\mathbb{Z}_6$.) Create $a(x), b(x) \in A[x]$ with $deg(a(x)) = deg(b(x)) = 3$, and $a(x) \neq b(x)$. Compute $a(x) + b(x)$ and $a(x) - b(x)$.
4. Consider $A = {}^{\mathbb{Z}_2 \times \mathbb{Z}_4}/_S$ with $S = \{(0,0), (0,2)\}$ and the usual coset operations. (Don't forget to use $+_2$ and $\cdot_2$ or $+_4$ and $\cdot_4$ in appropriate coordinates in $\mathbb{Z}_2 \times \mathbb{Z}_4$.) Create $a(x), b(x) \in A[x]$ with $deg(a(x)) = deg(b(x)) = 3$ and $a(x) \neq b(x)$. Compute $a(x) + b(x)$ and $a(x) - b(x)$.
5. For $a(x) = 2 + 3x + 3x^2$ and $b(x) = 4 + x + 3x^2$ in $\mathbb{Z}_6[x]$, compute $a(x) + b(x)$ and $a(x)b(x)$. (Don't forget to use $+_6$ and $\cdot_6$ for elements of $\mathbb{Z}_6$.)
6. For $a(x) = 1 + 7x + x^2 + 9x^3$ and $b(x) = 2 + x + 6x^2$ in $\mathbb{Z}_{12}[x]$, compute $a(x) + b(x)$ and $a(x)b(x)$. (Don't forget to use $+_{12}$ and $\cdot_{12}$ for elements of $\mathbb{Z}_{12}$.)
7. For $a(x) = 0 + 4x + 5x^2$ and $b(x) = 1 + 2x + x^2 + 6x^3$ in $\mathbb{Z}_7[x]$, compute $a(x) + b(x)$ and $a(x)b(x)$. (Don't forget to use $+_7$ and $\cdot_7$ for elements of $\mathbb{Z}_7$.)
8. For $a(x) = 2 + x + 3x^2 + 2x^3$ and $b(x) = 1 + 2x + 2x^2 + x^3$ in $\mathbb{Z}_4[x]$, compute $a(x) + b(x)$ and $a(x)b(x)$. (Don't forget to use $+_4$ and $\cdot_4$ for elements of $\mathbb{Z}_4$.)
9. For $a(x) = 2 + x + x^2 + 2x^3$ and $b(x) = 0 + x + 2x^2 + x^3$ in $\mathbb{Z}_3[x]$, compute $a(x) + b(x)$ and $a(x)b(x)$. (Don't forget to use $+_3$ and $\cdot_3$ for elements of $\mathbb{Z}_3$.)
10. For $a(x) = 5 + x + 4x^2$ and $b(x) = 3 + 6x + 7x^2$ in $\mathbb{Z}_8[x]$, compute $a(x) + b(x)$ and $a(x)b(x)$. (Don't forget to use $+_8$ and $\cdot_8$ for elements of $\mathbb{Z}_8$.)
11. For $a(x) = 3 + 6x + 9x^2$ and $b(x) = 2 + 4x + 6x^2$ in $\mathbb{Z}_{10}[x]$, compute $a(x) + b(x)$ and $a(x)b(x)$. (Don't forget to use $+_{10}$ and $\cdot_{10}$ for elements of $\mathbb{Z}_{10}$.)
12. For $a(x) = 1 + x + x^2$ and $b(x) = 1 + x + x^2 + x^3$ in $\mathbb{Z}_2[x]$, compute $a(x) + b(x)$ and $a(x)b(x)$. (Don't forget to use $+_2$ and $\cdot_2$ for elements of $\mathbb{Z}_2$.)
13. For $a(x) = 4 + 2x + 4x^2$ and $b(x) = 4 + x + 3x^2 + 2x^3$ in $\mathbb{Z}_5[x]$, compute $a(x) + b(x)$ and $a(x)b(x)$. (Don't forget to use $+_5$ and $\cdot_5$ for elements of $\mathbb{Z}_5$.)
14. Suppose $A = \{0_A, 1_A\}$ is an integral domain. Find all of the polynomials in $A[x]$ with degree 3. How many did you find?
15. Suppose A is a commutative ring with unity and $|A| = 3$. How many polynomials have degree 2 in $A[x]$?
16. Prove Theorem 7.13.

17. Prove Theorem 7.15 by showing that the polynomial $0(x) = 0_A + 0_A x$ and $1(x) = 1_A + 0_A x$ are the zero and unity of $A[x]$ and that $-a(x)$ is the polynomial $c(x)$ where $c_i = -a_i$ for all i.
18. Suppose A is a commutative ring with unity and $a(x) \in A[x]$. Prove by induction that for any positive integer n the polynomial $n \bullet a(x) = c(x)$ has $c_i = n \bullet a_i$.
19. Suppose A is a commutative ring with unity and $a(x) \in A[x]$. Use the previous exercise to help prove that for any integer n, the polynomial $n \bullet a(x) = c(x)$ has $c_i = n \bullet a_i$.
20. Determine if $a(x) = 4 + 2x$ is a zero divisor in $\mathbb{Z}_6[x]$. Either prove that it is not a zero divisor or find $b(x) \in \mathbb{Z}_6[x]$ with $b(x) \neq 0(x)$ and $a(x)b(x) = 0(x)$. (Don't forget to use $+_6$ and $\cdot_6$ for elements of $\mathbb{Z}_6$.)
21. Determine if $a(x) = 6 + 2x + 4x^2$ is a zero divisor in $\mathbb{Z}_8[x]$. Either prove that it is not a zero divisor or find $b(x) \in \mathbb{Z}_8[x]$ with $b(x) \neq 0(x)$ and $a(x)b(x) = 0(x)$. (Don't forget to use $+_8$ and $\cdot_8$ for elements of $\mathbb{Z}_8$.)
22. Determine if $a(x) = 2 + 3x$ is a zero divisor in $\mathbb{Z}_6[x]$. Either prove that it is not a zero divisor or find $b(x) \in \mathbb{Z}_6[x]$ with $b(x) \neq 0(x)$ and $a(x)b(x) = 0(x)$. (Don't forget to use $+_6$ and $\cdot_6$ for elements of $\mathbb{Z}_6$.)
23. Let A be a commutative ring with unity and $a(x) \in A[x]$. Determine if the following statement is true or false. Either prove it or find a counterexample: If every coefficient of $a(x)$ is a zero divisor of A, then $a(x)$ is a zero divisor in $A[x]$.
24. Let A be a commutative ring with unity and $a(x) \in A[x]$. Determine if the following statement is true or false, then either prove it or find a counterexample: If $a_0 \neq 0_A$ is not a zero divisor of A, then $a(x)$ is not a zero divisor in $A[x]$.
25. Let A be the ring $(\mathbb{Z}_8, +_8, \cdot_8)$. We know $S = \{0, 4\}$ is a subring of A. Is the set $S[x] = \{a(x) \in A[x] : a_i \in S \text{ for all } i\}$ an ideal of $A[x]$? Either prove that it is true or find a counterexample showing it fails to be an ideal.
26. Prove: If A is a commutative ring with unity and S is an ideal of A then $S[x] = \{a(x) \in A[x] : a_i \in S \text{ for all } i\}$ is an ideal of $A[x]$.
27. Suppose A is a commutative ring with unity. Let $S = \{a(x) \in A[x] : a_0 \neq 0_A\}$. Determine if S is an ideal of $A[x]$. Either prove that it is an ideal or find a counterexample showing it fails to be an ideal.
28. Suppose A is a commutative ring with unity. Let $S = \{a(x) \in A[x] : deg(a(x)) < 3\}$. Determine if S is an ideal of $A[x]$. Either prove that it is an ideal or find a counterexample showing it fails to be an ideal.

Section 7.2 Properties of Polynomial Rings

29. Determine if the following statement is true or false. Either prove it is true or find a counterexample: If A is an integral domain and $a(x), b(x) \in A[x]$ are nonzero polynomials then $deg(a(x) + b(x)) = max\{deg(a(x)), deg(b(x))\}$.
30. If A is an integral domain with nonzero polynomials $a(x), b(x) \in A[x]$, where $deg(a(x)) = n$ and $deg(b(x)) = m$, we have repeatedly used that $d(x) = a(x)b(x)$

must have $d_{n+m} \neq 0_A$. If for some $j < n$ and $s < m$ we have $a_j \neq 0_A$ and $b_s \neq 0_A$, can we also guarantee that $d_{j+s} \neq 0_A$? Either prove that it is true or give a counterexample showing it can fail.

31. Prove Theorem 7.22.
32. Prove (i) of Theorem 7.24.
33. Complete the proof of Theorem 7.24 by showing that if $a(x)$, $b(x)$, $q_1(x)$, $q_2(x)$, $r_1(x)$, $r_2(x) \in K[x]$ with $a(x) = b(x)q_1(x) + r_1(x)$ and $a(x) = b(x)q_2(x) + r_2(x)$ then $q_1(x) = q_2(x)$ and $r_1(x) = r_2(x)$. Don't forget that $K[x]$ is an integral domain, $b(x) \neq 0(x)$, and $deg(r_i(x)) < deg(b(x))$ or $r_i(x) = 0(x)$.
34. For $b(x) = 0 + 3x + 2x^2 + x^3$ and $a(x) = 4 + 2x + 4x^2 + 6x^3 + 7x^4 + x^5 + 9x^6$ in $\mathbb{Q}[x]$, find $q(x), r(x) \in \mathbb{Q}[x]$ with $a(x) = b(x)q(x) + r(x)$ and $deg(r(x)) < 3$ or $r(x) = 0(x)$. (The ring $\mathbb{Q}$ has usual addition and multiplication.)
35. For $b(x) = \frac{1}{2} + 2x$ and $a(x) = 6 + 0x + x^2 + 2x^3 + 5x^4 + 4x^5$ in $\mathbb{Q}[x]$, find $q(x), r(x) \in \mathbb{Q}[x]$ with $a(x) = b(x)q(x) + r(x)$ and $deg(r(x)) < 1$ or $r(x) = 0(x)$. (The ring $\mathbb{Q}$ has usual addition and multiplication.)
36. For $b(x) = 1 + \frac{2}{3}x$ and $a(x) = 7 + 2x - 3x^2 + 0x^3 - 4x^4$ in $\mathbb{Q}[x]$, find $q(x), r(x) \in \mathbb{Q}[x]$ with $a(x) = b(x)q(x) + r(x)$ and $deg(r(x)) < 1$ or $r(x) = 0(x)$. (The ring $\mathbb{Q}$ has usual addition and multiplication.)
37. For $b(x) = 1 + 2x + 2x^2$ and $a(x) = 0 + x + x^2 + 2x^3 + 0x^4 + 2x^5 + x^6$ in $\mathbb{Z}_3[x]$, find $q(x), r(x) \in \mathbb{Z}_3[x]$ with $a(x) = b(x)q(x) + r(x)$ and $deg(r(x)) < 2$ or $r(x) = 0(x)$. (Don't forget to use $+_3$ and $\cdot_3$ for elements of $\mathbb{Z}_3$.)
38. For $b(x) = 1 + 3x + 4x^2 + 2x^3$ and $a(x) = 2 + x + x^2 + 4x^3 + 0x^4 + 2x^5 + 3x^6$ in $\mathbb{Z}_5[x]$, find $q(x), r(x) \in \mathbb{Z}_5[x]$ with $b(x) = a(x)q(x) + r(x)$ and $deg(r(x)) < 3$ or $r(x) = 0(x)$. (Don't forget to use $+_5$ and $\cdot_5$ for elements of $\mathbb{Z}_5$.)
39. For $b(x) = 0 + x + x^2$ and $a(x) = 1 + 0x + x^2 + 0x^3 + x^4 + x^5 + 0x^6 + x^7$ in $\mathbb{Z}_2[x]$, find $q(x), r(x) \in \mathbb{Z}_2[x]$ with $a(x) = b(x)q(x) + r(x)$ and $deg(r(x)) < 2$ or $r(x) = 0(x)$. (Don't forget to use $+_2$ and $\cdot_2$ for elements of $\mathbb{Z}_2$.)
40. For $b(x) = 4 + 5x$ and $a(x) = 2 + 3x + 4x^2 + x^3 + 2x^4$ in $\mathbb{Z}_7[x]$, find $q(x), r(x) \in \mathbb{Z}_7[x]$ with $a(x) = b(x)q(x) + r(x)$ and $deg(r(x)) < 1$ or $r(x) = 0(x)$. (Don't forget to use $+_7$ and $\cdot_7$ for elements of $\mathbb{Z}_7$.)
41. For $b(x) = 2 + 3x$ and $a(x) = 0 + x + 2x^2 + 5x^3 + 7x^4$ in $\mathbb{Z}_{10}[x]$ is it possible to find $q(x), r(x) \in \mathbb{Z}_{10}[x]$ with $a(x) = b(x)q(x) + r(x)$, $deg(q(x)) \leq 3$ and $deg(r(x)) < 1$ or $r(x) = 0(x)$? Find them or explain why it is not possible. (Don't forget to use $+_{10}$ and $\cdot_{10}$ for elements of $\mathbb{Z}_{10}$.)
42. For $b(x) = 0 + 3x + 2x^2$ and $a(x) = 5 + 3x + 2x^2 + x^3 + 6x^4$ both in $\mathbb{Z}_8[x]$ is it possible to find $q(x), r(x) \in \mathbb{Z}_8[x]$ with $a(x) = b(x)q(x) + r(x)$, $deg(q(x)) \leq 2$, and $deg(r(x)) < 2$ or $r(x) = 0(x)$? Find them or explain why it is not possible.
43. For $b(x) = 7 + 3x + 2x^2$ and $a(x) = 3 + 4x + 8x^2 + 7x^3 + 6x^4$ in $\mathbb{Z}_9[x]$ is it possible to find $q(x), r(x) \in \mathbb{Z}_9[x]$ with $a(x) = b(x)q(x) + r(x)$, $deg(q(x)) \leq 2$, and $deg(r(x)) < 2$ or $r(x) = 0(x)$? Find them or explain why it is not possible. (Don't forget to use $+_9$

and $\cdot_9$ for elements of $\mathbb{Z}_9$.)

44. For $b(x) = 3 + 4x$ and $a(x) = 0 + 7x + 2x^2 + 4x^3 + 6x^4 + 6x^5$ in $\mathbb{Z}_{12}[x]$ is it possible to find $q(x), r(x) \in \mathbb{Z}_{12}[x]$ with $a(x) = b(x)q(x) + r(x)$, $deg(q(x)) \leq 4$, and $deg(r(x)) < 1$ or $r(x) = 0(x)$? Find them or explain why it is not possible. (Don't forget to use $+_{12}$ and $\cdot_{12}$ for elements of $\mathbb{Z}_{12}$.)
45. For $b(x) = 3 + 2x + 2x^2$ and $a(x) = 0 + 4x + x^2 + 3x^3 + 2x^4 + 5x^5$ in $\mathbb{Z}_6[x]$ is it possible to find $q(x), r(x) \in \mathbb{Z}_6[x]$ with $a(x) = b(x)q(x) + r(x)$, $deg(q(x)) \leq 3$, and $deg(r(x)) < 2$ or $r(x) = 0(x)$? Find them or explain why it is not possible. (Don't forget to use $+_6$ and $\cdot_6$ for elements of $\mathbb{Z}_6$.)
46. For $b(x) = 2 + x + 2x^2$ and $a(x) = 0 + x + x^2 + x^3 + 3x^4 + 2x^5 + 2x^6$ in $\mathbb{Z}_4[x]$ is it possible to find $q(x), r(x) \in \mathbb{Z}_4[x]$ with $a(x) = b(x)q(x) + r(x)$, $deg(q(x)) \leq 4$ and $deg(r(x)) < 2$ or $r(x) = 0(x)$? Find them or explain why it is not possible. (Don't forget to use $+_4$ and $\cdot_4$ for elements of $\mathbb{Z}_4$.)
47. Let K be a field. Prove: The only units in $K[x]$ are the nonzero constant polynomials.
48. Complete the proof of Theorem 7.27 by showing $f(ab) = f(a)f(b)$.
49. Complete the proof of Theorem 7.28 by showing that $\overline{f}(a(x)+b(x)) = \overline{f}(a(x))+\overline{f}(b(x))$ for $a(x), b(x) \in A[x]$.
50. Prove Theorem 7.30.
51. Let A, K be commutative rings with unity and suppose that $f : A \to K$ is an isomorphism. Prove: For any $a(x) \in A[x]$ we have $deg(a(x)) = deg(\overline{f}(a(x))$ where $\overline{f}$ is defined in Theorem 7.28.
52. Let A be a commutative ring with unity. Prove that the function $f : A[x] \to A$ defined by $f(a(x)) = a_0$ is a homomorphism. What is the kernel of f?
53. Let A be a commutative ring with unity. Determine if the function defined by $f : A[x] \to A$ defined by $f(a(x)) = a_1$ is a homomorphism. Either prove that it is a homomorphism or give a counterexample showing that it fails to be a homomorphism.

Section 7.3 Polynomial Functions and Roots

54. Consider the polynomials $a(x) = 2 + 3x$ and $b(x) = 4 + x + 2x^2$ in $\mathbb{Z}_5[x]$. If these polynomials defined functions f and g from $\mathbb{Z}_5$ to $\mathbb{Z}_5$, are the functions equal? Justify your answer. (Don't forget to use $+_5$ and $\cdot_5$ for elements of $\mathbb{Z}_5$.)
55. Consider the polynomials $a(x) = 2 + 3x + x^2$ and $b(x) = 2 + x + 3x^2$ in $\mathbb{Z}_6[x]$. If these polynomials defined functions f and g from $\mathbb{Z}_6$ to $\mathbb{Z}_6$, are the functions equal? Justify your answer. (Don't forget to use $+_6$ and $\cdot_6$ for elements of $\mathbb{Z}_6$.)
56. Consider the polynomials $a(x) = 1+6x+3x^2+3x^3$ and $b(x) = 1+6x+6x^2+3x^3+6x^4$ in $\mathbb{Z}_9[x]$. If these polynomials defined functions f and g from $\mathbb{Z}_9$ to $\mathbb{Z}_9$, are the functions equal? Justify your answer. (Don't forget to use $+_9$ and $\cdot_9$ for elements of $\mathbb{Z}_9$.)
57. Consider the polynomials $a(x) = 2 + 4x$ and $b(x) = 2 + 0x + 4x^2$ in $\mathbb{Z}_8[x]$. If these polynomials defined functions f and g from $\mathbb{Z}_8$ to $\mathbb{Z}_8$, are the functions equal? Justify your answer. (Don't forget to use $+_8$ and $\cdot_8$ for elements of $\mathbb{Z}_8$.)

58. Consider the polynomials $a(x) = 6 + 4x + x^2$ and $b(x) = 6 + 0x + 3x^2 + 5x^3$ in $\mathbb{Z}_7[x]$. If these polynomials defined functions f and g from $\mathbb{Z}_7$ to $\mathbb{Z}_7$, are the functions equal? Justify your answer. (Don't forget to use $+_7$ and $\cdot_7$ for elements of $\mathbb{Z}_7$.)
59. For the polynomial $a(x) = 0 + 2x + 3x^2 + x^3$ in $\mathbb{Z}_6[x]$, calculate $a(c)$ for every $c \in \mathbb{Z}_6$. Are there any roots for $a(x)$ in $\mathbb{Z}_6$? (Don't forget to use $+_6$ and $\cdot_6$ for elements of $\mathbb{Z}_6$.)
60. For the polynomial $a(x) = 1 + 4x + 6x^2$ in $\mathbb{Z}_7[x]$, calculate $a(c)$ for every $c \in \mathbb{Z}_7$. Are there any roots for $a(x)$ in $\mathbb{Z}_7$? (Don't forget to use $+_7$ and $\cdot_7$ for elements of $\mathbb{Z}_7$.)
61. For the polynomial $a(x) = 4 + 3x + 5x^2$ in $\mathbb{Z}_8[x]$, calculate $a(c)$ for every $c \in \mathbb{Z}_8$. Are there any roots for $a(x)$ in $\mathbb{Z}_8$? (Don't forget to use $+_8$ and $\cdot_8$ for elements of $\mathbb{Z}_8$.)
62. For the polynomial $a(x) = 1 + x + 2x^2 + 3x^3 + 2x^4 + x^5$ in $\mathbb{Z}_4[x]$, calculate $a(c)$ for every $c \in \mathbb{Z}_4$. Are there any roots for $a(x)$ in $\mathbb{Z}_4$? (Don't forget to use $+_4$ and $\cdot_4$ for elements of $\mathbb{Z}_4$.)
63. For the polynomial $a(x) = 3 + 2x + 4x^2 + 4x^3$ in $\mathbb{Z}_6[x]$, calculate $a(c)$ for every $c \in \mathbb{Z}_6$. Are there any roots for $a(x)$ in $\mathbb{Z}_6$? (Don't forget to use $+_6$ and $\cdot_6$ for elements of $\mathbb{Z}_6$.)
64. For the polynomial $a(x) = 5 + 6x + x^2$ in $\mathbb{Z}_9[x]$, calculate $a(c)$ for every $c \in \mathbb{Z}_9$. Are there any roots for $a(x)$ in $\mathbb{Z}_9$? (Don't forget to use $+_9$ and $\cdot_9$ for elements of $\mathbb{Z}_9$.)
65. For the polynomial $a(x) = 8 + x + 7x^2 + x^3$ in $\mathbb{Z}_{10}[x]$, calculate $a(c)$ for every $c \in \mathbb{Z}_{10}$. Are there any roots for $a(x)$ in $\mathbb{Z}_{10}$? (Don't forget to use $+_{10}$ and $\cdot_{10}$ for elements of $\mathbb{Z}_{10}$.)
66. For the polynomial $a(x) = 2 + x + 2x^2 + x^3 + 0x^4 + 2x^5$ in $\mathbb{Z}_3[x]$, calculate $a(c)$ for every $c \in \mathbb{Z}_3$. Are there any roots for $a(x)$ in $\mathbb{Z}_3$? (Don't forget to use $+_3$ and $\cdot_3$ for elements of $\mathbb{Z}_3$.)
67. Suppose A and K are commutative rings with unity and $f : A \to K$ is a nonzero ring homomorphism. Prove: If $c \in A$ is a root for $a(x) \in A[x]$ then $f(c)$ is a root for $\overline{f}(a(x))$ in $K[x]$.
68. Complete the proof of Theorem 7.35 by showing that $h_c(a(x) + b(x)) = a(c) + b(c) = h_c(a(x)) + h_c(b(x))$ for $a(x), b(x) \in A[x]$.
69. Suppose A is an integral domain and $a(x) \in A[x]$ with $a(x) \neq 0(x)$. Prove: If $a_0 = 0_A$ then 0_A is a root of $a(x)$.
70. Define $S = \{a(x) \in \mathbb{Z}[x] : a(0) = 0 \text{ and } a(2) = 0\}$. Prove that S is an ideal of $\mathbb{Z}[x]$.
71. Let A be a commutative ring with unity and $c \in A[x]$. Define $S = \{a(x) \in A[x] : a(c) = 0_A\}$. Prove that S is an ideal of $A[x]$.

Projects for Chapter 7

Project 7.1

Consider the following polynomials in $\mathbb{Z}_6[x]$. (Don't forget to use $+_6$ and $\cdot_6$ for elements of $\mathbb{Z}_6$.)

$$a(x) = 5 + x + 3x^2 + 4x^3 \qquad b(x) = 0 + 2x + 4x^2 + 4x^3 + 3x^4$$
$$c(x) = 4 + 2x + 5x^2 + 3x^3$$

1. Find the polynomials $a(x) + b(x)$, $a(x) + c(x)$, and $b(x) + c(x)$ in $\mathbb{Z}_6[x]$.
2. Find the polynomials $a(x) - b(x)$, $c(x) - a(x)$, and $b(x) - c(x)$ in $\mathbb{Z}_6[x]$.
3. Find the polynomials $3 \bullet a(x)$, $8 \bullet b(x)$, and $-4 \bullet c(x)$ in $\mathbb{Z}_6[x]$.
4. Find the polynomials $a(x)b(x)$ and $b(x)c(x)$ in $\mathbb{Z}_6[x]$.

Now use the same $a(x), b(x), c(x)$ but as elements of $\mathbb{Z}_7[x]$. (Don't forget to use $+_7$ and $\cdot_7$ for elements of $\mathbb{Z}_7$.)

5. Find the polynomials $a(x) + b(x)$, $a(x) + c(x)$, and $b(x) + c(x)$ in $\mathbb{Z}_7[x]$.
6. Find the polynomials $a(x) - b(x)$, $c(x) - a(x)$, and $b(x) - c(x)$ in $\mathbb{Z}_7[x]$.
7. Find the polynomials $3 \bullet a(x)$, $8 \bullet b(x)$, and $-4 \bullet c(x)$ in $\mathbb{Z}_7[x]$.
8. Find the polynomials $a(x)b(x)$, and $b(x)c(x)$ in $\mathbb{Z}_7[x]$.

Project 7.2

Suppose we have a nontrivial finite commutative ring with unity, A. Let $|A| = n$ for some $n > 1$.

1. Prove: There are exactly $n - 1$ distinct polynomials of $A[x]$ whose degree is 0.
2. In the ring $\mathbb{Z}_3[x]$ list all of the polynomials with degree 1. How is the number of them related to $|\mathbb{Z}_3|$?
3. Including the zero polynomial, how many polynomials in $\mathbb{Z}_3[x]$ have degree less than 2? How is the number of them related to $|\mathbb{Z}_3|$?
4. Now list of the polynomials of $\mathbb{Z}_3[x]$ with degree 2. How many polynomials of degree less than 3 are in $\mathbb{Z}_3[x]$? How is it related to $|\mathbb{Z}_3|$?
5. Try to generalize the pattern you found to know how many polynomials in $\mathbb{Z}_3[x]$ have degree less than n for an arbitrary $n > 0$. Use the PMI to prove that your formula is correct for all $n > 0$ in $\mathbb{Z}_3[x]$.

Project 7.3

According to the Division Algorithm for Polynomials, if $a(x), b(x) \in K[x]$ where K is a field, there exist $q(x), r(x) \in K[x]$ so that $b(x) = a(x)q(x) + r(x)$ and either $deg(r(x)) < deg(a(x))$ or $r(x) = 0(x)$.

We find $q(x), r(x)$ by a process like *long division.* For example, in $\mathbb{Q}[x]$ using $b(x) = 2 + 4x + 0x^2 + 2x^3$ and $a(x) = 2 + 5x + 4x^3 + x^5$ we do the following steps. (The ring $\mathbb{Q}$ has usual addition and multiplication.)

(i) Beginning with the highest powers first, we need to determine what to multiply $2x^3$ by to get x^5 so that when we subtract the highest power cancels. In this case, we need to multiply by $\frac{1}{2}x^2$, so we multiply all of $a(x)$ by $\frac{1}{2}x^2$ and subtract the result from $b(x)$.

$$\begin{array}{r|l} & 1+0x+\frac{1}{2}x^2 \\ \hline 2+4x+0x^2+2x^3 & 2+5x+0x^2+4x^3+0x^4+x^5 \\ & \quad x^2+2x^3+0x^4+x^5 \\ \hline & 2+5x-1x^2+2x^3 \end{array}$$

(ii) Now we look at the remainder and repeat the process only needing to multiply $2x^3$ by 1:

$$\begin{array}{r|l} & 1+0x+\frac{1}{2}x^2 \\ \hline 2+4x+0x^2+2x^3 & 2+5x+0x^2+4x^3+0x^4+x^5 \\ & \quad x^2+2x^3+0x^4+x^5 \\ \hline & 2+5x-1x^2+2x^3 \\ & 2+4x+0x^2+2x^3 \\ \hline & 0+1x-1x^2 \end{array}.$$

Thus we have $q(x) = 1+0x+\frac{1}{2}x^2$ and $r(x) = 0+x-x^2$ with $b(x)q(x)+r(x) = 2+5x+0x^2+4x^3+0x^4+x^5 = a(x)$ and $deg(r(x)) < deg(b(x))$.

Consider the ring $\mathbb{Z}_7[x]$ and the polynomials shown below. Recall that $\mathbb{Z}_7$ is a field. (Don't forget to use $+_7$ and $\cdot_7$ for elements of $\mathbb{Z}_7$.)

$$b(x) = 2+x+3x^2 \qquad a(x) = 1+x+0x^2+5x^3+2x^4+0x^5+6x^6$$
$$c(x) = 0+5x+2x^2+0x^3+4x^4$$

1. Find polynomials $q(x), r(x) \in \mathbb{Z}_7[x]$ with $a(x) = b(x)q(x)+r(x)$ and $deg(r(x) < 2$ or $r(x) = 0(x)$.
2. Find polynomials $q(x), r(x) \in \mathbb{Z}_7[x]$ with $c(x) = b(x)q(x)+r(x)$ and $deg(r(x) < 2$ or $r(x) = 0(x)$.
3. Find polynomials $q(x), r(x) \in \mathbb{Z}_7[x]$ with $a(x) = c(x)q(x)+r(x)$ and $deg(r(x) < 4$ or $r(x) = 0(x)$.

   Suppose we now decided to try this in $\mathbb{Z}_8[x]$ instead, with $+_8$ and $\cdot_8$ for elements of $\mathbb{Z}_8$. Use the same $a(x), b(x), c(x)$ as in the previous parts of the problem.

4. Can you find polynomials $q(x), r(x) \in \mathbb{Z}_8[x]$ with $a(x) = b(x)q(x)+r(x)$ and $deg(r(x)) < 2$ or $r(x) = 0(x)$? Find them or explain why it is impossible.
5. Can you find polynomials $q(x), r(x) \in \mathbb{Z}_8[x]$ with $a(x) = c(x)q(x)+r(x)$ and $deg(r(x)) < 4$ or $r(x) = 0(x)$? Find them or explain why it is impossible.
6. What made one of the last two divisions possible but not the other?

Project 7.4

Let A denote an integral domain and $a(x) \in A[x]$. Define the new polynomial $a'(x)$ as follows:

If $deg(a(x) = n \geq 1$ then $a'(x) = a_1 + (2 \bullet a_2)x + \cdots + (n \bullet a_n)x^{n-1}$ and if $deg(a(x)) < 1$ then $a'(x) = 0(x)$. (This should look familiar!)

1. In the ring $\mathbb{Z}_7[x]$ with $a(x) = 2 + x + 3x^2$, $b(x) = 1 + x + 0x^2 + 5x^3 + 2x^4 + 6x^6$, $c(x) = 0 + 5x + 2x^2 + 4x^4$, find $a'(x), b'(x)$ and $c'(x)$. (Don't forget to use $+_7$ and $\cdot_7$ for elements of $\mathbb{Z}_7$.)
2. In the ring $\mathbb{Z}_6[x]$ with $a(x) = 2 + 2x + 5x^2$, $b(x) = 3 + 0x + 3x^2 + 4x^3 + 5x^4$, $c(x) = 5 + 0x + 4x^2 + 0x^3 + 0x^4 + 3x^5$, find $a'(x), b'(x)$ and $c'(x)$. (Don't forget to use $+_6$ and $\cdot_6$ for elements of $\mathbb{Z}_6$.)
3. For an integral domain A, explain why if $a(x) \in A[x]$ then $a'(x) \in A[x]$.
4. Prove: If A is an integral domain, $a(x) \in A[x]$, $deg(a(x)) > 1$, and $char(A) = 0$ then $deg(a'(x)) > 0$.
5. Why did we include $char(A) = 0$ in the previous problem? Find a polynomial $a(x) \in \mathbb{Z}_5[x]$ with $deg(a(x)) > 1$ for which $a'(x)$ is a nonzero constant polynomial. (Don't forget to use $+_5$ and $\cdot_5$ for elements of $\mathbb{Z}_5$.)
6. Is the function $f : \mathbb{Z}_5[x] \to \mathbb{Z}_5[x]$ defined by $f(a(x)) = a'(x)$ a homomorphism? Either prove it is true or find polynomials that show it fails to be a homomorphism. (Don't forget to use $+_5$ and $\cdot_5$ for elements of $\mathbb{Z}_5$.)
7. If we had $char(A) = 0$ instead, would the function, as defined in the previous problem, be a homomorphism? Determine if $f : \mathbb{Z}[x] \to \mathbb{Z}[x]$ defined by $f(a(x)) = a'(x)$ is a homomorphism. (The ring $\mathbb{Z}$ has usual addition and multiplication.)

Project 7.5

Let A denote an integral domain, and define $S = \{a(x) \in A[x] : a_0 = 0_A\}$. Thus S consists of all polynomials in $A[x]$ whose constant term is 0_A.

1. Prove that S is an ideal of $A[x]$.
2. Suppose for the moment that $A = \mathbb{Z}$ under usual addition and multiplication, and S is defined as above. Determine which of the following polynomials are in the same coset of S. Be sure to explain how you decided.

$$5 + 2x + 3x^2$$
$$3 + 0x + 0x^2 + 0x^3 + x^4 + 7x^5$$
$$2 + 0x + 0x^2 + 0x^3 + x^4 + 9x^5$$
$$2 + 0x + 0x^2 + 0x^3 + 0x^4 + 0x^5 + 0x^6 + 0x^7 + 0x^8 + 0x^9 + 4x^{10}$$
$$1 + 0x + 7x^2 + 0x^3 + 0x^4 + 0x^5 + 9x^6$$
$$5 + 0x + 2x^2 + 0x^3 + 0x^4 + 8x^5$$
$$1 + 2x + 3x^2$$
$$2 + 0x + 7x^2 + 0x^3 + 0x^4 + 0x^5 + 9x^6$$

3. From the work in the previous problem how do we know, by just looking at two polynomials, whether or not they are in the same coset of S? Describe what a polynomial of smallest degree would be in any coset.
4. This time using $A = \mathbb{Z}_3$ under $+_3$ and $\cdot_3$, with S as above, create the Cayley tables for the ring ${}^{A[x]}/_{S}$. Can you guess what the ring ${}^{A[x]}/_{S}$ will look like in general?

5. Let A be an arbitrary integral domain and S the ideal defined at the beginning of the project. Define: $f : A[x] \to A$ by $f(a(x)) = a_0$. Prove that f is an onto homomorphism, and $ker(f) = S$. Explain what we can conclude about ${}^{A[x]}/_{S}$.
6. What do the conclusions from the previous parts tells us about the ideal S since A is an integral domain?
7. If we begin with a *field* A and used the same ideal S, what could we conclude about S?

Project 7.6

Finding zero divisors in a polynomial ring.

1. Suppose $a(x) \in A[x]$ and $a(x) \neq 0(x)$, for a nontrivial commutative ring with unity, A. Prove: If $a(x)$ is a zero divisor in $A[x]$ and $deg(a(x)) = n$, then a_n is a zero divisor in A.
2. Suppose again that $a(x) \in A[x]$ and $a(x) \neq 0(x)$, for some nontrivial commutative ring with unity, A. If $a(x)$ is a zero divisor in $A[x]$ will it also guarantee that a_0 is a zero divisor in A? Consider $\mathbb{Z}_6[x]$ to help you decide, then either prove it is true or find a zero divisor $a(x) \in \mathbb{Z}_6[x]$ which has $a_0 = 0$. (Don't forget to use $+_6$ and $\cdot_6$ for elements of $\mathbb{Z}_6$.)
3. Suppose $a(x) \in A[x]$ and $a_0 \neq 0_A$, for some nontrivial commutative ring with unity, A. Prove: If $a(x)$ is a zero divisor in $A[x]$, then a_0 is a zero divisor in A.

   We now look at whether the converses of some of the previous statements are true, specifically if having a_n or a_0 as zero divisors guarantees that $a(x)$ is also a zero divisor. For the rest of this project we will concentrate on polynomials in the ring $\mathbb{Z}_{10}[x]$. (Don't forget to use $+_{10}$ and $\cdot_{10}$ for elements of $\mathbb{Z}_{10}$.)

4. Consider the polynomial, $c(x) = 2 + 5x$ where 2 and 5 are both zero divisors in $\mathbb{Z}_{10}$. Is there a nonzero constant polynomial $b(x) = b_0$ for which $c(x)b(x) = 0(x)$? Either find one or show that none make $c(x)b(x) = 0(x)$.
5. Again, using $c(x) = 2 + 5x$, is there a polynomial $b(x) = 0 + b_1 x$ with $b_1 \neq 0$ for which $c(x)b(x) = 0(x)$? Either find one or show that none make $c(x)b(x) = 0(x)$.
6. Again, with $c(x) = 2 + 5x$, consider a polynomial of the form $b(x) = b_0 + b_1 x$ with both $b_0 \neq 0, b_1 \neq 0$. Show a formula for the product $c(x)b(x)$ and determine what conditions on b_0, b_1 are needed to make $c(x)b(x) = 0(x)$. Is it possible?
7. Repeat the step above to see why there is no $b(x) = b_0 + b_1 x + b_2 x^2$ with $b_2 \neq 0$ for which $c(x)b(x) = 0(x)$ as well. You must consider the options when b_0 or b_1 are 0 as well.
8. Finally, assume there is $b(x) \in \mathbb{Z}_{10}[x]$ with $deg(b(x)) > 2$ and $(2+5x)b(x) = 0(x)$. Let m be the smallest nonnegative integer with $b_m \neq 0$. Explain why if $m \neq deg(b(x))$ then the coefficients b_m and b_{m+1} will guarantee that no such $b(x)$ exists. What if $m = deg(b(x))$? Is $2 + 5x$ a zero divisor?

Chapter 8

Factorization of Polynomials

We now turn our attention to factoring polynomials. We will again see many parallels between the rings $(\mathbb{Z}, +, \cdot)$ and $K[x]$ where K is a field. For the rest of the text, when we mention $\mathbb{Z}$, $\mathbb{Q}$, or $\mathbb{R}$ we will assume the operations are usual addition and multiplication. Similarly, for each $n > 1$, we assume $\mathbb{Z}_n$ uses $+_n$ and $\cdot_n$.

8.1 Factors and Irreducible Polynomials

For polynomials $a(x), b(x) \in A[x]$, where A is a commutative ring with unity, we defined the product $d(x) = a(x)b(x)$ in Chapter 7 (Definition 7.8). There is another word related to such a product $d(x)$, *factor*.

Definition 8.1 Let A be a commutative ring with unity and $a(x), d(x) \in A[x]$. We say that $a(x)$ is a ***factor*** of $d(x)$ if there exists a polynomial $b(x) \in A[x]$ with $d(x) = a(x)b(x)$.

Example 8.2 In $\mathbb{Z}_5[x]$, we know that $a(x) = 2 + x$ and $b(x) = 1 + 3x$ are factors of $d(x) = 2 + 2x + 3x^2$ since $d(x) = (2 + x)(1 + 3x)$. But we can also write $d(x) = 1(2 + 2x + 3x^2)$ or $d(x) = (4 + 2x)(3 + 4x)$, for example, so there are other factors.

Example 8.3 Consider the field $(\mathbb{Q}, +, \cdot)$, and $a(x) = 3+2x+x^2 \in \mathbb{Q}[x]$. Can we find a factor for $a(x)$? We need to find $b(x), c(x) \in \mathbb{Q}[x]$ so that $a(x) = b(x)c(x)$. Since $deg(b(x)) \leq 2$ and $deg(c(x)) \leq 2$ we can express the polynomials and product as shown below where $a_i, b_i \in \mathbb{Q}$.

$$b(x) = b_0 + b_1x + b_2x^2 \qquad c(x) = c_0 + c_1x + c_2x^2$$

$$a(x) = b_0c_0 + (b_0c_1 + b_1c_0)x + (b_0c_2 + b_1c_1 + b_2c_0)x^2 + (b_1c_2 + b_2c_1)x^3 + b_2c_2x^4$$

Finding solutions to the equations below (which must hold if $a(x) = b(x)c(x)$) will help us find factors of $a(x)$.

$$\begin{aligned} 3 &= c_0b_0 \\ 2 &= b_0c_1 + b_1c_0 \\ 1 &= b_0c_2 + b_1c_1 + b_2c_0 \\ 0 &= b_1c_2 + b_2c_1 \\ 0 &= b_2c_2 \end{aligned}$$

Notice that $c_0 = \frac{2}{3}$, $c_1 = 0 = c_2$, $b_0 = \frac{9}{2}$, $b_1 = 3$, and $b_2 = \frac{3}{2}$ satisfy these equations! Thus $c(x) = \frac{2}{3}$ and $b(x) = \frac{9}{2} + 3x + \frac{3}{2}x^2$ have $a(x) = b(x)c(x)$, and we have found factors $c(x)$ and $b(x)$ of $a(x)$.

However, if we choose *any* nonzero rational number for c_0 we can use the constant polynomial $c(x) = c_0$ with $b(x) = \frac{3}{c_0} + \frac{2}{c_0}x + \frac{1}{c_0}x^2$ to show $a(x) = b(x)c(x)$. Thus we can find infinitely many (but not necessarily all) factors of $a(x)$ this way. A generalization of this property is an exercise at the end of the chapter: For any field K and $a(x) \in K[x]$, every constant polynomial is a factor of $a(x)$. (Remember that $0(x)$ is not considered a constant polynomial since it does not have degree 0.)

If there is a constant polynomial $c(x) = c$ for which $a(x) = c(x)b(x)$ then $a(x)$ and $b(x)$ will share many properties, so we need a word to describe this situation.

Definition 8.4 Let A be an integral domain. Polynomials $a(x), b(x) \in A[x]$ are called ***associates*** if there is a nonzero element $c \in A$ so that the constant polynomial $c(x) = c$ has $a(x) = c(x)b(x)$.

We will frequently write $a(x) = cb(x)$ instead of first defining the constant polynomial $c(x) = c$.

In Example 8.3 we found $a(x) = cb(x)$ where $c = \frac{2}{3}$, thus $a(x)$ and $b(x)$ are associates. Notice also that $deg(a(x)) = deg(b(x))$, which is not a coincidence. In an exercise at the end of the chapter you will show that associate polynomials in $K[x]$ *must* have the same degree when K is a field. However, having the same degree will *not* guarantee that polynomials are associates. In $\mathbb{Q}[x]$, $a(x) = 2 + x + 3x^2$ and $b(x) = 2 - x - x^2$ have the same degree, but they are <u>not</u> associates.

Another property shared by associate polynomials is seen in the next theorem, whose proof is left as an exercise at the end of the chapter. Recall that $c \in A$ is a root of $a(x) \in A[x]$ if $a(c) = 0_A$.

Theorem 8.5

Let A be an integral domain and suppose $a(x), b(x) \in A[x]$ are associates. Then $c \in A$ is a root of $a(x)$ if and only if c is a root of $b(x)$.

Example 8.6 Consider the polynomial $a(x) = 1 + 3x \in \mathbb{Q}[x]$. Suppose there are polynomials $b(x), c(x) \in \mathbb{Q}[x]$ with $a(x) = b(x)c(x)$. Since $deg(a(x)) = deg(b(x)) + deg(c(x))$ we must have $deg(b(x)) = 0$ or $deg(b(x)) = 1$.

If $deg(b(x)) = 0$ then $b(x)$ is a constant polynomial, but if $deg(b(x)) = 1$ then $deg(c(x)) = 0$ and $c(x)$ is a constant polynomial. Thus $a(x), b(x)$ are associates or $b(x)$ is a constant polynomial. This shows us that the only factors of $a(x)$ are constant polynomials or associates of $a(x)$.

Note that in $\mathbb{Z}$ the only units are 1 and -1, while in $K[x]$ the constant polynomials are the units (an exercise in Chapter 7). Thus factoring with a constant polynomial is equivalent to writing $1 \cdot n = n$ or $(-1) \cdot (-n) = n$. We consider factors that are constant polynomials or associates of $d(x)$ as *trivial* factors and are interested in whether a polynomial has nontrivial factors, leading to the next definition.

Definition 8.7 Let A be an integral domain with $a(x) \in A[x]$ and $deg(a(x)) > 0$. We say that $a(x)$ is ***irreducible over A*** if every factor of $a(x)$ in $A[x]$ is either a constant polynomial or an associate of $a(x)$.

If instead a nonconstant factor of $a(x)$ which is not an associate of $a(x)$ exists in $A[x]$, we say that $a(x)$ is ***reducible over A***.

Most of the time we will be using polynomials in $K[x]$ where K is a field. In this case the definition of an irreducible polynomial in $K[x]$ parallels the definition of a "prime" in $\mathbb{Z}$. In $\mathbb{Z}$ we do not consider 1 to be prime, and similarly we do not consider a constant polynomial in $K[x]$ (a unit) to be irreducible.

Determining whether a polynomial $a(x) \in K[x]$ is reducible or irreducible will be critical to all we do in the rest of this book, and can be very difficult. To show that $a(x)$ is irreducible we must be able to guarantee that **no** factor can exist in $K[x]$, other than a constant polynomial or associate of $a(x)$. The next theorem allows us to consider an associate of $a(x)$ in order to find factors of $a(x)$; the proof is left as an exercise at the end of the chapter.

Theorem 8.8

Let K be a field and suppose $a(x), b(x) \in K[x]$ are associates. The polynomial $a(x)$ is irreducible over K if and only if $b(x)$ is irreducible over K.

The proof of the next theorem is virtually identical to the steps in Example 8.6, so is also left as an exercise at the end of the chapter.

Theorem 8.9

Let K be a field. Every polynomial in $K[x]$ of degree 1 is irreducible over K.

Example 8.10 Consider the field $\mathbb{Z}_5$. This is a finite field, with only $0, 1, 2, 3, 4$ as elements, but there are *infinitely many* polynomials in $\mathbb{Z}_5[x]$. Thus listing the irreducible polynomials is impossible. However, we can list the polynomials of degree 1, which are all irreducible over $\mathbb{Z}_5$ by Theorem 8.9. Thus, we already know 20 irreducible polynomials in $\mathbb{Z}_5[x]$.

$$\begin{array}{ccccc} 0+x & 1+x & 2+x & 3+x & 4+x \\ 0+2x & 1+2x & 2+2x & 3+2x & 4+2x \\ 0+3x & 1+3x & 2+3x & 3+3x & 4+3x \\ 0+4x & 1+4x & 2+4x & 3+4x & 4+4x \end{array}$$

In Example 6.24, we discovered that for every prime $p \in \mathbb{Z}$, $\langle p \rangle$ is a maximal ideal. For another parallel with $K[x]$, the next theorem tells us that irreducible polynomials generate maximal ideals as well. This theorem will actually allow us to create field extensions in Chapter 9.

Theorem 8.11

Suppose K is a field, and $p(x) \in K[x]$. If $p(x)$ is irreducible over K then $\langle p(x) \rangle$ is a maximal ideal of $K[x]$.

Proof Assume K is a field, $p(x) \in K[x]$, and $p(x)$ is irreducible over K. Let $S = \langle p(x) \rangle$, and we will show S is a maximal ideal. Assume instead there is an ideal T in $K[x]$ with $S \subset T \subset K[x]$. Notice that as $deg(p(x)) > 0$ and $p(x) \in S$ then $S \neq \{0(x)\}$ and $T \neq \{0(x)\}$.

By Theorem 7.26 T must be a principal ideal. Thus there exists $b(x) \in T$ with $T = \langle b(x) \rangle$ and $b(x) \neq 0(x)$. Now $p(x) \in T$ and so $p(x) = b(x)q(x)$ for some $q(x) \in K[x]$. But $p(x)$ is irreducible over K and $b(x)$ is a factor of $p(x)$, so either $b(x)$ is an associate of $p(x)$ or $b(x)$ is a constant polynomial.

Suppose first that $b(x)$ is a constant polynomial, $b(x) = b_0$ with $b_0 \neq 0_K$. Since K is a field, the polynomial $s(x) = b_0^{-1}$ is in $K[x]$ and $b(x)s(x) \in T$. But $b(x)s(x) = 1(x)$ so by Theorem 5.12 we have $T = K[x]$ contradicting the choice of T.

Thus $b(x)$ must be an associate of $p(x)$ instead, and there is a nonzero $c \in K$ with $p(x) = cb(x)$. K is a field, so we know $c^{-1} \in K$ which tells us $c^{-1}p(x) = b(x)$ and thus $b(x) \in S$. Since S is an ideal we now know that every element of T, of the form $b(x)w(x)$, is also in S and $T = S$ which again contradicts the choice of T. Every possibility has lead us to a contradiction, so no such T can exist, and $S = \langle p(x) \rangle$ is a maximal ideal of $K[x]$. □

We now have many examples of maximal ideals in polynomial rings. In $\mathbb{Z}_7[x]$ we know that $\langle 3 + 4x \rangle$ is a maximal ideal, and in $\mathbb{Q}[x]$ the ideal $\left\langle \frac{1}{2} + \frac{5}{3}x \right\rangle$ is maximal as well.

A direct consequence of the previous theorem and Theorem 6.23 gives another parallel between primes in $\mathbb{Z}$ and irreducible polynomials in $K[x]$; see Theorem 0.21. The proof of this theorem is an exercise at the end of the chapter.

Theorem 8.12

Let K be a field, and assume that $p(x) \in K[x]$ is irreducible over K. If $a(x), b(x) \in K[x]$ and $p(x)$ is a factor of the product $a(x)b(x)$, then $p(x)$ is a factor of at least one of $a(x)$ or $b(x)$.

Example 8.13 Consider the polynomial $a(x) = 3 + 2x + 3x^2 + x^3 + x^4$ in $\mathbb{Z}_5[x]$. You should verify that $a(x) = b(x)c(x)$ where $b(x) = 1 + x + 2x^2$ and $c(x) = 3 + 4x + 3x^2$. But notice that $a(x) = (4 + x)(2 + 0x + 2x^2 + x^3)$ as well which means that $p(x) = 4 + x$ is a factor of $a(x)$. By Theorem 8.9 we know $p(x)$ is irreducible over $\mathbb{Z}_5$ and so by Theorem 8.12 $p(x)$ must be a factor of either $b(x)$ or $c(x)$. Notice that $c(x) = (4 + x)(2 + 3x)$ in $\mathbb{Z}_5[x]$ so $p(x)$ is a factor of $c(x)$.

We end this section with a theorem about irreducible polynomials which involves the homomorphism $\overline{f}$ defined in Theorem 7.28 and uses Theorem 7.30 in the proof. One direction of the proof is left as an exercise at the end of the chapter.

Theorem 8.14

Let K and E be fields, and suppose that $f : K \to E$ is an *isomorphism*. The polynomial $p(x) \in K[x]$ is irreducible over K if and only if $\overline{f}(p(x))$ is irreducible over E.

Proof Let K and E be fields, $f : K \to E$ an isomorphism, and $p(x) \in K[x]$.

($\rightarrow$) Assume $p(x)$ is irreducible over K, and for a contradiction suppose $\overline{f}(p(x))$ is reducible over E. Thus there exist polynomials $q(x), r(x) \in E[x]$ with $\overline{f}(p(x)) = q(x)r(x)$, $deg(q(x)) > 0$, and $deg(r(x)) > 0$.

By Theorem 7.30 $\overline{f}$ is an isomorphism, and thus onto, so we must have $s(x), t(x) \in K[x]$ with $\overline{f}(s(x)) = q(x)$, and $\overline{f}(t(x)) = r(x)$. Also $deg(s(x)) = deg(q(x))$, and $deg(r(x)) = deg(t(x))$ (an exercise in Chapter 7). Since $\overline{f}$ is a homomorphism we have the following equalities:

$$\overline{f}(s(x)t(x)) = \overline{f}(s(x))\overline{f}(t(x)) = q(x)r(x) = \overline{f}(p(x)).$$

But $\overline{f}$ is one to one so $p(x) = s(x)t(x)$. Since $deg(s(x)) > 0$ and $deg(t(x)) > 0$ this contradicts that $p(x)$ is irreducible over K. Thus $\overline{f}(p(x))$ is irreducible over E.

($\leftarrow$) An exercise at the end of the chapter. □

8.2 Roots and Factors

The relationship between roots and factors of a polynomial over a field is one of the most useful tools in the difficult task of factoring polynomials.

Theorem 8.15

Let K be a field and $a(x) \in K[x]$ with $a(x) \neq 0(x)$. The element $c \in K$ is a root of $a(x)$ if and only if $b(x) = -c + x$ is a factor of $a(x)$.

Proof Assume that K is a field and $a(x) \in K[x]$ with $a(x) \neq 0(x)$.

($\rightarrow$) Suppose that $c \in K$ is a root of $a(x)$ and define $b(x) = -c + x$, which is also in $K[x]$. As $b(x) \neq 0(x)$, by Theorem 7.24 there exist $q(x), r(x) \in K[x]$ with $a(x) = b(x)q(x) + r(x)$ and $deg(r(x)) < deg(b(x))$ or $r(x) = 0(x)$. If $r(x) = 0(x)$ we then have $a(x) = b(x)q(x)$ so $b(x)$ is a factor of $a(x)$ as needed. Thus assume instead that $r(x) \neq 0(x)$.

Recall that the substitution function h_c defined in Theorem 7.35 is a ring homomorphism. Thus $h_c(a(x)) = h_c(b(x))h_c(q(x)) + h_c(r(x))$ which can also be written as $a(c) = q(c)b(c) + r(c)$. Since c is a root of $a(x)$ then $a(c) = 0_K$. Also $b(c) = -c + c = 0_K$ so $0_K = q(c)0_K + r(c)$, or $0_K = r(c)$.

Since $deg(b(x)) = 1$ and $r(x) \neq 0(x)$ then $r(x) = r_0$ for some nonzero $r_0 \in K$. By Definition 7.3 $r(c) = r_0$, but then $r(c) \neq 0_K$ giving us a contradiction. Thus $r(x) = 0(x)$, $a(x) = q(x)b(x)$, and $b(x)$ is a factor of $a(x)$.

($\leftarrow$) An exercise at the end of the chapter. □

Example 8.16 Consider $a(x) = 3-4x-2x^2+5x^3$ in $\mathbb{Q}[x]$. Notice that $a(-1) = 3+4-2-5 = 0$, so -1 is a root of $a(x)$. The previous theorem tells us that $1+x$ is a factor of $a(x)$. We can verify this by finding $a(x) = (1+x)(3-7x+5x^2)$.

An extension of Theorem 8.15 will be very useful in later chapters. The proof (by induction) uses Theorem 8.15 more than once and is an exercise at the end of the chapter.

Theorem 8.17

Suppose K is a field and $a(x) \in K[x]$ with $a(x) \neq 0(x)$. If the distinct elements $c_1, c_2, \ldots, c_n \in K$ are all roots of $a(x)$, then the product $b(x) = (-c_1+x)(-c_2+x)\cdots(-c_n+x)$ is a factor of $a(x)$.

Example 8.18 In the ring $\mathbb{Z}_5[x]$ define the following set:

$$S = \{a(x) \in \mathbb{Z}_5[x] : a(1) = 0 \text{ and } a(2) = 0\}.$$

S is an ideal (an exercise in Chapter 7) so by Theorem 7.26 S is a principal ideal. How do we find $p(x) \in \mathbb{Z}_5[x]$ with $S = \langle p(x) \rangle$?

The definition of S tells us that both 1 and 2 are roots of every polynomial in S, thus $(4+x)(3+x)$ must be a factor of $a(x)$ when $a(x) \in S$ by the previous theorem. Consider the polynomial $p(x) = (4+x)(3+x)$, and we will show that $S = \langle p(x) \rangle$.

Clearly, $p(x) \in S$ so $\langle p(x) \rangle \subseteq S$. Now suppose we have $a(x) \in S$. By Theorem 7.24 there exist $q(x), r(x) \in \mathbb{Z}_5[x]$ with $a(x) = p(x)q(x) + r(x)$ where $deg(r(x)) < 2$ or $r(x) = 0(x)$. If $r(x) = 0(x)$ then $a(x) \in \langle p(x) \rangle$ as needed, so assume $r(x) \neq 0(x)$. Thus $r(x) = r_0 + r_1 x$ with at least one of r_0 and r_1 nonzero.

Using substitution we must have $a(1) = p(1)q(1) + r(1)$ or $0 = 0q(1) + r(1)$, giving us $r(1) = 0$. Thus we know $r_0 + r_1 = 0$, or $r_0 = 4r_1$. Similarly, $r(2) = 0$ so $r_0 + 2r_1 = 0$. Now $4r_1 + 2r_1 = 0$ or $r_1 = 0$ so $r(x) = r_0$. But $r(1) = 0$ so we must have $r_0 = 0$ as well, a contradiction. Thus $r(x) = 0(x)$ and $a(x) \in \langle p(x) \rangle$. Thus $S \subseteq \langle p(x) \rangle$ and so $S = \langle p(x) \rangle$.

An immediate consequence of Theorem 8.17 is next, whose proof is an exercise at the end of the chapter.

Theorem 8.19

Let K be a field. If $c_1, c_2, \ldots, c_n \in K$ are distinct roots of the nonzero polynomial $a(x) \in K[x]$, then $deg(a(x)) \geq n$.

It is critical to have K a field (or at least an integral domain) for the previous theorem. This can be seen in Example 7.38 where a polynomial in $\mathbb{Z}_6[x]$ of degree 2 has 4 distinct roots.

The parallel between primes in $\mathbb{Z}$ and irreducible polynomials over a field gives us the next result, analogous to Theorem 0.20. Whenever we factor a polynomial over a field, our goal will be to factor it into all irreducible factors in this way.

Theorem 8.20

Suppose K is a field and $a(x) \in K[x]$. If $deg(a(x)) > 0$ then there exist a positive integer m and polynomials $b_1(x), b_2(x), \ldots, b_m(x) \in K[x]$ which are irreducible over K and $a(x) = b_1(x)b_2(x)\cdots b_m(x)$.

Proof Suppose K is a field and $a(x) \in K[x]$ with $deg(a(x)) > 0$. If $a(x)$ is irreducible then we simply use $b_1(x) = a(x)$, and the statement is complete with $m = 1$. Thus we suppose instead that $a(x)$ is reducible, but cannot be written as a product of irreducible polynomials in $K[x]$.

Define $B = \{m \in \mathbb{Z}^+ :$ there exists $c(x) \in K[x]$ with $deg(c(x)) = m$, but $c(x)$ cannot be written as a product of irreducible polynomials in $K[x]\}$.

Our assumption about $a(x)$ tells us $deg(a(x)) \in B$ so $B \neq \emptyset$. Since $\mathbb{Z}$ is an integral system (Theorem 6.34) we can find a least element in B, $j > 0$. Also by definition of B there is $d(x) \in K[x]$ so that $deg(d(x)) = j$ and $d(x)$ cannot be factored into irreducible polynomials over K.

As $d(x)$ is not a product of irreducible polynomials, $d(x)$ must be reducible. Thus there are $u(x), v(x) \in K[x]$ with $d(x) = u(x)v(x)$, $deg(u(x)) > 0$, and $deg(v(x)) > 0$. However, $deg(u(x)) < j$ and $deg(v(x)) < j$ and j is the least element of B, so $u(x)$ and $v(x)$ can be factored into irreducible polynomials over K as shown below where all of the $s_i(x)$ and $q_i(x)$ are irreducible over K.

$$u(x) = s_1(x)s_2(x)\cdots s_t(x) \qquad v(x) = q_1(x)q_2(x)\cdots q_h(x)$$

Thus we have $d(x) = s_1(x)s_2(x)\cdots s_t(x)q_1(x)q_2(x)\cdots q_h(x)$ factored into irreducible polynomials over K, a contradiction. Thus $a(x)$ can be factored into irreducible polynomials over K as we needed to show. □

Example 8.21 Consider the polynomial $b(x) = 1 + 2x + 4x^2 + x^3$ in $\mathbb{Z}_5[x]$. How do we factor this polynomial into irreducible factors? First, we look for roots:

$$b(0) = 1, b(1) = 3, b(2) = 4, b(3) = 0, b(4) = 2.$$

The only root found is $c = 3$, so we know that one factor is $p(x) = 2 + x$ (since $-3 = 2$ in $\mathbb{Z}_5$). After long division we find $b(x) = (2 + x)(3 + 2x + x^2)$. Since $p(x)$ is irreducible we need to focus on $q(x) = 3 + 2x + x^2$ to see if we can factor $b(x)$ more.

Again, we look for roots, but you can check that there are no roots of $q(x)$ in $\mathbb{Z}_5$. Does this guarantee that $q(x)$ is irreducible? Suppose we have a factor $c(x)$ of $q(x)$ which is not constant nor an associate of $q(x)$, then $deg(c(x)) = 1$. Thus $c(x) = c_0 + c_1x$ with $c_1 \neq 0$.

As $\mathbb{Z}_5$ is a field, c_1 has an inverse, and $c(x) = c_1\left(c_1^{-1}c_0 + x\right)$. Thus $c_1^{-1}c_0 + x$ is a factor of $q(x)$ and so by Theorem 8.15 $q(x)$ has $-c_1^{-1}c_0$ as a root. Since there is no root for $q(x)$ in $\mathbb{Z}_5$ this is a contradiction and $q(x)$ is irreducible over $\mathbb{Z}_5$. Thus $a(x) = (2 + x)(3 + 2x + x^2)$ is written as a product of irreducible polynomials in $\mathbb{Z}_5[x]$.

Within the previous example we find a fact that will be often used when we try to factor polynomials over a field. One direction of the proof is left as an exercise at the end of the chapter.

Theorem 8.22

Let K be a field and $a(x) \in K[x]$ with $deg(a(x)) = 2$ or $deg(a(x)) = 3$. The polynomial $a(x)$ is reducible over K if and only if $a(x)$ has a root in K.

Proof Suppose K is a field and $a(x) \in K[x]$ with $deg(a(x)) = 2$ or $deg(a(x)) = 3$.

$(\rightarrow)$ Suppose that $a(x)$ is reducible over K. Thus there are $b(x), c(x) \in K[x]$ with $a(x) = b(x)c(x)$, $deg(b(x)) > 0$, and $deg(c(x)) > 0$. Since K is a field we know from Theorem 7.20 that $deg(a(x)) = deg(b(x)) + deg(c(x))$, and because $deg(a(x)) = 2$ or $deg(a(x)) = 3$, one of $b(x)$ or $c(x)$ must have degree 1.

Without loss of generality assume that $deg(b(x)) = 1$, so we can write $b(x) = b_0 + b_1x$ with $b_1 \neq 0_K$. Since $b(x) = b_1(b_1^{-1}b_0 + x)$ then $u = -b_1^{-1}b_0$ is a root of $b(x)$ in K and $b(u) = 0_K$. Thus $a(u) = b(u)c(u) = 0_K$, and $a(x)$ has a root in K as we needed to show.

$(\leftarrow)$ This is an exercise at the end of the chapter. □

Example 8.23 Consider the polynomial $a(x) \in \mathbb{Q}[x]$ with $a(x) = \frac{3}{2} + \frac{9}{2}x + \frac{7}{2}x^2 + x^3$. To factor $a(x)$ into irreducible factors, notice that $a\left(-\frac{1}{2}\right) = 0$, and so we know that $\frac{1}{2} + x$ must be a factor. Also by Theorem 8.9, $\frac{1}{2} + x$ is irreducible and so far $a(x) = \left(\frac{1}{2} + x\right)(3 + 3x + x^2)$.

Let $b(x) = 3 + 3x + x^2$. Thus $deg(b(x)) = 2$ so $b(x)$ is irreducible unless it has a root in $\mathbb{Q}$. Now $b(x)$ is a quadratic polynomial, so by the quadratic formula the only possible roots must be u and v below, which are not in $\mathbb{Q}$ since $\sqrt{-3}$ is not a rational number.

$$u = \frac{-3 + \sqrt{3^2 - 4(1)(3)}}{2(1)} = \frac{-3 + \sqrt{-3}}{2} \qquad v = \frac{-3 - \sqrt{3^2 - 4(1)(3)}}{2(1)} = \frac{-3 - \sqrt{-3}}{2}$$

Hence $b(x)$ has no roots in $\mathbb{Q}$ and by Theorem 8.22 $b(x)$ is irreducible. Thus $a(x) = \left(\frac{1}{2} + x\right)\left(3 + 3x + x^2\right)$ shows $a(x)$ factored into irreducible polynomials over $\mathbb{Q}$.

In the previous example we could have found $a(x) = (1 + 2x)\left(\frac{3}{2} + \frac{3}{2}x + \frac{1}{2}x^2\right)$, and so the irreducible polynomials needed to factor $a(x)$ are not actually unique, but are "unique up to associates."

Definition 8.24 Let K be a field and $a(x) \in K[x]$. Suppose $a(x) \neq 0(x)$, with $deg(a(x)) = n$. The polynomial $a(x)$ is ***monic*** if $a_n = 1_K$.

If we always choose an associate that is monic we have a way to uniquely factor (up to the order of the factors, of course) any polynomial in $K[x]$ into the product of a constant polynomial and some number of irreducible monic polynomials. For this reason we will often work with monic polynomials in the rest of the text.

Example 8.25 Consider the polynomial $a(x) = 3 + 2x + 3x^2 + x^3 + x^4$ in $\mathbb{Z}_5[x]$ that was discussed in Example 8.1. We found that $a(x) = b(x)c(x)$ where $b(x) = 1 + x + 2x^2$ and $c(x) = 3 + 4x + 3x^2 = (4 + x)(2 + 3x)$. But we want monic associates for each irreducible factor so we notice that $2 + 3x = 3(4 + x)$. Thus $c(x) = 3(4 + x)(4 + x)$ is a product of a constant polynomial and two irreducible polynomials by Theorem 8.9.

To complete the factorization of $a(x)$ we now concentrate on $b(x) = 1 + x + 2x^2$. You should verify that $b(x)$ has no roots in $\mathbb{Z}_5$, so by Theorem 8.22, $b(x)$ is irreducible over $\mathbb{Z}_5$. Now we need the monic associate for $b(x)$. Now since $2 \cdot_5 3 = 1$ then $b(x) = 2(3 + 3x + x^2)$ and so $a(x)$ is written as a product of a constant and monic irreducible polynomials over $\mathbb{Z}_5$ below.

$$a(x) = [3(4 + x)(4 + x)][2(3 + 3x + x^2)] = 1(4 + x)(4 + x)(3 + 3x + x^2)$$

In Example 8.25 we saw that the polynomial $c(x) = 3 + 4x + x^2$ has degree 2, but only 1 root. Since $c(x) = 3(4 + x)(4 + x)$ has $4 + x$ as a factor more than once we think of 1 as a root "twice." This leads us to define the notion of multiple or repeated roots.

Definition 8.26 Let K be a field and $a(x) \in K$ with $a(x) \neq 0(x)$. Suppose $c \in K$ is a root of $a(x)$. If there is an integer $m > 0$ for which the polynomial $b(x) = (-c + x)^m$ is a factor of $a(x)$ but $d(x) = (-c + x)^{m+1}$ is not a factor of $a(x)$, then we say that c is a root of $a(x)$ **with multiplicity** m.

Again, consider Example 8.25 where $b(x) = (4 + x)^2$ is a factor of $a(x) = 3 + 2x + 3x^2 + x^3 + x^4 \in \mathbb{Z}_5[x]$. To see that $d(x) = (4 + x)^3$ is not a factor of $a(x)$, notice that $d(x) = 4 + 3x + 2x^2 + x^3$, and try to divide $a(x)$ by $d(x)$ as in Example 8.25.

$$\begin{array}{r|l} & \quad 4 + x \\ \hline 4 + 3x + 2x^2 + x^3 & 3 + 2x + 3x^2 + x^3 + x^4 \\ & \quad\;\; \underline{4x + 3x^2 + 2x^3 + x^4} \\ & 3 + 3x + 0x^2 + 4x^3 \\ & \underline{1 + 2x + 3x^2 + 4x^3} \\ & 2 + x + 2x^2 \end{array}$$

The remainder is not $0(x)$ so $d(x)$ does not divide $a(x)$, and $d(x)$ is not a factor of $a(x)$. Thus the root $c = 1$ has multiplicity 2 for $a(x)$.

A straightforward modification of Theorem 8.19 is next, whose proof is an exercise at the end of the chapter. It was left until now since it is the key to the proof of Theorem 8.28. Don't forget that a constant polynomial cannot have a root.

Theorem 8.27

Let K be a field and $a(x) \in K[x]$ with $a(x) \neq 0(x)$. If $deg(a(x)) = n$ then there can be at most n distinct roots of $a(x)$ in K.

The next theorem illustrates an interesting point about polynomials and functions. Recall that polynomials must have the same degree and coefficients to be equal, but functions are considered equal if the answers are the same for every value plugged in. We saw in Example 7.32 two different polynomials in $\mathbb{Z}_7[x]$ which acted exactly the same when considered as functions.

Theorem 8.28

Let K be an <u>infinite</u> field. If $a(x), b(x) \in K[x]$, and $a(x) \neq b(x)$, then there must exist some $c \in K$ for which $a(c) \neq b(c)$.

Proof Let K be an infinite field, and suppose we have $a(x), b(x) \in K[x]$ and $a(x) \neq b(x)$. Let $d(x) = a(x) - b(x)$, then as $a(x) \neq b(x)$ we know that $d(x) \neq 0(x)$. Thus $deg(d(x)) = n$ for some nonnegative integer n.

If $n = 0$ then $d(x)$ is a constant polynomial so has no roots in K. Hence $d(0_K) \neq 0_K$ and so $a(0_K) \neq b(0_K)$ as needed. Assume now that $deg(d(x)) > 0$, so by Theorem 8.27 $d(x)$ can have at most n distinct roots in K. Since K is an infinite field there are *more* than n elements in K, so at least one of them, c, is not a root of $d(x)$. Thus $d(c) \neq 0_K$ and so again $a(c) \neq b(c)$ as needed. □

Thus the distinction between polynomial functions from K to K and polynomials over K becomes somewhat blurred when K is an infinite field.

8.3 Factorization over $\mathbb{Q}$

As we have seen in the previous sections it is relatively simple to determine if a polynomial over a finite field K has a root; simply substitute in each element of K to see if any result in 0_K. However, over an infinite field such as $\mathbb{Q}$ we cannot substitute in every member of the field so we need other strategies.

In this section, we will concentrate on the irreducibility of polynomials over $\mathbb{Q}[x]$. Although $\mathbb{Z}$ is not a field, $\mathbb{Z}[x]$ has an important role to play as seen in the next theorem.

Theorem 8.29

If $a(x) \in \mathbb{Q}[x]$ with $a(x) \neq 0(x)$ then there is a polynomial $b(x) \in \mathbb{Z}[x]$ with $deg(a(x)) = deg(b(x))$ which has exactly the same rational roots as $a(x)$.

Proof Suppose $a(x) \in \mathbb{Q}[x]$ and $a(x) \neq 0(x)$. If $deg(a(x)) = 0$ then $a(x)$ has no roots at all, so $b(x) = 1(x)$ has exactly the same roots and $deg(a(x)) = deg(b(x))$.

Let $deg(a(x)) = n$ where $n > 0$. The coefficients of $a(x)$ can be expressed as the fractions below with $u_i, w_i \in \mathbb{Z}$ and $w_i \neq 0$. If $a_i = 0$ we use $u_i = 0$, $w_i = 1$.

$$a_0 = \frac{u_0}{w_0}, \qquad a_1 = \frac{u_1}{w_1}, \qquad \cdots, \qquad a_n = \frac{u_n}{w_n}$$

Now the integers $w_0, w_1, \ldots, w_n$ are nonzero so $s = w_0 w_1 \cdots w_n$ is a nonzero integer, since $\mathbb{Z}$ is an integral domain. Let $b(x) = sa(x)$, and notice that for each i, $b_i = sa_i = (w_0 w_1 \cdots w_{i-1}, w_{i+1}, \cdots w_n)(u_i)$ as the w_i in the denominator of a_i is cancelled. Thus $b_i \in \mathbb{Z}$ for each i and $b(x) \in \mathbb{Z}[x]$. Now $b(x) \in \mathbb{Q}[x]$ as well so by Theorem 8.5 we know $b(x)$ and $a(x)$ have the same roots (from $\mathbb{Q}$) since $b(x) = sa(x)$. □

Example 8.30 Consider the polynomial $a(x) = -\frac{1}{8} - \frac{5}{8}x - \frac{1}{4}x^2 + x^3$ in $\mathbb{Q}[x]$. Using the method described in the previous theorem we multiply each coefficient by $s = (8)(8)(4)(1) = 256$ to get $b(x) = -32 - 160x - 64x^2 + 256x^3$. Clearly, we have $b(x) \in \mathbb{Z}[x]$.

The following list shows that $a(x)$ and $b(x)$ both have roots $1, -\frac{1}{2}, -\frac{1}{4}$. However, by Theorem 8.27 since each has degree 3 they have no more than 3 distinct roots, and so both $a(x)$ and $b(x)$ have exactly the same rational roots.

$$\begin{array}{ll} a(1) = -\frac{1}{8} - \frac{5}{8} - \frac{1}{4} + 1 = 0 & b(1) = -32 - 160 - 64 + 256 = 0 \\ a(-\frac{1}{2}) = -\frac{1}{8} + \frac{5}{16} - \frac{1}{16} - \frac{1}{8} = 0 & b(-\frac{1}{2}) = -32 + 80 - 16 - 32 = 0 \\ a(-\frac{1}{4}) = -\frac{1}{8} + \frac{5}{32} - \frac{1}{64} - \frac{1}{64} = 0 & b(-\frac{1}{4}) = -32 + 40 - 4 - 4 = 0 \end{array}$$

We will use Theorem 8.29 to help find roots of a polynomial in $\mathbb{Q}[x]$ by first writing its associate in $\mathbb{Z}[x]$. Once we have the associate in $\mathbb{Z}[x]$, the next theorem gives us a way to create a finite list of the only possible rational roots for the original polynomial over $\mathbb{Q}[x]$, eliminating the problem of infinitely many possibilities over the field $\mathbb{Q}$.

Theorem 8.31

The Rational Roots Theorem Let $a(x) \in \mathbb{Z}[x]$ with $a(x) \neq 0(x)$ and $deg(a(x)) = n$. If the rational number $\frac{s}{t}$ ($s, t \in \mathbb{Z}$ with no common prime factors and $t \neq 0$) is a root of $a(x)$ then s must evenly divide a_0 and t must evenly divide a_n.

Proof Suppose $a(x) \in \mathbb{Z}[x]$ with $a(x) \neq 0(x)$ and $deg(a(x)) = n$. Thus $a(x) = \sum_{i=0}^{n} a_i x^i$ and $a_n \neq 0$. Assume that the rational number $\frac{s}{t}$ is a root of $a(x)$ with $s, t \in \mathbb{Z}$ and no common prime factor. Thus $a(\frac{s}{t}) = 0$ and so $0 = \sum_{i=0}^{n} a_i (\frac{s}{t})^i$. Multiplying by the integer t^n we have $0 = \sum_{i=0}^{n} a_i t^{n-i} s^i$ and so subtracting one term gives us the equation below.

$$-a_0 t^n = \sum_{i=1}^{n} a_i t^{n-i} s^i = s \left(\sum_{i=1}^{n} a_i t^{n-i} s^{i-1} \right)$$

Notice that the right side of the equation is divisible by s, so the left must also be divisible by s. By Theorem 0.20 we can write $s = c p_{j_1}^{u_1} p_{j_2}^{u_2} \cdots p_{j_m}^{u_m}$ as a product of distinct primes and $c = 1$ or $c = -1$. Thus for each prime factor p_{j_i} of s, $p_{j_i}^{u_i}$ must also divide $-a_0 t^n$. However, s and t have no common prime factor so no power of p_{j_i} can divide t^n. Hence $p_{j_i}^{u_i}$ must divide a_0. We can repeat this for each prime factor of s showing that s must evenly divide a_0.

Similarly, from $0 = \sum_{i=0}^{n} a_i t^{n-i} s^i$, we could subtract a different term to find the next equation.

$$-a_n s^n = \sum_{i=0}^{n-1} a_i t^{n-i} s^i = t\left(\sum_{i=0}^{n-1} a_i t^{n-i-1} s^i\right)$$

Notice that every prime factor of t evenly divides the right side and so must also divide the left. Again, s and t have no common factors so t must divide a_n. □

Notice in the previous theorem if $a_0 = 0$ then any integer s will divide a_0, and so the theorem does not narrow down the list of possible roots much. However, this is not a problem since if $a_0 = 0$ we already know that 0 is a root of our polynomial (an exercise in Chapter 7).

Example 8.32 Consider the polynomial $a(x) = \frac{1}{3} + 2x + \frac{1}{2}x^2 + \frac{3}{4}x^3$ in $\mathbb{Q}[x]$. Multiplying by 12 we get the associate $b(x) = 4 + 24x + 6x^2 + 9x^3$ with exactly the same roots as $a(x)$. Now applying the Rational Roots Theorem (8.31) we see that the only possible rational roots are of the form $\frac{s}{t}$ where s evenly divides 4 and t evenly divides 9. The only choices for s are then $1, -1, 2, -2, 4, -4$, while the only choices for t are $1, -1, 3, -3, 9, -9$. This means the only possible roots for $b(x)$ are among the following list.

$$\begin{array}{ccccccccc} 1 & -1 & \frac{1}{3} & -\frac{1}{3} & \frac{1}{9} & -\frac{1}{9} & 2 & -2 & \frac{2}{3} \\ -\frac{2}{3} & \frac{2}{9} & -\frac{2}{9} & 4 & -4 & \frac{4}{3} & -\frac{4}{3} & \frac{4}{9} & -\frac{4}{9} \end{array}$$

While we still need to check 18 values for roots, it is much better than the infinite list of possibilities from $\mathbb{Q}$. We can also narrow down the possibilities by realizing that adding four positive rational numbers can never give us 0, so any roots would need to be negative. This cuts our list in half to only nine possibilities. You should check to see that none of these are roots of $b(x)$.

Thus we can see that $a(x)$ has no roots in $\mathbb{Q}$. Notice that if $a(x)$ is reducible then since $deg(a(x)) = 3$ there would be a root in $\mathbb{Q}$ by Theorem 8.22. Thus $a(x)$ is irreducible over $\mathbb{Q}$.

Even more interesting is the fact below that if $a(x) \in \mathbb{Z}[x]$ and we can factor $a(x)$ in $\mathbb{Q}[x]$, we could already have factored it in $\mathbb{Z}[x]$ as well.

Theorem 8.33

If $a(x) \in \mathbb{Z}[x]$ and $a(x) = b(x)c(x)$ with $b(x), c(x) \in \mathbb{Q}[x]$, $deg(b(x)) > 0$, and $deg(c(x)) > 0$, then there exist polynomials $u(x), w(x) \in \mathbb{Z}[x]$ with $a(x) = u(x)w(x)$, $deg(u(x)) > 0$, and $deg(w(x)) > 0$.

Proof Assume we have $a(x) \in \mathbb{Z}[x]$ and $a(x) = b(x)c(x)$ with $b(x), c(x) \in \mathbb{Q}[x]$, $deg(b(x)) > 0$, and $deg(c(x)) > 0$. Using the process described in Theorem 8.29 we can find $q(x), s(x) \in \mathbb{Z}[x]$ where $q(x) = mb(x)$ and $s(x) = nc(x)$ for some $m, n \in \mathbb{Z}$. Thus $mn \in \mathbb{Z}$, and we have the next equation.

$$q(x)s(x) = (mb(x))(nc(x)) = (mn)(b(x)c(x)) = (mn)(a(x))$$

Let $k = mn$. If $k < 0$ we could have used $q(x) = (-m)b(x)$ or $s(x) = (-n)c(x)$ in the equation above to have a positive k with $q(x)s(x) = ka(x)$. Thus we will assume that $k > 0$ in the rest of the proof.

Now k can be written as a product of primes $k = p_{j_1}^{i_1} p_{j_2}^{i_2} \cdots p_{j_m}^{i_m}$. Notice that if p_1 evenly divides *every* coefficient of $q(x)$ then we can write $q(x) = p_1 e(x)$ for some $e(x) \in \mathbb{Z}[x]$. But $p_{j_1}^{i_1} p_{j_2}^{i_2} \cdots p_{j_m}^{i_m} a(x) = p_1 e(x)s(x)$ allows us to cancel p_1 from both sides of the equation and reduce the value of k. (Similar steps could be used if p_1 divided every coefficient of $s(x)$.) Hence if we can repeat this for every prime dividing k, with one of the newest revisions of $q(x)$ or $s(x)$ we will eventually be able to cancel all of k and have $u(x), w(x) \in \mathbb{Z}[x]$ with $a(x) = u(x)w(x)$ as needed. So suppose instead that $p \in \mathbb{Z}$ is a prime dividing k, but does not divide some coefficient of $q(x)$ and does not divide some coefficient of $s(x)$.

Let r be the least integer with $0 \leq r \leq deg(q(x))$ for which p does not divide q_r, and let g be the least integer with $0 \leq g \leq deg(s(x))$ for which p does not divide s_g. Recall that for $v(x) = q(x)s(x)$ we have v_{r+g} as shown below. Also by subtracting we find an expression for the coefficient $q_r s_g$.

$$v_{r+g} = \sum_{i+j=r+g} q_i s_j = (q_r s_g) + \sum_{i+j=r+g}^{i \neq r, j \neq g} q_i s_j$$

$$v_{r+g} - \sum_{i+j=r+g}^{i \neq r, j \neq g} q_i s_j = q_r s_g$$

When $i < r$ then p divides q_i and so p divides $q_i s_j$. Also, if $i > r$, then $j < g$ so p divides s_j and p divides $q_i s_j$ again. But by assumption p divides v_{r+g} so p divides the left side of the expression for $q_r s_g$. Hence p divides $q_r s_g$ as well. This contradicts our assumptions on r and g, so p must divide every coefficient of either $q(x)$ or $s(x)$, and we have $a(x) = u(x)w(x)$ for some $u(x), w(x) \in \mathbb{Z}[x]$. □

Example 8.34 Consider the polynomial $a(x) = -1 - 4x + 15x^2 + 18x^3 \in \mathbb{Z}[x]$. To find factors we first look for roots in $\mathbb{Q}$. By Theorem 8.31 the possible rational roots are $\frac{s}{t}$ where s evenly divides -1 and t evenly divides 18. Thus $s = 1, -1$, and $t = 1, 2, 3, 6, 9, 18$ (and the negative of each). Thus we have the only possible roots shown next.

$$\frac{s}{t} = 1, \frac{1}{2}, \frac{1}{3}, \frac{1}{6}, \frac{1}{9}, \frac{1}{18} \text{ (or the negative of each)}$$

Now $a\left(\frac{1}{3}\right) = 0$ so we have a root of $a(x)$, and we know that $\left(-\frac{1}{3} + x\right)$ is also a factor of $a(x)$.

$$\begin{array}{r|l}
 & 3 + 21x + 18x^2 \\
\hline
-\frac{1}{3} + x & -1 - 4x + 15x^2 + 18x^3 \\
 & \qquad\quad \underline{-6x^2 + 18x^3} \\
 & -1 - 4x + 21x^2 \\
 & \quad \underline{-7x + 21x^2} \\
 & -1 + 3x \\
 & \underline{-1 + 3x} \\
 & 0
\end{array}$$

So we have $a(x) = (3 + 21x + 18x^2)\left(-\frac{1}{3} + x\right)$. In order to make each factor in $\mathbb{Z}[x]$ we would need to use $3a(x) = (3+21x+18x^2)(-1+3x)$. But notice that just as described in the theorem, every coefficient of the first factor is divisible by 3. Thus we have $a(x) = (1 + 7x + 6x^2)(-1 + 3x)$ with factors in $\mathbb{Z}[x]$. Continue factoring $a(x)$ into irreducible polynomials.

Theorem 8.31 tells us that any polynomial in $\mathbb{Z}[x]$ that is reducible as a polynomial in $\mathbb{Q}[x]$ must have been reducible in $\mathbb{Z}[x]$ already.

Similarly, being able to determine if a polynomial with integer coefficients is *irreducible* in $\mathbb{Z}[x]$ will also tell us it is irreducible as a polynomial of $\mathbb{Q}[x]$, even though $\mathbb{Z}$ is not a field! The next theorem helps determine irreducibility for many polynomials in $\mathbb{Z}[x]$.

Theorem 8.35

Eisenstein's Criterion Suppose $a(x) \in \mathbb{Z}[x]$ and $deg(a(x)) = n$ with $n > 0$. If there exists a prime number p which divides coefficients $a_0, a_1, \ldots, a_{n-1}$ but not a_n, and p^2 does not divide a_0, then $a(x)$ is irreducible over $\mathbb{Q}$.

Proof Suppose $a(x) \in \mathbb{Z}[x]$ and $deg(a(x)) = n$ with $n > 0$, so $a_n \neq 0$. Assume there is a prime p which divides coefficients $a_0, a_1, \ldots, a_{n-1}$ but not a_n, and p^2 does not divide a_0. (Note if $a_i = 0$ then p divides a_i since $a_i = 0 \cdot p$.) In order to show that $a(x)$ is irreducible over $\mathbb{Q}$, suppose instead that $a(x) = b(x)c(x)$ with nonconstant $b(x), c(x) \in \mathbb{Q}[x]$. Then by Theorem 8.33 there are nonconstant $u(x), w(x) \in \mathbb{Z}[x]$ with $a(x) = u(x)w(x)$.

Let $deg(u(x)) = m$ and $deg(w(x)) = s$, thus by Theorem 7.20 $m + s = n$. We know $a_0 = u_0 w_0$ so since p divides a_0 by our assumption, then p must divide one of u_0 or w_0. However, p^2 does not divide a_0, so p divides exactly one of u_0 or w_0. We will assume that

p divides u_0 but not w_0.

Also, p does not divide $a_n = u_m w_s$ and thus p cannot divide either of u_m or w_s. Since p divides u_0 but does not divide u_m, choose the smallest $0 < k \le m$ for which p does not divide u_k. With steps similar to those in Theorem 8.33 we write the expression below for $u_k w_0$.

$$a_k = \sum_{j+t=k} u_j w_t = u_k w_0 + \sum_{j+t=k}^{j \neq k} u_j w_t$$

$$a_k - \sum_{j+t=k}^{j \neq k} u_j w_t = u_k w_0$$

As $k < n$, p divides a_k. But also for each $j < k$, p divides u_j and hence p divides $u_j w_t$ when $j \neq k$. So p must also divide $u_k w_0$ as it divides the left side of the expression for $u_k w_0$. Since p does not divide u_k and does not divide w_0 we have a contradiction (and similarly if we had assumed p divides w_0 but not u_0). Thus $a(x)$ must be irreducible over $\mathbb{Q}$. □

Example 8.36 Consider the polynomial $a(x) = 2 + 6x + 10x^2 + 8x^3 + 4x^4 + x^5 \in \mathbb{Z}[x]$. The prime number 2 divides each of 2, 6, 10, 8, and 4, but 2 does not divide 1. Also 2^2 does not divide 2, so by Eisensteins Criterion (Theorem 8.35) we know $a(x)$ is irreducible over $\mathbb{Q}$.

Similarly, the polynomial $b(x) = \frac{3}{10} + \frac{3}{5}x + \frac{3}{10}x^2 + \frac{6}{5}x^3 + \frac{9}{10}x^4 + \frac{1}{2}x^5$ has associate $c(x) = 3 + 6x + 3x^2 + 12x^2 + 9x^4 + 5x^5$ in which 3 divides 3, 6, 3, 12, and 9, but not 5, and 9 does not divide 3. Thus $c(x)$ is irreducible over $\mathbb{Q}$ and so $a(x)$ is irreducible over $\mathbb{Q}$ as well.

Unfortunately, *Eisenstein's Criterion* is not an "if and only if" statement. It is possible to have a polynomial that does not satisfy the hypotheses but is still irreducible (such as $4 + x^2$) and so it is important to use the theorem correctly.

Another useful theorem allows us to use a finite ring to help determine if a polynomial with integer coefficients is irreducible as well. Recall from Theorem 4.41 that for each integer $n > 1$ we have a homomorphism $f_n : \mathbb{Z} \to \mathbb{Z}_n$ defined by $f_n(a) = a(\text{mod} n)$ for each $a \in \mathbb{Z}$. By Theorem 7.28 we can extend f_n to a homomorphism $\overline{f_n}$ from $\mathbb{Z}[x]$ to $\mathbb{Z}_n[x]$.

Theorem 8.37

Suppose $a(x) \in \mathbb{Z}[x]$ is a *monic* polynomial and $deg(a(x)) = k$ with $k > 0$. If there exists $n > 1$ so that $\overline{f_n}(a(x))$ is irreducible in $\mathbb{Z}_n[x]$ then $a(x)$ is also irreducible in $\mathbb{Z}[x]$.

Proof Suppose $a(x) \in \mathbb{Z}[x]$ is a monic polynomial and $deg(a(x)) = k$ with $k > 0$. Assume there exists $n > 1$ with $\overline{f_n}(a(x))$ irreducible in $\mathbb{Z}_n[x]$. Since we know $a_k = 1$ and for

$j > k, a_j = 0$ then $f_n(a_k) = 1(\text{mod}n) = 1$ and $f_n(a_j) = f(0) = 0$ for $j > k$. Thus $deg(\overline{f_n}(a(x))) = k$ as well.

We want to show that $a(x)$ is irreducible in $\mathbb{Z}[x]$, so assume instead that $a(x)$ is reducible in $\mathbb{Z}[x]$ and thus there are nonconstant polynomials $b(x), c(x) \in \mathbb{Z}[x]$ with $a(x) = b(x)c(x)$. Now $0 < deg(b(x)) < k$ and $0 < deg(c(x)) < k$ so let $s = deg(b(x))$ and $t = deg(c(x))$. Also since $a_k = 1$ we must have b_s and c_t both equal to 1 or -1.

As $b_s = 1$ or $b_s = -1$ then $f_n(b_s) \neq 0$ and so $deg(\overline{f_n}(b(x))) = s$. Similarly, we know $deg(\overline{f_n}(c(x))) = t$. As $\overline{f_n}$ is a homomorphism we know that the following must hold:

$$\overline{f_n}(a(x)) = \overline{f_n}(b(x))\overline{f_n}(c(x)).$$

Thus we have factored $\overline{f_n}(a(x))$ in $\mathbb{Z}_n[x]$ into a product of nonconstant polynomials contradicting that $\overline{f_n}(a(x))$ is irreducible. Hence $a(x)$ is irreducible in $\mathbb{Z}[x]$. □

Example 8.38 Let $a(x) = 7 + 5x + 6x^2 + x^3$ in $\mathbb{Z}[x]$. Using $f_5 : \mathbb{Z} \to \mathbb{Z}_5$ we see $\overline{f_5}(a(x)) = 2 + 0x + x^2 + x^3$. In $\mathbb{Z}_5[x]$ we now consider the polynomial $b(x) = 2 + 0x + x^2 + x^3$ to see if it is irreducible over $\mathbb{Z}_5$.

Since $\mathbb{Z}_5$ is a field then by Theorem 8.22 since $deg(b(x)) = 3$ then either $b(x)$ is irreducible over $\mathbb{Z}_5$ or $b(x)$ has a root in $\mathbb{Z}_5$. But $b(0) = 2, b(1) = 4, b(2) = 4, b(3) = 3$, and $b(4) = 2$, so $b(x)$ has no roots in $\mathbb{Z}_5$ and is irreducible over $\mathbb{Z}_5$. Therefore $a(x)$ is irreducible over $\mathbb{Z}$ (and $\mathbb{Q}$).

Once again we must be careful to only use the theorem under the correct hypotheses! It is possible to have $a(x) \in \mathbb{Z}[x]$ irreducible but for some n, $\overline{f_n}(a(x))$ reducible in $\mathbb{Z}_n[x]$. We saw that $a(x) = 7 + 5x + 6x^2 + x^3$ is irreducible over $\mathbb{Z}$ in Example 8.38. However, using $n = 7$ we find $\overline{f_7}(a(x)) = 0 + 5x + 6x^2 + x^3 = x(5 + 6x + x^2)$, which is reducible. So having $\overline{f_7}(a(x))$ reducible over $\mathbb{Z}_7$ does <u>not</u> tell us if $a(x)$ is reducible over $\mathbb{Z}$.

Irreducible polynomials are the foundation for field extensions and Galois Theory, the final topics for this book. Thus we will use many of the theorems in Chapters 7 and 8 as we reach toward our final goal of solvability of polynomials.

Exercises for Chapter 8

Section 8.1 Factors and Irreducible Polynomials

1. Prove: If K is a field and $a(x) \in K[x]$ then every constant polynomial in $K[x]$ is a factor of $a(x)$. (Remember that $0(x)$ is not a constant polynomial.)

2. Show that having K a field in the previous problem is critical by finding a constant polynomial in $\mathbb{Z}[x]$ which is not a factor of $3 + 2x + 5x^2 + 4x^3$
3. Prove: If K is an integral domain, $a(x), b(x) \in K[x]$ are associates, and $a(x) \neq 0(x)$, then $deg(a(x)) = deg(b(x))$.
4. Find nonzero polynomials $a(x), b(x) \in \mathbb{Z}_{10}[x]$ which are associates but $deg(a(x)) \neq deg(b(x))$.
5. Suppose K is a field and $a(x) \in K[x]$ with $a(x) \neq 0(x)$. Prove: If $b(x), c(x) \in K[x]$ with $a(x) = b(x)c(x)$ and $deg(c(x)) \neq 0$, then $deg(b(x)) < deg(a(x))$.
6. In $\mathbb{Z}_{11}[x]$, determine if $a(x) = 4x + 9x^2 + 8x^3$ and $b(x) = 8x + 7x^2 + 9x^3$ are associates. Either find the correct constant polynomial $c(x)$ with $a(x) = c(x)b(x)$ or prove that no such $c(x)$ can exist.
7. In $\mathbb{Z}_3[x]$, determine if $a(x) = 2 + x + 2x^2 + 2x^3 + x^4$ and $b(x) = 1 + 2x + x^2 + x^3 + x^4$ are associates. Either find the correct constant polynomial $c(x)$ with $a(x) = c(x)b(x)$ or prove that no such $c(x)$ can exist.
8. In $\mathbb{Z}_5[x]$, determine if $a(x) = 2+3x+x^2+4x^3+3x^4$ and $b(x) = 1+4x+3x^2+3x^3+4x^4$ are associates. Either find the correct constant polynomial $c(x)$ with $a(x) = c(x)b(x)$ or prove that no such $c(x)$ can exist.
9. In $\mathbb{Q}[x]$, determine if $a(x) = 1+2x+4x^2-6x^3-x^4$ and $b(x) = -4-8x-16x^2+24x^3+4x^4$ are associates. Either find the correct constant polynomial $c(x)$ with $a(x) = c(x)b(x)$ or prove that no such $c(x)$ can exist.
10. In $\mathbb{Z}_7[x]$, determine if $a(x) = 6 + 2x + 5x^2 + x^3$ and $b(x) = 2 + 3x + 4x^2 + 5x^3$ are associates. Either find the correct constant polynomial $c(x)$ with $a(x) = c(x)b(x)$ or prove that no such $c(x)$ can exist.
11. In $\mathbb{Z}_{13}$, determine if $a(x) = 11 + 7x + 8x^2 + 4x^3$ and $b(x) = 10 + 6x + 7x^2 + 3x^3$ are associates. Either find the correct constant polynomial $c(x)$ with $a(x) = c(x)b(x)$ or prove that no such $c(x)$ can exist.
12. Prove Theorem 8.5.
13. Show that it is necessary for A to be an integral domain in Theorem 8.5 by finding nonconstant $a(x), b(x) \in \mathbb{Z}_6[x]$ which are associates but $a(x)$ and $b(x)$ do not have the same roots.
14. Show that in $\mathbb{Z}_7[x]$ the associate polynomials $a(x) = 2+x+2x^2$ and $b(x) = 4+2x+4x^2$ have the same roots. Be sure to verify that $a(x)$ and $b(x)$ are associates.
15. Show that in $\mathbb{Z}_3[x]$ the associate polynomials $a(x) = 1+x+x^2$ and $b(x) = 2+2x+2x^2$ have the same roots. Be sure to verify that $a(x)$ and $b(x)$ are associates.
16. Show that in $\mathbb{Z}_{11}[x]$ the associate polynomials $a(x) = 6+2x+3x^2$ and $b(x) = 2+8x+x^2$ have the same roots. Be sure to verify that $a(x)$ and $b(x)$ are associates.
17. Show that in $\mathbb{Z}_5[x]$ the associate polynomials $a(x) = 2 + 0x + x^2 + 3x^3$ and $b(x) = 1 + 0x + 3x^3 + 4x^3$ have the same roots. Be sure to verify that $a(x)$ and $b(x)$ are associates.
18. Show that in $\mathbb{Z}_7[x]$ the associate polynomials $a(x) = 2 + 6x + 5x^2 + 3x^3$ and $b(x) =$

$1+3x+6x^2+5x^3$ have the same roots. Be sure to verify that $a(x)$ and $b(x)$ are associates.

19. Find two nonconstant polynomials $a(x), b(x) \in \mathbb{Z}_5[x]$ which have exactly the same roots in $\mathbb{Z}_5$ but are not associates.
20. Find two nonconstant polynomials $a(x), b(x) \in \mathbb{Z}_3[x]$ which have exactly the same roots in $\mathbb{Z}_3$ but are not associates.
21. Find two nonconstant polynomials $a(x), b(x) \in \mathbb{Z}_7[x]$ which have exactly the same roots in $\mathbb{Z}_7$ but are not associates.
22. Prove Theorem 8.8.
23. Prove Theorem 8.9.
24. Show that it was necessary to have a field K in Theorem 8.9, by finding $a(x) \in \mathbb{Z}_8[x]$ with $deg(a(x)) = 1$ but $a(x)$ is not irreducible.
25. Prove Theorem 8.12.
26. Consider $a(x) = 6+6x+x^2+5x^3$, $b(x) = 2+x+x^2$ and $c(x) = 3+0x+2x^2+3x^3$ in $\mathbb{Z}_7[x]$. Show that $b(x)$ is a factor of the product $a(x)c(x)$ by finding $q(x) \in \mathbb{Z}_7[x]$ with $b(x)q(x) = a(x)c(x)$, then show that $b(x)$ is a factor of one of $a(x)$ and $c(x)$ by finding $r(x) \in \mathbb{Z}_7[x]$ with either $a(x) = b(x)r(x)$ or $c(x) = b(x)r(x)$.
27. Consider $a(x) = 1+4x+x^2+3x^3$, $b(x) = 4+3x$ and $c(x) = 2+2x+0x^2+3x^3$ in $\mathbb{Z}_5[x]$. Show that $b(x)$ is a factor of the product $a(x)c(x)$ by finding $q(x) \in \mathbb{Z}_5[x]$ with $b(x)q(x) = a(x)c(x)$, then show that $b(x)$ is a factor of one of $a(x)$ and $c(x)$ by finding $r(x) \in \mathbb{Z}_5[x]$ with either $a(x) = b(x)r(x)$ or $c(x) = b(x)r(x)$.
28. Show that the assumption of $p(x)$ irreducible in Theorem 8.12 was needed, by finding nonconstant polynomials $a(x), b(x), c(x) \in \mathbb{Z}_5[x]$ so that $b(x)$ is a factor of $a(x)c(x)$ but $b(x)$ is not a factor of either $a(x)$ or $c(x)$.
29. Complete the proof of Theorem 8.14.

Section 8.2 Roots and Factors

30. Complete the proof of Theorem 8.15.
31. Use the PMI to prove Theorem 8.17, for any $n \geq 1$. In the inductive step when you have $a(x) = (-c_1+x)\cdots(-c_k+x)q(x)$ be sure to show why $q(c_{k+1})$ must equal 0_K.
32. Find all of the roots $c_1, c_2, \cdots c_n$ of $a(x) = 4+6x+5x^2+4x^3$ in $\mathbb{Z}_7$, and factor $a(x) = (-c_1+x)(-c_2+x)\cdots(-c_n+x)q(x)$ in $\mathbb{Z}_{11}[x]$ as Theorem 8.17 tells us we can.
33. Find all of the roots $c_1, c_2, \cdots c_n$ of $a(x) = 1+2x+3x^2+3x^3+3x^4+2x^5$ in $\mathbb{Z}_5$, and factor $a(x) = (-c_1+x)(-c_2+x)\cdots(-c_n+x)q(x)$ in $\mathbb{Z}_5[x]$ as Theorem 8.17 tells us we can.
34. Find all of the roots $c_1, c_2, \cdots c_n$ of $a(x) = 10+7x+5x^2+9x^3+0x^4+3x^5$ in $\mathbb{Z}_{11}$, and factor $a(x) = (-c_1+x)(-c_2+x)\cdots(-c_n+x)q(x)$ in $\mathbb{Z}_{11}[x]$ as Theorem 8.17 tells us we can.
35. Find all of the roots $c_1, c_2, \cdots c_n$ of $a(x) = 1+3x+x^2+3x^3+0x^4+x^5+x^6$ in $\mathbb{Z}_7$, and factor $a(x) = (-c_1+x)(-c_2+x)\cdots(-c_n+x)q(x)$ in $\mathbb{Z}_7[x]$ as Theorem 8.17 tells us we can.

36. Find all of the roots $c_1, c_2, \cdots c_n$ of $a(x) = 6 + 0x + x^2 + x^3 + 6x^4 + x^5$ in $\mathbb{Z}_{11}$, and factor $a(x) = (-c_1 + x)(-c_2 + x) \cdots (-c_n + x)q(x)$ in $\mathbb{Z}_{11}[x]$ as Theorem 8.17 tells us we can.
37. Prove Theorem 8.19.
38. Complete the proof of Theorem 8.22.
39. Factor $a(x) = 4 + 4x + 4x^2 + 2x^3 + x^4$ into a product of irreducible polynomials in $\mathbb{Z}_5[x]$. Be sure to verify that the factors are irreducible over $\mathbb{Z}_5$.
40. Factor $a(x) = 2 + 9x + 2x^2 + 9x^3 + 7x^4 + x^5$ into a product of irreducible polynomials in $\mathbb{Z}_{11}[x]$. Be sure to verify that the factors are irreducible over $\mathbb{Z}_{11}$.
41. Factor $a(x) = 2 + 2x + 2x^2 + 0x^3 + x^4 + x^5 + x^6$ into a product of irreducible polynomials in $\mathbb{Z}_3[x]$. Be sure to verify that the factors are irreducible over $\mathbb{Z}_3$.
42. Factor $a(x) = 3 + 2x + 6x^2 + 4x^3 + 5x^4 + x^5$ into a product of irreducible polynomials in $\mathbb{Z}_7[x]$. Be sure to verify that the factors are irreducible over $\mathbb{Z}_7$.
43. Factor $a(x) = 7 + 3x + 9x^2 + x^3 + 7x^4 + 12x^5$ into a product of irreducible polynomials in $\mathbb{Z}_{13}[x]$. Be sure to verify that the factors are irreducible over $\mathbb{Z}_{13}$.
44. Factor $a(x) = 1 + x + 6x^2 + 3x^3 + 2x^4$ into a product of irreducible polynomials in $\mathbb{Z}_7[x]$. Be sure to verify that the factors are irreducible over $\mathbb{Z}_7$.
45. Factor $a(x) = 4 + x + x^2 + 4x^3 + 3x^4$ into a product of irreducible polynomials in $\mathbb{Z}_5[x]$. Be sure to verify that the factors are irreducible over $\mathbb{Z}_5$.
46. Prove Theorem 8.27.

Section 8.3 Factorization over $\mathbb{Q}$

47. Let $a(x) = \frac{1}{3} + x + \frac{2}{3}x^2 + 3x^3 + \frac{1}{2}x^4$ in $\mathbb{Q}[x]$. Find an associate of $a(x)$ in $\mathbb{Z}[x]$ then determine the possible roots for $a(x)$ in $\mathbb{Q}$.
48. For the possible roots of $a(x) = \frac{1}{3} + x + \frac{2}{3}x^2 + 3x^3 + \frac{1}{2}x^4$ found in the previous problem determine which actually are roots.
49. Let $b(x) = -\frac{1}{6} - \frac{4}{6}x - \frac{8}{3}x^2 - \frac{3}{2}x^3 + 5x^4$ in $\mathbb{Q}[x]$. Find an associate of $b(x)$ in $\mathbb{Z}[x]$ then determine the possible roots for $b(x)$ in $\mathbb{Q}$.
50. For the possible roots of $b(x) = -\frac{1}{6} - \frac{4}{6}x - \frac{8}{3}x^2 - \frac{3}{2}x^3 + 5x^4$ found in the previous problem determine which actually are roots.
51. Let $c(x) = -\frac{2}{5} - \frac{1}{2}x + \frac{103}{20}x^2 + \frac{11}{4}x^3 - x^4$ in $\mathbb{Q}[x]$. Find an associate of $c(x)$ in $\mathbb{Z}[x]$ then determine the possible roots for $c(x)$ in $\mathbb{Q}$.
52. For the possible roots of $c(x) = -\frac{2}{5} - \frac{1}{2}x + \frac{103}{20}x^2 + \frac{11}{4}x^3 - x^4$ found in the previous problem determine which actually are roots.
53. Let $d(x) = -\frac{1}{6} - \frac{1}{4}x + \frac{1}{2}x^2 + \frac{3}{4}x^3 + \frac{2}{3}x^4 + x^5$ in $\mathbb{Q}[x]$. Find an associate of $d(x)$ in $\mathbb{Z}[x]$ then determine the possible roots for $d(x)$ in $\mathbb{Q}$.
54. For the possible roots of $d(x) = -\frac{1}{6} - \frac{1}{4}x + \frac{1}{2}x^2 + \frac{3}{4}x^3 + \frac{2}{3}x^4 + x^5$ found in the previous problem determine which actually are roots.
55. Let $q(x) = -\frac{3}{49} - \frac{2}{7}x + x^2 - \frac{3}{49}x^3 - \frac{2}{7}x^4 + x^5$ in $\mathbb{Q}[x]$. Find an associate of $q(x)$ in $\mathbb{Z}[x]$ then determine the possible roots for $q(x)$ in $\mathbb{Q}$.
56. For the possible roots of $q(x) = -\frac{3}{49} - \frac{2}{7}x + x^2 - \frac{3}{49}x^3 - \frac{2}{7}x^4 + x^5$ found in the previous

problem determine which actually are roots.

57. Let $s(x) = \frac{2}{5} + x + \frac{5}{3}x + \frac{19}{15}x^2 - \frac{1}{15}x^3 + x^4$ in $\mathbb{Q}[x]$. Find an associate of $s(x)$ in $\mathbb{Z}[x]$ then determine the possible roots for $s(x)$ in $\mathbb{Q}$.
58. For the possible roots of $s(x) = \frac{2}{5} + x + \frac{5}{3}x + \frac{19}{15}x^2 - \frac{1}{15}x^3 + x^4$ found in the previous problem determine which actually are roots.
59. Use Eisenstein's Criterion to show that the polynomial $a(x) = \frac{2}{5} + \frac{8}{15}x + \frac{2}{3}x^2 + \frac{4}{5}x^3 + \frac{2}{15}x^4 + \frac{4}{15}x^5 + \frac{1}{3}x^6$ is irreducible over $\mathbb{Q}$.
60. Use Eisenstein's Criterion to show that the polynomial $a(x) = \frac{1}{4} + \frac{3}{4}x + \frac{1}{2}x^2 + \frac{1}{4}x^3 + \frac{7}{4}x^4 + \frac{9}{4}x^5 + \frac{1}{4}x^6 + \frac{1}{28}x^7$ is irreducible over $\mathbb{Q}$.
61. Use Eisenstein's Criterion to show that the polynomial $a(x) = \frac{1}{2} + x + \frac{3}{2}x^2 + \frac{1}{2}x^3 + 2x^4 + x^5 + \frac{5}{6}x^6$ is irreducible over $\mathbb{Q}$.
62. Use Eisenstein's Criterion to show that the polynomial $a(x) = \frac{3}{2} + \frac{9}{2}x + 0x^2 + \frac{3}{2}x^3 + 3x^4 + \frac{3}{5}x^5$ is irreducible over $\mathbb{Q}$.
63. Use the function h defined in Project 8.3 to help show that the polynomial $a(x) = 1 + 3x + 6x^2 + 9x^3 + 3x^4$ is irreducible over $\mathbb{Q}$.
64. Use the function h defined in Project 8.3 to help show that the polynomial $a(x) = \frac{5}{4} + \frac{1}{2}x + x^2 + x^3 + \frac{1}{2}x^4$ is irreducible over $\mathbb{Q}$.
65. Use the function h defined in Project 8.3 to help show that the polynomial $a(x) = \frac{1}{5} + \frac{1}{7}x + 0x^2 + \frac{3}{7}x^3 + \frac{2}{7}x^4 + \frac{1}{7}x^5$ is irreducible over $\mathbb{Q}$.
66. Use Theorem 8.37 to prove that $a(x) = 56 + 36x + 29x^2 + x^3$ is irreducible over $\mathbb{Q}$.
67. Use Theorem 8.37 to help prove that $a(x) = 11 + 8x + 16x^2 + 7x^3 + x^4$ is irreducible over $\mathbb{Q}$.
68. Use Theorem 8.37 to help prove that $a(x) = \frac{15}{2} + 4x + 0x^2 + 0x^3 + \frac{1}{2}x^4$ is irreducible over $\mathbb{Q}$.
69. Factor $a(x) = -\frac{1}{12} + 0x + \frac{13}{12}x^2 + \frac{11}{12}x^3 + \frac{1}{12}x^4 + x^5$ into a product of irreducible polynomials in $\mathbb{Q}[x]$. Be sure to verify that the factors are irreducible over $\mathbb{Q}$.
70. Factor $a(x) = -\frac{40}{3} - \frac{50}{3}x - \frac{25}{3}x^2 - \frac{50}{3}x^3 + \frac{7}{3}x^4 - \frac{10}{3}x^5 + x^6$ into a product of irreducible polynomials in $\mathbb{Q}[x]$. Be sure to verify that the factors are irreducible over $\mathbb{Q}$.
71. Factor $a(x) = \frac{1}{6} + \frac{5}{2}x + 11x^2 + 13x^3 + \frac{13}{2}x^4 + 3x^5$ into a product of irreducible polynomials in $\mathbb{Q}[x]$. Be sure to verify that the factors are irreducible over $\mathbb{Q}$.

Projects for Chapter 8

Project 8.1

Let K be a field and $a(x), b(x) \in K[x]$. We say $b(x)$ is a factor of $a(x)$, if $a(x) = b(x)q(x)$ for some $q(x) \in K[x]$.

For example in $\mathbb{Z}_7[x]$ we see that $2+3x$ is a factor of $1+4x+2x^2$ since $1+4x+2x^2 = (2+3x)(4+3x)$.

1. Show that $2+x$ is a factor of both $2+x^3$ and $2+x^7$ in $\mathbb{Z}_3[x]$.
2. Show that in $\mathbb{Z}_5[x]$, the polynomial $2+x$ is <u>not</u> a factor of $2+x^3$.
3. Use PMI to prove that for every positive integer n the polynomial $2+x$ is a factor of $2+x^n$ in $\mathbb{Z}_3[x]$.
4. Show that $4+x$ is a factor of $4+x^3$ and $4+x^6$ in $\mathbb{Z}_5[x]$.
5. Use PMI to prove that for every positive integer n the polynomial $4+x$ is a factor of $4+x^n$ in $\mathbb{Z}_5[x]$.
6. Do you see a pattern? What would we use instead of 4 in the field $\mathbb{Z}_7$?

Project 8.2

We have seen that for polynomials of degree 1, 2, or 3 over a field K we only need to discover if they have roots to see if they are irreducible over K. However, if our polynomial $a(x)$ has degree 4, that is not enough. We also have to show that $a(x)$ cannot factor into quadratics. For the rest of this project let $a(x) = 2+0x+4x^2+0x^3+x^4$ in $\mathbb{Z}_5[x]$.

1. Show that $a(x)$ has no roots in $\mathbb{Z}_5$.

   Now we have to check and see if we can write $a(x) = (d+ex+jx^2)(u+vx+wx^2)$ for some $e,d,j,u,v,w \in \mathbb{Z}_5$.

2. Explain why if we can find $e,d,j,u,v,w \in \mathbb{Z}_5$ with $a(x) = (d+ex+jx^2)(u+vx+wx^2)$ we can also find $q,r,s,t, \in \mathbb{Z}_5$ with $a(x) = (q+rx+x^2)(s+tx+x^2)$.
3. Show that for $a(x) = (q+rx+x^2)(s+tx+x^2)$ we must have $qs=2$, $qt+rs=0$, $q+rt+s=4$, and $r+t=0$, in $\mathbb{Z}_5$.
4. Prove that the equations found in the previous problem imply that either $r=t=0$, or $q=s$.
5. Explain why there is no solution for $q,r,s,t, \in \mathbb{Z}_5$ that will make $a(x) = (q+rx+x^2)(s+tx+x^2)$.
6. What can we conclude about $a(x)$?

Project 8.3

Let K be a field and define $V = \{a(x) \in K[x] : a_0 \neq 0\}$.

1. Use the field $\mathbb{Z}_7$ to show that V is **not** a subring of $K[x]$, since it is not closed under subtraction. It will simply be a useful set for the next few problems.
2. Prove: V is closed under multiplication.
3. Suppose $a(x) \in V$. Prove: If $b(x) \in K[x]$ and $b(x)$ is a factor of $a(x)$ then $b(x) \in V$.

   Define the function $h : V \to V$ by $h(a_0 + a_1x + \cdots + a_nx^n) = a_n + a_{n-1}x + \cdots + a_0x^n$

when $deg(a(x)) = n$.

For example: in $\mathbb{Z}_7[x]$, $a(x) = 3 + 0x + 2x^2 + 0x^3 + 4x^4 + x^5$ is in V and $h(a(x)) = 1 + 4x + 0x^2 + 2x^3 + 0x^4 + 3x^5$.

This function reverses the order of the coefficients of the polynomial! It is trivial to see that h is onto since for any $a(x) \in V$ and $b(x) = h(a(x))$ we have $h(b(x)) = a(x)$. The function h is also easily one to one since if $c(x) = h(a(x)) = h(b(x))$ then $h(c(x)) = a(x)$ and $h(c(x)) = b(x)$ so $a(x) = b(x)$.

4. Find $h(a(x))$ for each of $a(x), b(x), c(x) \in \mathbb{Z}_7[x]$ defined below.
   $a(x) = 1 + 5x + 0x^2 + 6x^3$ $\qquad h(a(x)) =$
   $b(x) = 4 + 3x + 4x^2 + 0x^3 + 5x^4$ $\qquad h(b(x)) =$
   $c(x) = 3 + 2x + x^2 + 4x^3 + x^5 + 2x^6$ $\qquad h(c(x)) =$
5. Prove: If $a(x) \in V$ then $deg(a(x)) = deg(h(a(x)))$.
6. With simply a chase of indices it it straightforward to see that for any $a(x), b(x) \in V$ we have $h(a(x)b(x)) = h(a(x))h(b(x))$. We will not write out the steps but will assume it is true. Verify in $\mathbb{Z}_7[x]$ that $h(a(x)b(x)) = h(a(x))h(b(x))$ is true for $a(x) = 2 + x + 4x^2$ and $b(x) = 5 + 6x + 2x^2$.
7. Prove: If $a(x) \in V$ and $h(a(x))$ is irreducible in $K[x]$, then $a(x)$ is irreducible in $K[x]$ as well.
8. For $a(x) = 6 + 2x + x^2 + 4x^3$ in $\mathbb{Z}_7[x]$, verify that $a(x)$ and $h(a(x))$ are irreducible.

Project 8.4

A useful technique for working with polynomials is referred to as "substitution." Suppose K is a field. For a polynomial $a(x) \in K[x]$ of degree n, and $c \in K$, define the new polynomial $a(c + x)$ as shown below. For a constant polynomial $a(x) = a_0$ we say $a(c + x) = a_0$.

$$a(c + x) = a_0 + a_1(c + x) + a_2(c + x)^2 + \cdots + a_n(c + x)^n.$$

For example: In $\mathbb{Z}_5[x]$ with $a(x) = 2 + 3x + 4x^2$ and $c = 2$ we find:

$$\begin{aligned} a(2 + x) &= 2 + 3(2 + x) + 4(2 + x)^2 \\ &= 2 + 3(2 + x) + 4(4 + 4x + x^2) \\ &= 2 + 1 + 3x + 1 + x + 4x^2 \\ &= 4 + 4x + 4x^2 \end{aligned}$$

1. Explain why for any $c \in K$ and $a(x) \in K[x]$ we know $a(c + x) \in K[x]$.
2. We will not write out the steps (it is repeated use of the distributive laws), but you may use the following fact: If $a(x), b(x) \in K[x]$, $c \in K$, $q(x) = a(x) + b(x)$, and $d(x) = a(x)b(x)$ then $q(c + x) = a(c + x) + b(c + x)$ and $d(c + x) = a(c + x)b(c + x)$. Verify that this holds in $\mathbb{Z}_5[x]$ with $a(x) = 1 + 4x + x^2$, $b(x) = 2 + 2x$, and $c = 3$.
3. Prove by PMI on $n \geq 0$ that if $deg(a(x)) = n$ then $deg(a(c + x)) = n$.
4. Prove: If $a(x) \in K[x]$, $c \in K$, and $a(c + x)$ is irreducible over K then $a(x)$ is also

irreducible over K.

5. Try this technique in $\mathbb{Q}[x]$ with $a(x) = 21 + 24x + 11x^2 + 4x^3 + x^4$ and $c = -1$. Why do we know that $a(-1 + x)$ is irreducible over $\mathbb{Q}$?

Project 8.5

Consider the ring $\mathbb{Z}[x]$, and suppose we have $a(x) \in \mathbb{Z}[x]$. The rational roots theorem will be very useful here.

1. Prove: If $a(x)$ is monic and has a root $c \in \mathbb{Q}$, then $c \in \mathbb{Z}$.
2. Does the previous statement still hold if we do not assume that $a(x)$ is monic? Give a proof or a counterexample to support your statement.
3. Prove that $a(x) = 2 - nx + 0x^2 + x^3$ is reducible in $\mathbb{Q}[x]$ when n is -1, 3, or 5.
4. Prove that $a(x) = 2 - nx + 0x^2 + x^3$ is irreducible in $\mathbb{Q}[x]$ whenever $n \in \mathbb{Z}$ and $n \neq -1, 3, 5$.
5. Is $b(x) = 2 - nx + 0x^2 + 2x^3$ irreducible whenever $n \in \mathbb{Z}$ and $n \neq -1, 3, 5$ as well? Show that it is or find an $n \neq -1, 3, 5$ where $b(x)$ is reducible.
6. Find which integer values of n make $b(x) = 2 - nx + 0x^2 + 2x^3$ reducible in $\mathbb{Q}[x]$, and show that $b(x)$ is irreducible for the other $n \in \mathbb{Z}$.

Project 8.6

The polynomials of the form $-1 + x^n$ in $\mathbb{Q}$ will be very useful, so we will learn more about them in this project.

1. Prove: For all $n > 0$ the polynomial $a(x) = -1 + x^n$ always has $-1 + x$ as a factor in $\mathbb{Q}[x]$.
2. For any $n > 1$ we can write $a(x) = -1 + x^n$ as $(-1 + x)p(x)$ for some $p(x) \in \mathbb{Q}[x]$. Find the correct polynomial $p(x)$ for each of $n = 2, 3, 4, 5$
3. For $n = 2$ and $n = 3$ prove that appropriate $p(x)$ is irreducible over $\mathbb{Q}$.
4. Prove that for $n = 4$ the polynomial $p(x)$ is reducible by factoring it in $\mathbb{Q}[x]$.
5. Let $n > 1$ and $a(x) = -1 + x^n = (-1 + x)p(x)$. Explain why $a(1 + x) = xp(1 + x)$.
6. From the binomial formula we know that $q(x) = (1 + x)^n = 1 + nx + q_2x^2 + \cdots + q_{n-2}x^{n-2} + nx^{n-1} + x^n$ where n evenly divides each of $q_2, \ldots, q_{n-2}$. Also notice that for $a(x) = -1 + x^n$ we have $a(1 + x) = -1 + (1 + x)^n = -1 + q(x)$. Hence

$$\begin{aligned} a(1+x) &= -1 + (1 + nx + q_2x^2 + \cdots + q_{n-2}x^{n-2} + nx^{n-1} + x^n) \\ &= nx + q_2x^2 + \cdots q_{n-2}x^{n-2} + nx^{n-1} + x^n \\ &= x(n + q_2x + \cdots q_{n-2}x^{n-3} + nx^{n-2} + x^{n-1}) \end{aligned}$$

Thus we know $p(1 + x) = n + q_2x + \cdots + nx^{n-2} + x^{n-1}$. Why do we know that $p(1 + x)$ is irreducible when n is prime? What can we conclude about $p(x)$ when n is prime?

Project 8.7

Consider the polynomial $a(x) = 144 + 24x + 12x^2 + 4x^3$ in $\mathbb{Q}[x]$. Since the degree is 3, if $a(x)$ is reducible in $\mathbb{Q}[x]$ then it must have a root in $\mathbb{Q}$.

1. Use the rational roots theorem to show there are more than 20 possible roots of $a(x)$ in $\mathbb{Q}$. (Do not check them to see if they are roots!)
2. Consider now the polynomial $b(x) = 36 + 6x + 3x^2 + x^3$. Explain why we know that if $b(x)$ is irreducible over $\mathbb{Q}$ then $a(x)$ is also irreducible. How many possible rational roots does $b(x)$ have? (Do not check them.)

   Recall from Theorem 8.37, if $\overline{f_5}(b(x))$ is irreducible in $\mathbb{Z}_5$ then we will know that $a(x)$ is irreducible in $\mathbb{Z}[x]$.

3. Find $q(x) = \overline{f_5}(b(x))$, and determine if $q(x)$ has roots in $\mathbb{Z}_5$.
4. What can we now conclude about $b(x)$ and $a(x)$?
5. Repeat this same process for the polynomial $d(x) = -240 + 90x + 150x^2 + 10x^3$ using $\mathbb{Z}_7$ to see that $d(x)$ is irreducible in $\mathbb{Q}[x]$

Chapter 9

Extension Fields

The focus for the rest of this book is on fields. We have seen many fields such as $\mathbb{Z}_p$ where p is prime, and of course, $\mathbb{Q}$, $\mathbb{R}$, or the complex numbers $\mathbb{C}$. But are there others? Roots of polynomials will be the driving force behind finding other fields, and the properties of extension fields will lead us to the final topic of the text, solvability of polynomials.

9.1 Extension Field

When we looked at the structures of groups and rings it was natural to look for substructures, subgroups, and subrings. Subfields will again have an important role, but in addition we consider the idea of an extension field, a field that contains the original.

Definition 9.1 Suppose that K and E are fields with $K \subseteq E$. If for all $a, b \in K$ we have $a +_K b = a +_E b$ and $a \cdot_K b = a \cdot_E b$, then K is a subfield of E or E is an ***extension field*** of K.

Note: If we have E an extension field of K then $0_K = 0_E$ (since K is a subring of E), thus we use them interchangeably from now on. Also, if $a(x) \in K[x]$, then $a(x) \in E[x]$.

$(\mathbb{R}, +, \cdot)$ is easily an extension field of the rational numbers $\mathbb{Q}$. It is not as obvious, however, if there is an extension field of $(\mathbb{Z}_5, +_5, \cdot_5)$. Also, as the set of rational numbers, $\mathbb{Q}$, is countable but $\mathbb{R}$ is uncountable, is there a field K between them, i.e., $\mathbb{Q} \subset K \subset \mathbb{R}$? A first step is to describe different "types" of elements in an extension field.

Definition 9.2 Suppose E is an extension field of K, and $c \in E$.

(i) If there exists $a(x) \in K[x]$ with $a(x) \neq 0(x)$ and $a(c) = 0_E$, then c is ***algebraic over K***.
(ii) If for every nonzero $a(x) \in K[x]$ we have $a(c) \neq 0_E$, then c is ***transcendental over K***.

Example 9.3 Consider the fields $\mathbb{Q} \subset \mathbb{R}$. The polynomial $a(x) = -13 + x^2$ in $\mathbb{Q}[x]$ has root $c = \sqrt{13}$ with $c \in \mathbb{R}$. Thus $\sqrt{13}$ is algebraic over $\mathbb{Q}$. Similarly, for any positive rational number q, we have $b(x) = -q + x^2$ in $\mathbb{Q}[x]$, and $c = \sqrt{q}$ is a root of $b(x)$. Thus $\sqrt{q}$ is algebraic over $\mathbb{Q}$.

It is much more difficult to show that a number $c \in \mathbb{R}$ is transcendental over $\mathbb{Q}$, since we must be able to verify that *no* polynomial in $\mathbb{Q}[x]$ can have it as a root. Most students learn that e and π are transcendental over $\mathbb{Q}$ in a calculus course.

Since our final goal is solvability of polynomials, we will concentrate on algebraic elements over a field (more about transcendental elements can be found in [1] or [9]). First, we state a theorem that we will frequently reference; the proof of each part is an exercise at the end of the chapter.

Theorem 9.4

Suppose E is an extension field of K, $a(x) \in K[x]$, and there is $c \in E$ with $a(c) = 0_E$.

(i) If $deg(a(x)) = 1$ then $c \in K$.
(ii) If $a(x)$ is irreducible over K and $deg(a(x)) > 1$ then $c \notin K$.

When we determine if a polynomial is irreducible (or reducible) we always say which field we are trying to reduce it over. For example, by Eisenstein's Criterion (Theorem 8.35) the polynomial $a(x) = -13 + x^2$ is irreducible over $\mathbb{Q}$. But the same polynomial considered as an element of $\mathbb{R}[x]$ is reducible over $\mathbb{R}$ into $(\sqrt{13} + x)(-\sqrt{13} + x)$. Thus a polynomial can be irreducible over one field, but over an extension field it might become reducible. This idea will help us find many fields K with $\mathbb{Q} \subset K \subseteq \mathbb{R}$, where $a(x)$ is reducible over K. The existence of such a field is in the next theorem, which references theorems from Chapters 5, 6, 7, and 8!

Theorem 9.5

Suppose K is a field, E is an extension field of K and $c \in E$. If c is algebraic over K, then there exist a field $K(c)$ ("***K adjoin c***") with:

(i) $K \subseteq K(c) \subseteq E$.
(ii) $c \in K(c)$.
(iii) For any subfield S of E with $K \subseteq S$ and $c \in S$ we have $K(c) \subseteq S$.

Proof Suppose K is a field, E is an extension field of K, and $c \in E$ is algebraic over K. Define the function $f_c : K[x] \to E$ by $f_c(a(x)) = a(c)$ for every $a(x) \in K[x]$. The proof that f_c is a homomorphism is an exercise at the end of the chapter. Let $T = ker(f_c)$, then by Theorem 5.31 T is an ideal of $K[x]$. Also, by Theorem 7.26, we know T is a principal ideal so there is a polynomial $b(x) \in K[x]$ with $T = \langle b(x) \rangle$.

Since we assumed c is algebraic over K there is some nonzero polynomial $a(x) \in K[x]$ with $a(c) = 0_E$. Thus $f_c(a(x)) = 0_E$ which tells us $a(x) \in T$, $T \neq \{0(x)\}$, and $b(x) \neq 0(x)$. In an exercise at the end of the chapter, you will be asked to explain why $deg(b(x)) > 0$ so $b(x)$ is not a constant polynomial. Our goal is to show that $b(x)$ is irreducible over K, and $f_c(K[x])$ is the field we hope to find.

Suppose instead that $b(x)$ is reducible over K, then there are nonconstant polynomials $t(x), d(x) \in K[x]$ with $b(x) = t(x)d(x)$. Thus $deg(t(x)) < deg(b(x))$ and $deg(d(x)) < deg(b(x))$ as shown in an exercise of Chapter 8. Since f_c is a homomorphism, $b(x) \in ker(f_c)$, and E has no zero divisors, the following equation shows that either $0_E = t(c)$ or $0_E = d(c)$:

$$t(c)d(c) = f_c(t(x))f_c(d(x)) = f_c(b(x)) = 0_E.$$

Thus either $t(x) \in T$ or $d(x) \in T$. But for any $q(x) \in T$ we have $q(x) = b(x)s(x)$ for some $s(x) \in K[x]$ and so $deg(q(x)) \geq deg(b(x))$. Hence $d(x), t(x) \notin T$ since $deg(t(x)) < deg(b(x))$ and $deg(d(x)) < deg(b(x))$. By this contradiction we know $b(x)$ is irreducible over K, and so by Theorem 8.11 T is a maximal ideal of $K[x]$. By Theorem 6.22 we can conclude that ${}^{K[x]}/_{T}$ is a field.

Define $K(c) = f_c(K[x])$, then by Theorem 5.34 (FHT) $K(c) \cong {}^{K[x]}/_{T}$. This tells us that $K(c)$ is a field, and since $f_c(K[x]) \subseteq E$ then $K(c) \subseteq E$. Also for each constant polynomial $p(x) = p_0$ with $p_0 \in K$ we see that $f_c(p_0) = p_0$ so we have $K \subseteq K(c) \subseteq E$ and (i) is complete.

(ii) and (iii) are left as as exercises at the end of the chapter. Don't forget that $K(c) = f_c(K[x])$.

□

Example 9.6 Let $c = \sqrt{13}$ for this example. Since $p(x) = -13 + x^2$ is irreducible over $\mathbb{Q}$ we know $c \notin \mathbb{Q}$ by Theorem 9.4. Define the function $f_c : \mathbb{Q}[x] \to \mathbb{R}$ as in the previous proof with $f_c(a(x)) = a(c)$. We know that $\mathbb{Q}\left(\sqrt{13}\right) = \mathbb{Q}(c) = f_c\left(\mathbb{Q}[x]\right)$ is a field by Theorem 9.5, but it is not yet clear whether $\mathbb{Q}(c)$ is simply the same as $\mathbb{R}$. In Example 9.14 we will see why $\sqrt{5} \in \mathbb{R}$ but $\sqrt{5} \notin \mathbb{Q}(c)$ and thus $\mathbb{Q} \subset \mathbb{Q}(c) \subset \mathbb{R}$.

So far we have only considered a subfield K of a field E and created extensions of K between K and E, but how can we create an extension of K if we do not know of an extension E that exists? For example, how can we find an extension field for $\mathbb{Z}_5$? We use the same ideas, but things are a little bit more interesting.

Theorem 9.7

Let K be a field and assume $a(x) \in K[x]$ is irreducible over K. Then there exists a field E so that E is an extension field of K and $a(x)$ has a root in E.

Proof Let K be a field and assume $a(x) \in K[x]$ is irreducible over K. Since constant polynomials cannot be irreducible, we know $deg(a(x)) > 0$. Let T be the principal ideal of $K[x]$ generated by $a(x)$, so $T = \langle a(x) \rangle$. T is a maximal ideal by Theorem 8.11 and so by Theorem 6.22 ${}^{K[x]}/_{T}$ is a field as we have seen before. Define $E = {}^{K[x]}/_{T}$.

The tricky part of this proof is understanding how E is an extension field of K, and that E has a root for $a(x)$. Consider an element $u \in K$. We know there is a constant polynomial $u(x) = u \in K[x]$ and thus the coset $u(x) + T$ (or $u + T$) is an element of E.

$$\text{Define } f : K \to {}^{K[x]}/_{T} \text{ by } f(w) = w + T \text{ for each } w \in K.$$

The proof that f is a one to one homomorphism is left as an exercise at the end of the chapter. Since f is one to one we know $ker(f) = \{0_K\}$ so by Theorem 5.34 we know that $f(K) \cong {}^{K}/_{\{0_K\}}$. However, ${}^{K}/_{\{0_K\}} \cong K$ (an exercise in Chapter 5) so we know $f(K) \cong K$. Now we have an isomorphic **copy** of K inside of E, and each element $u \in K$ is identified with the coset $u + K$. This allows us to consider $K \subseteq E$, and so E is an extension field of K.

Finally, we need to show that there is a root for $a(x)$ in E. In the polynomial $a(x)$ we replace each coefficient a_i by the coset $a_i + T$ as described above to create the polynomial $\overline{a}(x)$ in ${}^{K[x]}/_{T}$.

$$a(x) = a_0 + a_1 x + \cdots + a_n x^n$$

$$\overline{a}(x) = (a_0 + T) + (a_1 + T)x + \cdots + (a_n + T)x^n$$

But the polynomial $0_K + x$ is in $K[x]$, so we have the coset $x+T \in {}^{K[x]}/_T$ or $x+T \in E$. Let $c = x+T$, and we will show that $\overline{a}(c) = 0_E$ using coset rules.

$$\begin{aligned} \overline{a}(c) &= (a_0+T)+(a_1+T)(c)+\cdots+(a_n+T)(c)^n \\ &= (a_0+T)+(a_1+T)(x+T)+\cdots+(a_n+T)(x+T)^n \\ &= (a_0+T)+(a_1x+T)+\cdots+(a_nx^n+T) \\ &= (a_0+a_1x+\cdots+a_nx^n)+T \end{aligned}$$

This gives us $\overline{a}(c) = a(x)+T$ but $a(x) \in T$, so $\overline{a}(c) = 0(x)+T = 0_E$. Thus $c = x+T$ is a root of $\overline{a}(x)$ and $c \in E$. We therefore conclude that the field $E = {}^{K[x]}/_T$ has a root for the original polynomial $a(x)$. □

Although this seems strange, we can use it to find a field extension for $\mathbb{Z}_5$.

Example 9.8 Let $a(x) = 2+x^2 \in \mathbb{Z}_5[x]$. Since $a(0) = 2, a(1) = 3, a(2) = 1, a(3) = 1$, and $a(4) = 3$ then $a(x)$ has no root in $\mathbb{Z}_5$. Now we know $a(x)$ is irreducible by Theorem 8.22. Thus Theorem 9.7 tells us there is a field, call it E, which is an extension of $\mathbb{Z}_5$ and contains a root for $a(x)$, which we can call c. Now with field extension E, and root $c \in E$, we can use Theorem 9.5 to find $\mathbb{Z}_5(c) = f_c(\mathbb{Z}_5[x]) = \{b(c) : b(x) \in \mathbb{Z}_5[x]\}$. This allows us to begin a list of elements in $\mathbb{Z}(c)$ as seen below.

$$\begin{gathered} 0,1,2,3,4,0+c,1+c,2+c,3+c,4+c,2c,3c,4c,1+2c,1+3c,1+4c, \\ 2+2c,2+3c,2+4c,3+2c,3+3c,3+4c,4+2c,4+3c,4+4c \end{gathered}$$

We have 25 elements so far, but are there more in $\mathbb{Z}_5(c)$? In the next section we will see why these are in fact the <u>only</u> elements of $\mathbb{Z}_5(c)$. Notice that since c is a root of $a(x) = 2+x^2$, then $2+c^2 = 0_E$. Thus we know $c^2 = 3$ since both $\mathbb{Z}_5$ and $\mathbb{Z}_5(c)$ use $+_5$. This helps us calculate products in $\mathbb{Z}_5(c)$ as shown below.

$$(3c)\cdot(4c) = (3 \cdot_5 4)c^2 = 2c^2 = 2(3) = 1$$

$$(1+c)(2+3c) = 2+3c+2c+3c^2 = 2+3c^2 = 2+3(3) = 1$$

9.2 Minimum Polynomial

In order to create an extension field of the form $K(c)$ we needed an irreducible polynomial in $K[x]$ for which c is a root. But there could be more than one such polynomial, so how do we choose the one to use?

Theorem 9.9

If K is a field, E is an extension field of K, and $c \in E$ is algebraic over K, then there is a ***unique monic*** polynomial $p(x) \in K[x]$ that is irreducible over K and has c as a root.

Proof Suppose K is a field, E is an extension of K, and $c \in E$ is algebraic over K. By Definition 9.2 there is a nonzero polynomial $b(x) \in K[x]$ which has c as a root. Using the familiar homomorphism $f_c : K[x] \to E$ we know $T = ker(f_c)$ is the principal ideal generated by some $q(x) \in K[x]$ with $q(c) = 0_E$, so $T = \langle q(x) \rangle$.

The fact that $q(x)$ is irreducible over K is left as an exercise at the end of the chapter. Thus $q(x)$ is irreducible over K, and has c as a root, but may not be monic. Since K is a field, $q(x)$ has a monic associate, $p(x) = uq(x)$ for some $u \in K$, which is also irreducible by Theorem 8.8. But since $p(x) = uq(x)$ then $p(c) = uq(c) = 0_E$ so c is a root of $p(x)$. In an exercise at the end of the chapter you will show that with $p(x)$ and $q(x)$ nonzero associates we also know that $T = \langle p(x) \rangle$. To complete the proof we must now show $p(x)$ is unique.

Suppose we have any monic polynomial $t(x) \in K[x]$, which is irreducible over K and has c as a root. Since $t(c) = 0_E$ then by $t(x) \in ker(f_c)$, or $t(x) \in \langle p(x) \rangle$, we know $t(x) = p(x)w(x)$ for some $w(x) \in K[x]$. As $deg(p(x)) > 0$ by $p(x)$ irreducible, we must have $deg(w(x)) = 0$ or it will contradict that $t(x)$ is irreducible over K. Thus $deg(t(x)) = deg(p(x)) = n$ for some $n > 0$. Now $t_n = p_n w_0$, but both $p(x)$ and $t(x)$ are monic so $1_K = 1_K w_0$ or $w_0 = 1_K$. Hence $t(x) = p(x)$, and we have seen that $p(x)$ is unique. □

> **Definition 9.10** Let K be a field, E an extension field of K, and $c \in E$ algebraic over K. The unique monic polynomial $p(x) \in K[x]$ that is irreducible over K and has c as a root is called ***the minimum polynomial*** for c over K.

Example 9.11 Consider the field $\mathbb{Q}$ and the element $u = \frac{2}{\sqrt[3]{5}}$ in the extension field $\mathbb{R}$. The steps below show that u is a root of $a(x) = -8 - 8x + 5x^3 + 5x^4$.

$$\begin{aligned} a(u) &= -8 - 8\left(\tfrac{2}{\sqrt[3]{5}}\right) + 5\left(\tfrac{2}{\sqrt[3]{5}}\right)^3 + 5\left(\tfrac{2}{\sqrt[3]{5}}\right)^4 \\ &= -8 - \tfrac{16}{\sqrt[3]{5}} + 8 + \tfrac{16}{\sqrt[3]{5}} = 0 \end{aligned}$$

Thus we know $\frac{2}{\sqrt[3]{5}}$ is algebraic over $\mathbb{Q}$, but this polynomial is reducible since -1 is also a root, so how do we find the minimum polynomial? From $u = \frac{2}{\sqrt[3]{5}}$, cube each side and $u^3 = \frac{8}{5}$, therefore $-\frac{8}{5} + u^3 = 0$. Thus u is a root of $p(x) = -\frac{8}{5} + x^3$, which is monic and has u as a root, but we need to determine if $p(x)$ is irreducible over $\mathbb{Q}$.

Consider the associate $q(x) = -8 + 5x^3$ which has $q_0 \neq 0$. Using the homomorphism h defined in Project 8.3 we find $h(q(x)) = 5 + 0x + 0x^2 - 8x^3$. By Eisenstein's Criterion $h(q(x))$ is irreducible over $\mathbb{Q}$, and so by Project 8.3 we know $q(x)$ and $p(x)$ are irreducible over $\mathbb{Q}$. Thus we have found the minimum polynomial $p(x)$ for $u = \frac{2}{\sqrt[3]{5}}$ over $\mathbb{Q}$.

Two special properties of minimum polynomials were discovered in the proof of Theorem 9.9. The proofs are left as exercises at the end of the chapter.

Theorem 9.12

Suppose K is a field, E is an extension field of K, and $c \in E$ is algebraic over K with minimum polynomial $p(x) \in K[x]$.

(i) Using the homomorphism $f_c : K[x] \to E$ as defined in Theorem 9.5, $ker(f_c) = \langle p(x) \rangle$.
(ii) If $b(x) \in K[x]$ is a nonzero polynomial with $b(c) = 0_E$, then $b(x) = p(x)q(x)$ for some $q(x) \in K[x]$.

One of the keys to understanding an extension $K(c)$ is next.

Theorem 9.13

Suppose K is a field, E is an extension of K, and $c \in E$ is algebraic over K. If $p(x)$ is the minimum polynomial for c over K, and $deg(p(x)) = n$, then:

$$K(c) = \{a(c) : a(x) \in K[x] \text{ and either } a(x) = 0(x) \text{ or } deg(a(x)) < n\}.$$

Proof Suppose K is a field, E is an extension field of K, and $c \in E$ is algebraic over K. Let $p(x)$ be the minimum polynomial for c over K, and $deg(p(x)) = n$ with $n > 0$. Let $S = \{a(c) : a(x) \in K[x] \text{ and either } a(x) = 0(x) \text{ or } deg(a(x)) < n\}$, and we need to show that $K(c) = S$. Recall that we defined $K(c) = f_c(K[x])$ where $f_c : K[x] \to E$ was defined by $f_c(a(x)) = a(c)$.

($\supseteq$) Let $u \in S$. By definition of S, $u = a(c)$ for some $a(x) \in K[x]$. Thus $u \in f_c(K[x])$, so $u \in K(c)$. Hence $K(c) \supseteq S$.

($\subseteq$) Let $w \in K(c)$. Thus by definition of $K(c)$ we know $w = a(c)$ for some $a(x) \in K[x]$. However, we don't yet know $w \in S$ since we don't know if $a(x) = 0(x)$ or $deg(a(x)) < n$. Clearly, if $a(x) = 0(x)$ or if $deg(a(x)) < n$ then $w \in S$, so assume $deg(a(x)) \geq n$.

Using the Division Algorithm (Theorem 7.24) we can find $q(x), r(x) \in K[x]$ with $a(x) = q(x)p(x) + r(x)$ and $r(x) = 0(x)$ or $deg(r(x)) < n$. Now $a(c) = q(c)p(c) + r(c)$, but c is a root of $p(x)$ so $p(c) = 0_E$. Thus $w = a(c) = 0_E + r(c)$. This tells us that $w = r(c)$ and $r(x) = 0(x)$ or $deg(r(x)) < n$. Thus $w \in S$, $K(c) \subseteq S$, and $K(c) = S$, as we needed to show. $\square$

This answers the question we had in Example 9.8 about how many elements there should be in $\mathbb{Z}_5(c)$ where c is a root of $p(x) = 2 + x^2$. Since $p(x)$ is the minimum polynomial for c over $\mathbb{Z}_5$ with $deg(p(x)) = 2$ then the elements of $\mathbb{Z}_5(c)$ are found by substituting c into polynomials of degree 0 or 1 (and $0(x)$, of course), giving us exactly the 25 elements we found in Example 9.8.

Example 9.14 In Example 9.6 we wondered if $\sqrt{5} \in \mathbb{Q}(\sqrt{13})$. We can answer this with help from Theorem 9.13. Recall that we had $p(x) = -13 + x^2$ in $\mathbb{Q}[x]$, a monic irreducible polynomial with $\sqrt{13}$ as a root. Thus $p(x)$ is the minimum polynomial for $\sqrt{13}$ so by Theorem 9.13 if $\sqrt{5} \in \mathbb{Q}(\sqrt{13})$ then we could find $a_0, a_1 \in \mathbb{Q}$ with $\sqrt{5} = a_0 + a_1\sqrt{13}$. Note that as $-5 + x^2$ is irreducible over $\mathbb{Q}$ by Eisenstein's Criterion we know $\sqrt{5} \notin \mathbb{Q}$. Thus $a_1 \neq 0$.

Squaring both sides of $\sqrt{5} = a_0 + a_1\sqrt{13}$ we find $5 = a_0^2 + 2a_0a_1\sqrt{13} + 13a_1^2$. If $a_0 \neq 0$ then we would have $\sqrt{13}$ as a root of the polynomial $d(x) = (-5 + a_0^2 + 13a_1^2) + (2a_0a_1)x$ in $\mathbb{Q}[x]$. With $deg(d(x)) = 1$ and Theorem 9.4 we would then know $\sqrt{13} \in \mathbb{Q}$. Since this is not true we can conclude that $a_0 = 0$. However, $\sqrt{5} = a_1\sqrt{13}$ gives us $5 = 13(a_1)^2$. Thus $a_1 \in \mathbb{Q}$ is a root of the polynomial $q(x) = -5 + 13x^2$. But by Eisenstein's Criterion $q(x)$ is irreducible over $\mathbb{Q}$, a contradiction. Hence we can say $\sqrt{5} \notin \mathbb{Q}(\sqrt{13})$.

Example 9.15 Consider the field $\mathbb{Z}_3$ and $p(x) = 2 + 2x + x^2$ in $\mathbb{Z}_3[x]$. Since $p(0) = 2, p(1) = 2$, and $p(2) = 1$ there is no root for $p(x)$ in $\mathbb{Z}_3$. As $deg(p(x)) = 2$ then $p(x)$ is irreducible over $\mathbb{Z}_3$ and monic. By Theorem 9.7 there is an extension of $\mathbb{Z}_3$ which has a root, c, for $p(x)$, the minimum polynomial for c over $\mathbb{Z}_3$. We know from Theorem 9.13 that $\mathbb{Z}_3(c) = \{u(c) : u(x) \in \mathbb{Z}_3[x]$ and either $u(x) = 0(x)$ or $deg(u(x)) < 2\}$. Thus we only need to list the polynomials in $\mathbb{Z}_3[x]$ of degree less than 2, using c instead of x. This gives us exactly nine elements in $\mathbb{Z}_3(c)$, $0, 1, 2, 0 + c, 1 + c, 2 + c, 0 + 2c, 1 + 2c, 2 + 2c$.

Let's look more closely at the operations in this field. Recall that addition is done by coefficients since we add the polynomials then plug in c. Thus adding $(1 + 2c) + (2 + 2c)$ is the same as taking $a(x) = 1 + 2x$ and $b(x) = 2 + 2x$, finding $a(x) + b(x) = 0 + x$ and plugging in c to get $(1 + 2c) + (2 + 2c) = (1 +_3 2) + (2 +_3 2)c = 0 + 1c = c$. Other examples are below. Be sure you see how the answers were found.

$$2 + (2 + c) = 1 + c \qquad (2 + 2c) + (0 + c) = 2 \qquad (1 + c) + (2 + 2c) = 0$$

In order to calculate products in $\mathbb{Z}(c)$ we take advantage of the fact that $0 = 2+2c+c^2$, so $c^2 = 1+c$. Thus to calculate $c(1+c)$ we first use the distributive law to find $c(1+c) = c+c^2$ then using $c^2 = 1+c$ we get $c(1+c) = c+(1+c) = 1+2c$. More examples are shown below. Be sure you can verify them.

$$(2 + c)(1 + c) = c \qquad (2c)(2 + 2c) = 1 + 2c$$

An exercise at the end of the chapter asks you to complete the Cayley tables for the operations in this field.

9.3 Algebraic Extensions

Recall that we are working with algebraic elements over a field. The extension fields we create will all have a special property; they are *algebraic* extensions of the original field K as defined below.

> **Definition 9.16** Let K be a field and E an extension field of K. If every element of E is algebraic over K we say that E is an ***algebraic extension*** of K.

It should be clear that for any field K, K is an algebraic extension of itself. For any $c \in K$ the polynomial $b(x) = -c + x$ is in $K[x]$ and has c as a root. Thus c is algebraic over K. In this section we will show that for c algebraic over a field K, $K(c)$ is an algebraic extension of K. However, some extension fields may not be algebraic extensions; for example, $\mathbb{R}$ is **not** an algebraic extension of $\mathbb{Q}$ since π is not algebraic over $\mathbb{Q}$.

By definition each element of $K(c)$ can be written as $a_0 + a_1c + \cdots + a_nc^n$ where $a_i \in K$. This should remind you of a "linear combination" of the elements $1_E, c, c^2, \ldots, c^n$ with "scalars" from the field K. This parallel to having a *basis* in a vector space is exactly what we need. Similar to Definition 0.28 we define a basis for an extension field. (We only consider finite bases here, as all of our extensions will be finite extensions.)

Definition 9.17 Let K be a field, and E an extension field of K. A nonempty subset of E, $B = \{u_1, u_2, \ldots, u_m\}$, is called a ***basis for E over K*** when the following hold:

(i) For every element $s \in E$ there exist $a_1, a_2, \ldots, a_m \in K$ so that $s = a_1u_1 + a_2u_2 + \cdots + a_mu_m$ (*B spans E over K*).
(ii) If $a_1, a_2, \ldots, a_m \in K$ with $a_1u_1 + a_2u_2 + \ldots + a_mu_m = 0_E$ then $a_i = 0_K$ for all $i = 1, \ldots m$ (*B is independent over K*).

If there exist m elements of E that form a basis for E over K we say E is a ***finite extension*** of K of *degree* m, and write $[E : K] = m$.

Example 9.18 In Example 9.15 we found the elements of the extension field $\mathbb{Z}_3(c)$ where c is a root of the irreducible polynomial $p(x) = 2 + 2x + x^2$. $\mathbb{Z}_3(c) = \{0, 1, 2, c, 1 + c, 2 + c, 2c, 1 + 2c, 2 + 2c\}$. Then $B = \{1, c\}$ appears to be a basis for $\mathbb{Z}_3(c)$ over $\mathbb{Z}_3$. We will soon see that it is true.

We assume some Linear Algebra facts related to bases without proof. For details see [5].

Theorem 9.19

Let K be a field and E an extension field of K.

(i) Every basis for E over K has the same cardinality.
(ii) Every subset of E that spans E contains a basis for E over K.

Theorem 9.20

Suppose K is a field, c is algebraic over K with minimum polynomial $p(x)$, and $deg(p(x)) = n$. Then the set $B = \{1_K, c, c^2, \ldots, c^{n-1}\}$ is a basis for $K(c)$ over K and $[K(c) : K] = deg(p(x))$.

Proof Suppose K is a field, c is algebraic over K with minimum polynomial $p(x)$, and $deg(p(x)) = n$. Consider the set $B = \{1_K, c, c^2, \ldots, c^{n-1}\}$. By Theorem 9.13 for an element $y \in K(c)$, $y \neq 0_K$ there is $a(x) \in K[x]$ with $y = a(c)$ and $deg(a(x)) < n$. So there are $a_0, a_1, \ldots, a_{n-1} \in K$ with $y = a(c) = a_0 + a_1c + \ldots + a_{n-1}c^{n-1}$ and B spans $K(c)$ as needed. We only need to show that B is independent over K to complete the proof.

Suppose we have $a_0, a_1, \ldots, a_{n-1} \in K$ with $a_0 + a_1c + \ldots + a_{n-1}c^{n-1} = 0_E$. Since we want to prove that all of the a_i are equal to 0_K, we suppose instead that at least one of them is nonzero. Now we have a nonzero polynomial in $K[x]$, $a(x) = a_0 + a_1x + \ldots + a_{n-1}x^{n-1}$, with c as a root, so by Theorem 9.12 $a(x) = p(x)q(x)$ for some $q(x) \in K[x]$. But then $deg(a(x)) \geq deg(p(x))$ which is a contradiction since $deg(p(x)) = n$. Hence $a(x) = 0(x)$ or $a_i = 0_K$ for all i, and B is independent over K. Thus B is a basis for $K(c)$ over K and $[K(c) : K] = deg(p(x))$. □

The previous theorem helps us find the *number of elements* in a finite extension of a finite field. Suppose $[K(c) : K] = n$ and $|K| = q$, then every element of $K(c)$ is of the form $a_0 + a_1c + \cdots + a_{n-1}c^{n-1}$. Each of the elements $a_0, a_1, \ldots, a_{n-1} \in K$ can be one of exactly q choices. Notice that no two choices can lead to the same element of $K(c)$. Suppose $u(c) = u_0 + u_1c + \cdots + u_{n-1}c^{n-1}$ and $w(c) = w_0 + w_1c + \cdots + w_{n-1}c^{n-1}$ with $u(c) = w(c)$, but for some $i < n$ we have $u_i \neq w_i$. Then $u(c) - w(c) = 0_K$ and c is a root of the polynomial $q(x) = (u_0 - w_0) + (u_1 - w_1)x + \cdots + (u_{n-1} - w_{n-1})x^{n-1}$ of degree less than n which is impossible. Thus each element of the form $a_0 + a_1c + \cdots + a_{n-1}c^{n-1}$ is different. Thus we have $|K(c)| = q^n$.

Notice that in Example 9.18 we have $[\mathbb{Z}_3(c) : \mathbb{Z}_3] = 2$ and $|\mathbb{Z}_3| = 3$. Thus we should have 3^2 elements in $\mathbb{Z}_3(c)$, and we found exactly nine of them. Similarly with $\mathbb{Z}_5$ and irreducible polynomial $p(x) = 2 + x^2$, $\mathbb{Z}_5(c)$ should have 5^2 elements, exactly as we found in Example 9.8.

The next theorem guarantees us that $K(c)$ is always an algebraic extension of K when c is algebraic over K.

Theorem 9.21

Let K be a field and E an extension of K with $[E : K] = n$ for some $n > 0$. Then E is an *algebraic* extension of K.

Proof Suppose K is a field and E is an extension of K with $[E : K] = n$ and $n > 0$. We must show that every element $y \in E$ is algebraic over K, i.e., the root of a nonzero polynomial over K.

Let $y \in E$, then the set $\{1_K, y, y^2, \ldots, y^n\}$ cannot be independent over K or there would be a basis with more than n elements, which is impossible. Thus there is a linear combination over K, $a_0 1_K + a_1y + \cdots + a_ny^n = 0_E$ where not all a_i are 0_K. Thus y is a root of the nonzero polynomial $a(x) = a_0 + a_1x + \ldots + a_nx^n$ in $K[x]$, and y is algebraic over K. Hence every element of E is algebraic over K. □

We end this section with a property of finite extensions that will be extremely important. The proof is simply a chase of indices.

Theorem 9.22

Suppose that K is a field and L is a finite extension of K. If E is a finite extension of L, then E is a also finite extension of K and $[E : K] = [E : L][L : K]$.

Proof Suppose that K is a field, L is a finite extension of K with $[L : K] = t$, and E is a finite extension of L with $[E : L] = m$. We know by definition that $K \subseteq L \subseteq E$, and since the operations of K agree with those in L which agree with those in E, then E is an extension field of K. Suppose we have a basis $B = \{u_1, u_2, \ldots, u_t\}$ for L over K, and a basis $A = \{w_1, w_2, \ldots, w_m\}$ for E over L.

It is straightforward to see that set $C = \{u_j w_i \in E : i = 1 \ldots m, j = 1 \ldots t\}$ spans E over K. Notice that C contains exactly tm elements, thus if we show C is linearly independent over K then we have C a basis and $[E : K] = tm = [E : L][L : K]$.

To show that C is linearly independent over K, suppose that there is a combination equal to 0_E, $\sum b_{ij} u_j w_i = 0_E$, and each $b_{ij} \in K$ with $1 \leq i \leq m$ and $1 \leq j \leq t$. By rearranging the sum, using distributivity, associativity, and commutativity in E, we have the equation below.

$$0_E = \left(\sum_{j=1}^{t} b_{1j} u_j\right) w_1 + \cdots + \left(\sum_{j=1}^{t} b_{mj} u_j\right) w_m \text{ and for each } i, \sum_{j=1}^{t} b_{ij} u_j \in L$$

But since A is a basis for E over L, this guarantees that for each i, $\sum_{j=1}^{t} b_{ij} u_j = 0_E$. Also for each i, the fact that B is a basis for L over K tells us that $b_{ij} = 0_E$ for all j. Hence For all i and j we know $b_{ij} = 0_E$ so C is linearly independent over K. Now we have a basis C of tm many elements therefore E is a finite extension of K. As $[L : K] = t$ and $[E : L] = m$, then $[E : K] = mt = [E : L][L : K]$ as needed. □

9.4 Root Field of a Polynomial

In Example 9.18 the field $\mathbb{Z}_3(c)$ was created, where c is a root of $p(x) = 2 + 2x + x^2$. Having c as a root tells us that $2c + x$ is a factor, and $p(x) = (2c + x)((2 + c) + x)$ in $\mathbb{Z}_3(c)[x]$. Thus $p(x)$ factors into two polynomials of degree 1, and $\mathbb{Z}_3(c)$ contains both roots, c and $1 + 2c$, for $p(x)$.

Definition 9.23 Let K be a field and $a(x) \in K[x]$ have $deg(a(x)) > 0$. The ***root field for a(x) over K*** is a field extension E of K with the following properties:

(i) In $E[x]$, $a(x)$ can be factored into a product of polynomials of degree 1.
(ii) For any extension of K, L, which satisfies (i), we have $K \subseteq E \subseteq L$.

In Example 9.18 the field $\mathbb{Z}_3(c)$ contains every root for $p(x)$. Any other field that contains c and is an extension of $\mathbb{Z}_3$ must also contain $\mathbb{Z}_3(c)$ by Theorem 9.5. Thus $\mathbb{Z}_3(c)$ is the root field of $p(x)$ over $\mathbb{Z}_3$.

Example 9.24 Consider $p(x) = -5 + x^3 \in \mathbb{Q}[x]$. By Eisenstein's Criterion we know that $p(x)$ is irreducible over $\mathbb{Q}$. Clearly, $\sqrt[3]{5}$ is a root of $p(x)$ so we can factor $p(x)$ over $\mathbb{Q}\left(\sqrt[3]{5}\right)$.

$$
\begin{array}{r@{}l}
 & \sqrt[3]{25} + \sqrt[3]{5}x + x^2 \\
\cline{2-2}
-\sqrt[3]{5} + x \,\big) & -5 + 0x + 0x^2 + x^3 \\
 & \qquad\qquad -\sqrt[3]{5}x^2 + x^3 \\
\cline{2-2}
 & -5 + 0x + \sqrt[3]{5}x^2 \\
 & \quad -\sqrt[3]{25}x + \sqrt[3]{5}x^2 \\
\cline{2-2}
 & -5 + \sqrt[3]{25}x \\
 & -5 + \sqrt[3]{25}x \\
\cline{2-2}
 & 0
\end{array}
$$

Thus $p(x) = \left(-\sqrt[3]{5} + x\right)\left(\sqrt[3]{25} + \sqrt[3]{5}x + x^2\right)$. Using the *quadratic formula* we find the roots of $\sqrt[3]{25} + \sqrt[3]{5}x + x^2$ as $c = \frac{-\sqrt[3]{5}}{2}(1 + \sqrt{-3})$ and $d = \frac{-\sqrt[3]{5}}{2}(1 - \sqrt{-3})$. Hence we need to see if $c, d \in \mathbb{Q}\left(\sqrt[3]{5}\right)$.

Suppose $c \in \mathbb{Q}\left(\sqrt[3]{5}\right)$, then we can conclude $\sqrt{-3} \in \mathbb{Q}\left(\sqrt[3]{5}\right)$ by the equation below. A similar equation would be used if we assumed $d \in \mathbb{Q}\left(\sqrt[3]{5}\right)$.

$$-1 - \frac{2}{\sqrt[3]{5}}c = -1 - \frac{2}{\sqrt[3]{5}}\left(\frac{-\sqrt[3]{5}}{2}(1 + \sqrt{-3})\right) = -1 + (1 + \sqrt{-3}) = \sqrt{-3}$$

Let $E = \mathbb{Q}\left(\sqrt[3]{5}\right)$ and $L = \mathbb{Q}\left(\sqrt{-3}\right)$. Since $\sqrt{-3} \in \mathbb{Q}\left(\sqrt[3]{5}\right)$ we then have $\mathbb{Q} \subseteq L \subseteq E$. Thus by Theorem 9.22 we know $[E : \mathbb{Q}] = [E : L][L : \mathbb{Q}]$. Since $p(x)$ is irreducible we know $[E : \mathbb{Q}] = 3$, but $[L : \mathbb{Q}] = 2$ since $3 + x^2$ is the minimum polynomial for $\sqrt{-3}$ over $\mathbb{Q}$. It is impossible to have $3 = [E : L](2)$, so we know $\sqrt{-3} \notin \mathbb{Q}\left(\sqrt[3]{5}\right)$. Hence $\mathbb{Q}\left(\sqrt[3]{5}\right)$ is **not** the root field for $p(x) = -5 + x^3$ over $\mathbb{Q}$.

A critical piece of notation that will help us find a root field is an iterated extension. So far we have found extension fields by taking a single c algebraic over K and creating $K(c)$. We repeat this process to create an iterated extension as defined below.

Definition 9.25 Let K be a field and c_1, c_2 algebraic over K. Let $L = K(c_1)$, then the field $K(c_1, c_2) = L(c_2)$ is called an ***iterated extension*** of K.

Recall that the basis for $L = K(c_1)$ over K is of the form $\{1_K, c_1, c_1^2, \ldots, c_1^m\}$. Also, the basis for $L(c_2)$ over $K(c_1)$, is of the form $\{1_K, c_2, c_2^2, \ldots, c_2^n\}$. Thus from the proof of Theorem 9.22, the field $K(c_1, c_2)$ has a basis consisting of all elements of the form $(c_1)^{i_1}(c_2)^{i_2}$ where $0 \leq i_1 \leq m$ and $0 \leq i_2 \leq n$.

We can repeat the process to create an extension $K(c_1, c_2, \ldots, c_n)$ as needed, whose basis is the set of all elements of the form $c_1^{i_1} c_2^{i_2} \cdots c_n^{i_n}$ for appropriate i_j. This brings us to the next theorem.

Theorem 9.26

Let K be a field and $a(x) \in K[x]$ with $deg(a(x)) > 0$. If E is the root field of $a(x)$ over K, and the elements $c_1, c_2, \ldots, c_n \in E$ are all of the distinct roots of $a(x)$, then $E = K(c_1, c_2, \ldots, c_n)$.

Proof Let K be a field and $a(x) \in K[x]$ with $deg(a(x)) > 0$. Assume E is the root field of $a(x)$ over K, and the elements $c_1, c_2, \ldots, c_n \in E$ are all of the distinct roots of $a(x)$. Since E is the root field of $a(x)$, by Definition 9.23 we must have $E \subseteq K(c_1, c_2, \ldots, c_n)$.

Now suppose we have $u \in K(c_1, c_2, \ldots, c_n)$. We know u is a finite sum of the form $\sum a_j(c_1^{i_1} c_2^{i_2} \cdots c_n^{i_n})$ with $a_j \in K$. Since $c_j \in E$ for each $1 \leq j \leq n$, then each of $c_j^{i_j} \in E$ as well. Hence by E a field, each $c_1^{i_1} c_2^{i_2} \cdots c_n^{i_n} \in E$, and so $a_j(c_1^{i_1} c_2^{i_2} \cdots c_n^{i_n}) \in E$. Finally, as E is closed under addition we have $u \in E$ as needed, so $K(c_1, c_2, \ldots, c_n) \subseteq E$. Therefore $E = K(c_1, c_2, \ldots, c_n)$. □

Example 9.27 To create the root field for $p(x) = -5 + x^3$, we use the three roots found in Example 9.24, so by Definition 9.23 the root field is $\mathbb{Q}\left(\sqrt[3]{5},\ c,\ d\right)$. Since we showed $\sqrt{-3} \in \mathbb{Q}(\sqrt[3]{5},\ c)$ then easily $d = c + \sqrt[3]{5}(\sqrt{-3})$ so $d \in \mathbb{Q}\left(\sqrt[3]{5},\ c\right)$. Thus the root field for $p(x)$ is $E = \mathbb{Q}\left(\sqrt[3]{5},\ c\right)$ with basis $B = \{1, \sqrt[3]{5}, (\sqrt[3]{5})^2, c, \sqrt[3]{5}c, (\sqrt[3]{5})^2 c\}$ and $[E : \mathbb{Q}] = 6$.

The notation here can be cumbersome, since c is complicated, so we often try to find a simpler way to denote a field such as $E = \mathbb{Q}\left(\sqrt[3]{5},\, c\right)$. The following steps show that $E = \mathbb{Q}\left(\sqrt[3]{5}, \sqrt{-3}\right)$ as well, so the basis elements can be written as $\{1, \sqrt[3]{5}, (\sqrt[3]{5})^2, \sqrt{-3}, \sqrt[3]{5}\sqrt{-3}, (\sqrt[3]{5})^2\sqrt{-3}\}$.

($\subseteq$) Clearly, $\mathbb{Q} \subseteq \mathbb{Q}\left(\sqrt[3]{5}, \sqrt{-3}\right)$, and also $\sqrt[3]{5} \in \mathbb{Q}\left(\sqrt[3]{5}, \sqrt{-3}\right)$. But notice that $c = \frac{-\sqrt[3]{5}}{2}(1+\sqrt{-3}) \in \mathbb{Q}\left(\sqrt[3]{5}, \sqrt{-3}\right)$, and similarly $d = \frac{-\sqrt[3]{5}}{2}(1-\sqrt{-3}) \in \mathbb{Q}\left(\sqrt[3]{5}, \sqrt{-3}\right)$. Thus by definition of the root field, $E \subseteq \mathbb{Q}\left(\sqrt[3]{5}, \sqrt{-3}\right)$.

($\supseteq$)We showed in Example 9.24 that $\sqrt{-3} \in E$. Thus from $\sqrt[3]{5} \in E$ and E closed under products we find every basis element of $\mathbb{Q}\left(\sqrt[3]{5}, \sqrt{-3}\right)$ in E. But then an arbitrary element of $\mathbb{Q}\left(\sqrt[3]{5}, \sqrt{-3}\right)$ is a combination of these basis elements, so is also in E. Hence $\mathbb{Q}\left(\sqrt[3]{5}, \sqrt{-3}\right) \subseteq E$.

Notice that in the previous example we found the root field as $\mathbb{Q}\left(\sqrt[3]{5}, \sqrt{-3}\right)$, not just $\mathbb{Q}(s)$ for some $s \in E$. Could we have found such an s?

Definition 9.28 Suppose that E is an extension field of K with $c \in E$. A field extension $K(c)$ is called a ***simple extension*** of K.

We will show that every finite extension of a field is a simple extension, but the proof requires irreducible polynomials to have no multiple roots, i.e., to be separable.

Definition 9.29 Let K be a field and $p(x) \in K[x]$. We say $p(x)$ is ***separable*** if no irreducible factor of $p(x)$ has multiple roots in any extension field of K. Otherwise, we say $p(x)$ is ***inseparable***.

Example 9.30 Consider $p(x) = 2 + 2x + x^3 \in \mathbb{Z}_3[x]$. You can verify that $p(x)$ is irreducible over $\mathbb{Z}_3$ since it has no roots in $\mathbb{Z}_3$. Once we create an extension field with a root c for $p(x)$, we have a field $\mathbb{Z}_3(c)$ with exactly $3^3 = 27$ elements in it. The basis for this field is $\{1, c, c^2\}$. Now we can factor $p(x) = (-c + x)[(2 + c^2) + cx + x^2]$, and to see if $p(x)$ has multiple roots in $\mathbb{Z}_3(c)$ we need to see if c is again a root of $d(x) = (2+c^2)+cx+x^2$. The next calculations show us that $d(c) \neq 0$.

$$d(c) = (2 + c^2) + (c)c + c^2 = 2 + c^2 + c^2 + c^2 = 2$$

The final question we still need to ask is if we can write $d(x) = (-u + x)^2$ for u in an extension of $\mathbb{Z}(c)$. Suppose this is true, then $d(x) = (-u+x)^2 = u^2 - 2ux + x^2$. This requires that $2 + c^2 = u^2$ and $-2u = c$, or $u = c$ which is impossible since c is not a root of $d(x)$.

Thus $p(x)$ has no multiple roots in an extension of $\mathbb{Z}_3$, and is separable.

It will be critical later to know whether a polynomial is separable, so the next two theorems will help answer this for all of the fields we will consider.

Theorem 9.31

Let K be a field with $char(K) = 0$. Then every irreducible polynomial in $K[x]$ is separable.

Proof Let K be a field with $char(K) = 0$, and suppose that $p(x) \in K[x]$ is irreducible over K. If $p(x)$ is not monic we can choose a monic associate that is also irreducible with the same roots, so we will assume $p(x)$ is monic. In order to show $p(x)$ is separable suppose instead that $p(x)$ is inseparable.

Thus, there is an extension E of K which has a root of $p(x)$ that is a multiple root, say $c \in E$, so in $E[x]$ we can factor $p(x) = (-c + x)^2 q(x)$. In Project 7.4, we defined the polynomial $p'(x)$ in $K[x]$. It is straightforward to show that using $r(x) = (-c+x)^2$, we have $p'(x) = r(x)q'(x) + r'(x)q(x)$. But $r'(x) = 2 \bullet (-c + x)$ and so $p'(x) = (-c + x)^2 q'(x) + 2 \bullet (-c + x)q(x)$. As $-c + x$ is a factor of $p'(x)$ then c is a root of both $p(x)$ and $p'(x)$.

Since $p(x)$ is irreducible over K and monic it is the minimum polynomial for c over K. Also $p'(x) \in K[x]$ has c as a root, thus either $p'(x) = 0(x)$ or $deg(p'(x)) \geq deg(p(x))$. As $deg(p'(x)) < deg(p(x))$ by definition of $p'(x)$, then $p'(x) = 0(x)$. Now $deg(p(x)) = n$ for some $n > 0$, and $p_n = 1_K$. Thus the term $(n \bullet 1)x^{n-1} \neq 0_K$, a contradiction. Hence $p(x)$ is separable. $\square$

The previous theorem applies to fields such as $\mathbb{Q}$ and $\mathbb{R}$, but not to $\mathbb{Z}_5$. The same fact is true over a finite field as we will see next, but the proof relies on a few interesting properties of finite fields. You will be guided through the proof of the following theorem in Project 9.2.

Theorem 9.32

Let K be a finite field with $char(K) = q$ for some prime q.

(i) For any polynomial $b(x) = b_0 + b_1 x + \cdots + b_t x^t \in K[x]$ we have $(b(x))^q = b_0^q + b_1^q x^q + \cdots + b_t^q x^{qt}$.

(ii) For any element $s \in K$ there is $r \in K$ with $s = r^q$.

Theorem 9.33

Let K be a finite field. Then every irreducible polynomial in $K[x]$ is separable.

Proof Suppose we have a finite field K and $p(x) \in K[x]$ is irreducible over K. Since K is a finite field, we know from Theorems 6.12 and 6.9 that $char(K)$ must be a prime number q. As in Theorem 9.31 assume that $p(x)$ is inseparable.

Using the same steps as the proof of Theorem 9.31, we can find $p'(x)$ which must again be equal to $0(x)$. However, over the finite field K this is not an immediate contradiction, since $q \bullet 1_K = 0_K$. From any term of the form $p_{mq}x^{mq}$ we get the following equation:

$$mq \bullet p_{mq}x^{mq-1} = (mq \bullet 1_K)p_{mq}x^{mq-1} = 0_K x^{mq-1}.$$

But if q does not divide j, then $j \bullet p_j \neq 0_K$ unless $p_j = 0_K$ already. Thus to have $p'(x) = 0(x)$ we must have $p(x) = p_0 + p_q x^q + p_{2q}x^{2q} + \cdots + p_{mq}x^{mq}$ for some $m \geq 1$. By Theorem 9.32 we can find $a_i \in K$ with $(p_{iq}) = (a_i)^q$.

$$\text{Define } a(x) = a_0 + a_1 x + a_2 x^2 + \cdots + a_m x^m$$

Then $a(x) \in K[x]$, and as $p(x)$ is nonconstant we have $deg(a(x)) = m > 0$. But by Theorem 9.32 we know $p(x) = (a(x))^q$, contradicting that $p(x)$ is irreducible over K. This contradiction tells us that $p(x)$ is separable.

□

Unfortunately, irreducible polynomials need not be separable over an infinite field of nonzero characteristic (see [9] for an example). All extension fields we discuss in the rest of the text will extend fields that either have characteristic 0 or are finite fields and thus all of our irreducible polynomials will be separable.

The final theorem for this section shows us every finite extension of a field of characteristic 0 is a simple extension. This also true for finite extensions of finite fields, but the proof is significantly more complicated and will not be presented here. (See [9] for details.)

Theorem 9.34

Let K be a field of characteristic 0, and E a finite extension of K. Then E is a simple extension of K, meaning there is some $c \in E$ with $E = K(c)$.

Proof Let K be a field of characteristic 0 and E a finite extension of K. In an exercise at the end of the chapter you will show that $E = K(u_1, u_2, \ldots, u_m)$ for some $u_1, u_2, \ldots, u_m \in E$. Once we show that $K(u_1, u_2)$ can be written as $K(c_1)$ for some $c_1 \in K(u_1, u_2)$, this process is repeated at most $n-1$ times to get $K(c_1, u_2) = K(c_2)$, $K(c_2, u_3) = K(c_3)$, ..., and finally, $E = K(c_{m-1}, u_m) = K(c_m)$.

Consider the finite extension $L = K(u_1, u_2)$. By Theorem 9.21 both u_1, u_2 are algebraic over K. Thus for u_1 there is a minimum polynomial $s(x) \in K[x]$ and a minimum polynomial $r(x) \in K[x]$ for u_2. Each polynomial has all of its roots in some extension of K. Denote the roots of $s(x)$ by $s_1, s_2, \ldots, s_t$, the roots of $r(x)$ by $r_1, r_2 \ldots, r_n$ and assume $u_1 = s_1$, $r_1 = u_2$. Since both of $s(x)$ and $r(x)$ are irreducible over K, they are separable by Theorem 9.31. Thus $s_i \neq s_j$ and $r_i \neq r_j$ for any $i \neq j$, and so $u_2 - r_j \neq 0_K$ when $j \neq 1$.

Since K is an infinite field by Theorem 6.9, there must exist an element $b \in K$ so that for any $i \neq 1$ and $j \neq 1$, $b \neq (u_2 - r_j)^{-1}(s_i - u_1)$. Define $c_1 = u_1 + bu_2$, then $c_1 \in L$. Now we must show $L = K(c_1)$. Since $c_1 \in L$ then $K(c_1) \subseteq L$ already. If $u_2 \in K(c_1)$, then as $u_1 = c_1 - bu_2 \in K(c_1)$ and then $u_1 \in K(c_1)$. This would tell us that $L \subseteq K(c_1)$ and thus $K(c_1) = L$ as needed. Thus we only need to show $u_2 \in K(c_1)$ to complete the proof.

Assume for a contradiction that $u_2 \notin K(c_1)$. Let $h(x) \in K(c_1)[x]$ where $h(x) = s(c_1 - bx)$. Then $h(u_2) = s(u_1) = 0_K$, but for $j \neq 1$, $h(r_j) = s(c_1 - br_j) \neq 0_K$. Now, $j \neq 1$, each r_j is not a root of $h(x)$ so only u_2 is a root of both $r(x)$ and $h(x)$. The minimum polynomial $p(x)$ for u_2 over $K(c_1)$ must have $deg(p(x)) > 1$ and so by $p(x)$ separable it has another root w. But $p(x)$ is a factor of both $r(x)$ and $h(x)$ by Theorem 9.12, making w also a root of both $h(x)$ and $r(x)$, a contradiction. Hence $u_2 \in K(c_1)$ and therefore as discussed above $L = K(c_1)$ completing, the proof that E is a simple extension of K. $\square$

Example 9.35 Consider the field $\mathbb{Q}$ and the extension $\mathbb{Q}\left(\sqrt[3]{5}, \sqrt{-3}\right)$ seen in Example 9.27. How do we make this a simple extension? Using the same procedure described above, look at the minimum polynomials $s(x) = -5 + x^3$ and $r(x) = 3 + x^2$ and each of their roots. Then $b = 1$ satisfies the requirement $b \neq (u_2 - r_j)^{-1}(s_i - u_1)$ from the previous theorem, so we have $\mathbb{Q}\left(\sqrt[3]{5}, \sqrt{-3}\right) = \mathbb{Q}\left(\sqrt[3]{5} + \sqrt{-3}\right)$.

In Chapter 10, we will use all that we have learned about field extensions to discuss *isomorphisms* between field extensions, leading us to Galois Theory.

Exercises for Chapter 9

Section 9.1 Extension Fields

1. Prove (i) of Theorem 9.4.
2. Prove (ii) of Theorem 9.4.
3. Let $p \in \mathbb{Z}$ be prime. Prove $\sqrt{p} \notin \mathbb{Q}$.
4. Let $p \in \mathbb{Z}$ be prime. Prove $\sqrt[3]{p} \notin \mathbb{Q}$.
5. If $p \in \mathbb{Z}$ is prime, can we also say that for any integer $n > 1$ that $\sqrt[n]{p} \notin \mathbb{Q}$? Prove it or find a counterexample.

6. Let $m, n \in \mathbb{Z}$ where $m > 0$ and $n > 0$. Prove: If there is a prime p which divides n but does not divides m and p^2 does not divide n, then $\sqrt{\frac{n}{m}} \notin \mathbb{Q}$.
7. Suppose K is a field and E is an extension of K. Define $g : K \to E$ by $g(a) = a$ for each $a \in K$. Prove that g is a one to one homomorphism.
8. Suppose K is a field and E is an extension of K. Define $g : K \to E$ by $g(a) = a$ for each $a \in K$. Use Theorems 7.35 and 7.28 to prove that the function f_c defined in Theorem 9.5 is a homomorphism.
9. In the proof of Theorem 9.5, explain why $deg(b(x)) > 0$ for $ker(f_c) = T = \langle b(x) \rangle$.
10. In the proof of Theorem 9.5, explain why $c \in f_c(K[x])$. What polynomial will help?
11. Prove that the field $K(c)$, defined in the proof of Theorem 9.5, satisfies (iii) of the theorem. Don't forget that $K(c) = f_c(K[x])$. Use it to help show an arbitrary $u \in K(c)$ is also in S.
12. In the proof of Theorem 9.7 show that the function $f : K \to {}^{K[x]}/_{T}$ defined by $f(w) = w + T$ for each $w \in K$ is a one to one homomorphism. Don't forget that T cannot contain a nonzero constant polynomial since $deg(a(x)) > 0$.

Section 9.2 Minimum Polynomial

13. In the proof of Theorem 9.9 the polynomial $q(x) \in K[x]$ was defined as $ker(f_c) = \langle q(x) \rangle$. Prove that $q(x)$ is irreducible over K.
14. Find the minimum polynomial for $u = \sqrt[3]{10}$ over $\mathbb{Q}$. Be sure to prove your polynomial is irreducible.
15. Find the minimum polynomial for $u = 7 + \sqrt{2}$ over $\mathbb{Q}$. Be sure to prove your polynomial is irreducible.
16. Find the minimum polynomial for $6 + 2i$ over $\mathbb{Q}$. Be sure to prove your polynomial is irreducible.
17. Find the minimum polynomial for $2 - \sqrt[3]{7}$ over $\mathbb{Q}$. Be sure to prove your polynomial is irreducible.
18. Find the minimum polynomial for $\sqrt{2} + \sqrt{7}$ over $\mathbb{Q}$. Be sure to prove your polynomial is irreducible.
19. Find the minimum polynomial for $\sqrt[4]{2}i$ over $\mathbb{Q}$. Be sure to prove your polynomial is irreducible.
20. Suppose E is an extension field of K and $p(x), q(x) \in K[x]$ are nonzero polynomials. Prove: If $p(x) = uq(x)$ for some $u \in K$ then $\langle p(x) \rangle = \langle q(x) \rangle$.
21. Prove (i) of Theorem 9.12.
22. Explain how (ii) follows from (i) in Theorem 9.12.
23. Show that it was necessary to have the minimum polynomial for c, $p(x)$, in the statement of (ii) in Theorem 9.12 by finding two polynomials $a(x), b(x) \in \mathbb{Q}[x]$ which both have 3 as a root but neither polynomial divides the other.
24. Show that $\sqrt{-7}$ is a root of $a(x) = -49 + x^4$, then show that the minimum polynomial for $\sqrt{-7}$ over $\mathbb{Q}$ is a factor of $a(x)$.

25. Show that $-\frac{1}{2}+\frac{\sqrt{-11}}{2}$ is a root of $a(x)=3+x+4x^2+x^3+x^4$, then show that the minimum polynomial for $-\frac{1}{2}+\frac{\sqrt{-11}}{2}$ over $\mathbb{Q}$ is a factor of $a(x)$.
26. Show that $1+i$ is a root of $a(x)=4-4x+2x^2+2x^3-2x^4+x^5$, then show that the minimum polynomial for $1+i$ over $\mathbb{Q}$ is a factor of $a(x)$.
27. Show that $2+\sqrt[3]{2}$ is a root of $a(x)=-10+2x+6x^2-5x^3+x^4$, then show that the minimum polynomial for $\sqrt{-7}$ over $\mathbb{Q}$ is a factor of $a(x)$.
28. Consider the polynomial $a(x)=3+6x+3x^2+6x^3+x^4$ in $\mathbb{Q}[x]$. Prove $a(x)$ is irreducible over $\mathbb{Q}$, then if u is a root of $a(x)$ in an extension field of $\mathbb{Q}$, describe carefully the elements of $\mathbb{Q}(u)$.
29. Consider the polynomial $a(x)=\frac{1}{2}+3x-\frac{3}{2}x^2+\frac{1}{2}x^3$ in $\mathbb{Q}[x]$. Prove $a(x)$ is irreducible over $\mathbb{Q}$, then if u is a root of $a(x)$ in an extension field of $\mathbb{Q}$, describe carefully the elements of $\mathbb{Q}(u)$.
30. Consider the polynomial $a(x)=\frac{2}{3}-\frac{4}{3}x+\frac{5}{3}x^2$ in $\mathbb{Q}[x]$. Prove $a(x)$ is irreducible over $\mathbb{Q}$, then if u is a root of $a(x)$ in an extension field of $\mathbb{Q}$, describe carefully the elements of $\mathbb{Q}(u)$.
31. Consider the polynomial $a(x)=5+3x+4x^2+6x^3+x^4$ in $\mathbb{Q}[x]$. Prove $a(x)$ is irreducible over $\mathbb{Q}$ (try Theorem 8.37 to help), then if u is a root of $a(x)$ in an extension field of $\mathbb{Q}$, describe carefully the elements of $\mathbb{Q}(u)$.
32. Prove that $p(x)=2+2x+2x^2+x^3$ is irreducible over $\mathbb{Z}_3$, and for a root c of $p(x)$ list the elements of the field $\mathbb{Z}_3(c)$.
33. Prove that $p(x)=2+4x+x^2$ is irreducible over $\mathbb{Z}_5$, and for a root c of $p(x)$ list the elements of the field $\mathbb{Z}_5(c)$.
34. Prove that $p(x)=2+0x+x^2$ is irreducible over $\mathbb{Z}_7$, and for a root c of $p(x)$ list the elements of the field $\mathbb{Z}_7(c)$.
35. Prove that $p(x)=2+x+x^2$ is irreducible over $\mathbb{Z}_5$, and for a root c of $p(x)$ list the elements of the field $\mathbb{Z}_5(c)$.
36. Prove that $p(x)=1+0x+x^2+x^3$ is irreducible over $\mathbb{Z}_2$, and for a root c of $p(x)$ list the elements of the field $\mathbb{Z}_2(c)$.
37. Prove that $p(x)=1+x+x^2+x^3+x^4$ is irreducible over $\mathbb{Z}_2$, and for a root c of $p(x)$ list the elements of the field $\mathbb{Z}_2(c)$.
38. Let c be a root $p(x)=4+3x+x^2$ in an extension of $\mathbb{Z}_5$. Show that $p(x)$ is irreducible over $\mathbb{Z}_5$, and calculate the following elements in $\mathbb{Z}_5(c)$.
    $(3+2c)+(3+4c)=$
    $(2+3c)(4+2c)=$
    $(2+4c)^{-1}=$
39. Let c be a root $p(x)=6+6x+x^2$ in an extension of $\mathbb{Z}_7$. Show that $p(x)$ is irreducible over $\mathbb{Z}_7$, and calculate the following elements in $\mathbb{Z}_7(c)$.
    $(4+5c)+(6+3c)=$
    $(2+6c)(4+2c)=$
    $(6+4c)^{-1}=$

40. Let c be a root $p(x) = 10 + 9x + x^2$ in an extension of $\mathbb{Z}_{11}$. Show that $p(x)$ is irreducible over $\mathbb{Z}_{11}$, and calculate the following elements in $\mathbb{Z}_{11}(c)$.
    $(6 + 10c) + (9 + 8c) =$
    $(7 + 8c)(6 + 3c) =$
    $(5 + 9c)^{-1} =$
41. Let c be a root $p(x) = 8 + 11x + x^2$ in an extension of $\mathbb{Z}_{13}$. Show that $p(x)$ is irreducible over $\mathbb{Z}_{13}$, and calculate the following elements in $\mathbb{Z}_{13}(c)$.
    $(10 + 12c) + (11 + 8c) =$
    $(9 + 2c)(5 + 4c) =$
    $(6 + 8c)^{-1} =$
42. Let c be a root $p(x) = 2 + 3x + 0x^2 + x^3$ in an extension of $\mathbb{Z}_5$. Show that $p(x)$ is irreducible over $\mathbb{Z}_5$, and calculate the following elements in $\mathbb{Z}_5(c)$.
    $(2 + 3c + c^2) + (3 + 4c + 3c^2) =$
    $(2 + 4c)(1 + 3c + c^2) =$
    $(4 + 4c + c^2)^{-1} =$
43. Let c be a root $p(x) = 5 + 3x + 2x^2 + x^3$ in an extension of $\mathbb{Z}_7$. Show that $p(x)$ is irreducible over $\mathbb{Z}_7$, and calculate the following elements in $\mathbb{Z}_7(c)$.
    $(4 + 6c + 3c^2) + (4 + 4c + 5c^2) =$
    $(2 + 6c)(4 + 2c + c^2) =$
    $(6 + 4c + c^2)^{-1} =$
44. Let c be a root $p(x) = 6 + 3x + 5x^2 + x^3$ in an extension of $\mathbb{Z}_{11}$. Show that $p(x)$ is irreducible over $\mathbb{Z}_{11}$, and calculate the following elements in $\mathbb{Z}_{11}(c)$.
    $(10 + 9c + 4c^2) + (5 + 4c + 6c^2) =$
    $(5 + c)(1 + c + c^2) =$
    $(1 + 3c + c^2)^{-1} =$
45. We know $p(x) = 1 + x + x^2 + x^3 + x^4$ is irreducible over $\mathbb{Z}_2$ by a previous exercise. Let c be a root of $p(x)$ in an extension field of $\mathbb{Z}_2$ and calculate the following elements in $\mathbb{Z}_2(c)$.
    $(1 + c + 0c^2 + c^3) + (1 + 0c + c^2 + c^3) =$
    $(1 + c + c^2)(1 + 0c + 0c^2 + c^3) =$
    $(1 + 0c + c^2 + c^3)^{-1} =$
46. Find the complete addition and multiplication tables for the field $\mathbb{Z}_2(c)$ where c is a root of the polynomial $p(x) = 1 + x + x^2$ which is irreducible over $\mathbb{Z}_2$.
47. Using the result of the previous exercise, find the other root of $p(x)$ which is in $\mathbb{Z}_2(c)$.
48. Find the complete addition and multiplication tables for the field $\mathbb{Z}_3(c)$ where c is a root of the polynomial $p(x) = 2 + 2x + x^2$ which is irreducible over $\mathbb{Z}_3$. (See Example 9.15.)
49. Using the result of the previous exercise, find the other root of $p(x)$ which is in $\mathbb{Z}_3(c)$.
50. Find the complete addition and multiplication tables for the field $\mathbb{Z}_2(c)$ where c is a root of the polynomial $p(x) = 1 + x + 0x^2 + x^3$ which is irreducible over $\mathbb{Z}_2$.
51. Using the result of the previous exercise, determine if there are other roots of $p(x)$ in

$\mathbb{Z}_2(c)$.

52. Find the complete addition and multiplication tables for the field $\mathbb{Z}_3(c)$ where c is a root of the polynomial $p(x) = 2 + x + x^2$ which is irreducible over $\mathbb{Z}_3$.
53. Using the result of the previous exercise, find the other root of $p(x)$ which is in $\mathbb{Z}_3(c)$.

Sections 9.3 and 9.4 Algebraic Extensions and Roots

54. Suppose K is a field and E is a finite extension of K. If L is an extension field of K and $K \subseteq L \subseteq E$, use Theorem 9.19 to explain why E is a finite extension of L.
55. Suppose K is a field and c is algebraic over K. Prove $[K(c) : K] = 1$ if and only if $c \in K$.
56. Suppose K is a field and E is an extension field of K with $[E : K] = 3$. If $c \in E$ then the minimum polynomial for c over K must have degree 1 or 3.
57. Suppose K is a field, E is an extension field of K, and n is a positive integer. Prove: If $[E : K] = n$ and $c \in E$ then the degree of the minimum polynomial for c over K must divide n.
58. Suppose K is a field and E is an extension field of K with $[E : K] = n$ for some positive integer n. Prove: If $n = 2^k$ for some positive integer k and $c \in E$ then the minimum polynomial for c cannot be $1 + x + x^2 + x^3$.
59. Prove that $\sqrt[5]{2} \notin \mathbb{Q}(\sqrt[3]{2})$ using Theorem 9.22.
60. Prove that $2i \notin \mathbb{Q}(\sqrt[3]{2})$ using Theorem 9.22.
61. Prove that $\sqrt[3]{2} \notin \mathbb{Q}(\sqrt[4]{17})$ using Theorem 9.22.
62. Prove that $\sqrt[5]{7} \notin \mathbb{Q}(\sqrt[3]{7})$ using Theorem 9.22.
63. Prove that $\sqrt{2+i} \notin \mathbb{Q}(\sqrt[3]{2})$ using Theorem 9.22.
64. Suppose that K is a field and c is algebraic over K with $[K(c) : K] = 2$. Prove: $K(c)$ is the root field for the minimum polynomial of c over K.
65. Suppose K is a field and E is a finite extension of K. Prove: There exist $u_1, u_2, \ldots, u_m \in E$ with $E = K(u_1, u_2, \ldots, u_m)$.
66. Find the root field E for $a(x) = \frac{5}{4} + 6x + \frac{11}{2}x^2 + 2x^3 + \frac{1}{4}x^4$ over $\mathbb{Q}$.
67. Prove that $\sqrt{2} \in \mathbb{Q}(\sqrt{2} + \sqrt{3})$ and use it to show $[\mathbb{Q}(\sqrt{2} + \sqrt{3}) : \mathbb{Q}] = 4$.

Projects for Chapter 9

Project 9.1

Consider the polynomial $a(x) = -1 + x^n$ in $\mathbb{Q}[x]$ for some integer $n > 0$. Any root c of $a(x)$ must satisfy $c^n = 1$, and thus is called an nth root of unity. We know that 1 is always an nth root of unity, but are there others?

For each positive integer n define the following complex number:

$$\omega_n = \cos\left(\tfrac{2\pi}{n}\right) + i\sin\left(\tfrac{2\pi}{n}\right).$$

Clearly, $\omega_1 = \cos(2\pi) + i\sin(2\pi) = 1 + i(0) = 1$.

1. Using your amazing store of trigonometric knowledge, calculate $\omega_2, \omega_3, \omega_4$. Don't leave "cos" or "sin" in them, and don't write decimal approximations. (Does a right triangle with side lengths 1, 2, $\sqrt{3}$ remind you of anything?)
2. Calculate the complex numbers $(\omega_3)^2$, $(\omega_4)^2$, $(\omega_4)^3$. Remember that for two complex numbers $a+bi$ and $c+di$ we find $(a+bi)(c+di) = (ac-bd)+(ad+bc)i$.
3. Prove that $(\omega_2)^2 = 1$, $(\omega_3)^3 = 1$, and $(\omega_4)^4 = 1$. Hence ω_2, ω_3, and ω_4 are second, third, and fourth roots of unity respectively.

   Trigonometric identities show that for any $0 \le k \le n$ we have $(\omega_n)^k = \cos\left(\frac{2k\pi}{n}\right) + i\sin\left(\frac{2k\pi}{n}\right)$, so you may assume this fact for the rest of the project.

4. Use power rules to explain why for any n and $1 \le k \le n$, we know $(\omega_n)^k$ is also an nth root of unity.

   In Project 8.6, we saw that for each n, $-1+x^n = (-1+x)p(x)$ where $p(x) = 1 + x + x^2 + \cdots + x^{n-1}$ is irreducible when n is prime.

5. Find $[\mathbb{Q}(\omega_2) : \mathbb{Q}]$, $[\mathbb{Q}(\omega_3) : \mathbb{Q}]$, and $[\mathbb{Q}(\omega_4) : \mathbb{Q}]$.

Project 9.2

This project assists with the proof of Theorem 9.32. Let K be a finite field with $char(K) = q$ for some prime q. We will show the following facts:

(i) For any polynomial $b(x) = b_0 + b_1x + \cdots + b_tx^t \in K[x]$ we have $(b(x))^q = b_0^q + b_1^q x^q + \cdots + b_t^q x^{qt}$.
(ii) For any element $s \in K$ there is $r \in K$ with $s = r^q$.

Assume we have $b(x) \in K[x]$.

1. Show that if $b(x) = 0(x)$ or $b(x)$ is a constant polynomial, then (i) holds for $b(x)$.

   Now assume $deg(b(x) > 0$. We will prove (i) by induction on the degree of $b(x)$. Let $P(n)$ be the statement: If $deg(b(x)) = n$ then $(b(x))^q = b_0^q + b_1^q x^q + \cdots + b_n^q x^{qn}$.

2. Use Theorem 4.30 to show that $P(1)$ holds. (Look back at Exercise 37 of Chapter 6).

   Now assume that $P(k)$ is true for some k. Thus if $deg(b(x)) = k$ we know $(b(x))^q = b_0^q + b_1^q x^q + \cdots + b_k^q x^{qk}$.

3. Consider a polynomial of degree $k+1$, $b(x) = b_0 + b_1x + \cdots + b_{k+1}x^{k+1}$. Use $u = b_0 + b_1x + \cdots + b_kx^k$, $w = b_{k+1}x^{k+1}$, Theorem 4.30, and $P(k)$ to show that $P(k+1)$ is true.

4. What can we conclude from the previous questions?
5. Use Theorem 6.17 to prove (ii), from the list of facts at the beginning of the project.

Project 9.3

In this project we will completely describe an extension field of $\mathbb{Z}_3$, including its Cayley tables.

1. Show that $p(x) = 1 + 0x + x^2$ is irreducible over $\mathbb{Z}_3$.
2. List the nine elements of $\mathbb{Z}_3(c)$ where c is a root of $p(x) = 1 + 0x + x^2$.

   To create the addition table for our new field we need to remember that for the polynomials in $a(x), b(x) \in \mathbb{Z}_3[x]$, with $d(x) = a(x) + b(x)$ we have $d(c) = a(c) + b(c)$.

3. Compute the addition table for $\mathbb{Z}_3(c)$, and determine $char(\mathbb{Z}_3(c))$. Notice that the characteristic is *not* the same as the size of the field.

   Multiplication is a bit trickier in $\mathbb{Z}_3(c)$ since if we multiply $2c \cdot c$ we get $2c^2$ which is not officially one of the elements of our field. However, remember that c is a root of $p(x)$, so $1 + c^2 = 0$. Subtracting in $\mathbb{Z}_3(c)$ gives us $c^2 = 2$, so we now have the correct answer for our multiplication, $2c \cdot c = 2c^2 = 2(2) = 1$.

4. Compute the multiplication table for our field $\mathbb{Z}_3(c)$.

Project 9.4

Consider the complex numbers $u = \sqrt{i+2}$ and $w = \sqrt{-i+2}$. We will use $\mathbb{Q}(i)$ to help us understand $\mathbb{Q}(u, w)$.

1. Find the minimum polynomial for i over $\mathbb{Q}$ and $[\mathbb{Q}(i) : \mathbb{Q}]$.
2. Carefully describe the elements of $\mathbb{Q}(i)$.
3. Explain why if $u \in \mathbb{Q}(i)$ then there exists $b \in \mathbb{Q}$ which is a root of $a(x) = -1 + 8x^2 + 4x^4$. Show that there cannot be a rational root of $a(x)$ and thus $u \notin \mathbb{Q}(i)$. You may assume similarly that $w \notin \mathbb{Q}(i)$.
4. Show that $\mathbb{Q}(u)$ contains i, then find the minimum polynomial for u over $\mathbb{Q}(i)$ and $[\mathbb{Q}(u) : \mathbb{Q}]$. (Be sure to explain why your polynomial is irreducible.)
5. Show that w is not in $\mathbb{Q}(u)$ by showing it cannot be written as $a + bu$ for $a, b \in \mathbb{Q}(i)$.
6. Find the minimum polynomial for w over $\mathbb{Q}(u)$ and $[\mathbb{Q}(u, w) : \mathbb{Q}(i)]$.
7. Explain why $\mathbb{Q}(u, w)$ is the root field for $(-(i + 2) + x^2)((-(-i + 2) + x^2)$ over $\mathbb{Q}(i)$, and find a basis for $\mathbb{Q}(u, w)$ over $\mathbb{Q}(i)$.

Project 9.5

Suppose K is a field. Recall that if $p(x) \in K[x]$ is irreducible with degree n and c is a root of $p(x)$ in some extension of K, then ${}^{K[x]}/_{\langle p(x) \rangle} \cong K(c)$, $[K(c) : K] = n$, and $K(c) = \{a(c) : a(x) \in K[x] \text{ and } deg(a(x)) < n \text{ or } a(x) = 0(x)\}$.

1. Explain why $|K| = q$ guarantees us $|K(c)| = q^n$.
2. Find a field with exactly eight elements. Be sure to verify any necessary claims, including irreducibility of your polynomial.
3. Suppose p is prime. Any reducible monic quadratic in $\mathbb{Z}_p[x]$ can be factored as $(a + x)(b+x)$ where $a, b \in \mathbb{Z}_p$. Being careful not to count the same polynomial twice because of the order of the factors, count how many reducible monic quadratic polynomials there are in $\mathbb{Z}_p[x]$.
4. Explain why there must always be an *irreducible* monic quadratic polynomial in $\mathbb{Z}_p[x]$.
5. Use all of the work above to prove that for any prime p there exists a field with exactly p^2 elements.

Project 9.6

Consider the field $\mathbb{Z}_7$.

1. Show that there is no $a \in \mathbb{Z}_7$ with $a^4 = 6$. (Thus 6 does not have a "fourth root" in $\mathbb{Z}_7$.) Does any element other than 0 or 1 have a fourth root in $\mathbb{Z}_7$?

   Let c be an element in an extension field of $\mathbb{Z}_7$ with $c^4 = 6$. Thus c is a root of $b(x) = 1 + 0x + 0x^2 + 0x^3 + x^4$.

2. Show that the minimum polynomial for c over $\mathbb{Z}_7$ is either $p(x) = 1 + 3x + x^2$ or $q(x) = 1+4x+x^2$, but we are not sure which. Be sure to show they are both irreducible over $\mathbb{Z}_7$.
3. Describe the elements of $\mathbb{Z}_7(c)$ completely. Can you determine how many elements it contains without listing all of them?
4. Prove that $b(x)$ has all four of its roots in $\mathbb{Z}_7(c)$ and factors into a product polynomials of degree 1 over $\mathbb{Z}_7(c)$.

Project 9.7

Suppose that K is a field and E an extension field of K. For this project assume $u \in E$ is algebraic over K with minimum polynomial $p(x) \in K[x]$. Thus ${}^{K[x]}/_{\langle p(x) \rangle} \cong K(u)$.

Recall from Project 8.4 that for any $a(x) \in K[x]$ and $c \in K$ we defined $a(c + x) \in K[x]$ by $a(c + x) = a_0 + a_1(c + x) + \cdots + a_n(c + x)^n$. A similar definition when $c \neq 0_K$ gives $a(cx) = a_0 + a_1(cx) + \cdots + a_n(cx)^n$.

1. Prove: If $c \in K$ then $c + u$ and cu are also algebraic over K. What polynomial is each

a root of?

Remember that if z is algebraic over K, $K(z)$ has the property that for any field L, if $K \subseteq L$ and $z \in L$ then $K(z) \subseteq L$.

2. Prove: If $c \in K$, then $K(u) = K(c+u)$. (Try subsets.)
3. Prove: If $c \in K$ and $c \neq 0_K$, then $K(u) = K(cu)$. (Try subsets.)
4. How does the work in this project tell us that $\mathbb{Z}_7[x]/\langle 1+x^2 \rangle \cong \mathbb{Z}_7[x]/\langle 3+6x+x^2 \rangle$?

Project 9.8

Recall that for any positive integer n we defined ω_n as below and proved that ω_n is a root of $-1+x^n$.

$$\omega_n = \cos\left(\frac{2\pi}{n}\right) + i\sin\left(\frac{2\pi}{n}\right)$$

Also, for any $1 \leq k \leq n$ we have $(\omega_n)^k = \cos\left(\frac{2k\pi}{n}\right) + i\sin\left(\frac{2k\pi}{n}\right)$.

1. Use the fact that $\cos(u) = 1$ only when u is an integer multiple of 2π to explain why for $1 \leq k < n$ we know $(\omega_n)^k \neq 1$.
2. Explain why for $k \neq j$ and $1 \leq j, k < n$ we must have $(\omega_n)^j \neq (\omega_n)^k$.
3. Prove: For any $n > 1$, $\mathbb{Q}(\omega_n)$ is the root field for $-1+x^n$.
4. Suppose now we have $a(x) = -s+x^n$ where $a(x) \in \mathbb{Q}[x]$. Show that for each $1 \leq k \leq n$ the element $\sqrt[n]{s}(\omega_n)^k$ is a root of $a(x)$.
5. Prove that $\mathbb{Q}(\sqrt[n]{s}, \omega_n)$ is the root field of $a(x) = -s+x^n$ when $s \in \mathbb{Q}$.

Chapter 10

Galois Theory

The quest to find roots of polynomials over a field, such as $\mathbb{Q}$ or $\mathbb{R}$, has a rich history, beginning with Mesopotamian solutions ([4]) of quadratics and continuing to this day with efforts to find more efficient algorithms. It even included contests between mathematicians to see who could solve the most equations, and of highest degree ([4])! In 1823, Niels Abel (for whom abelian groups are named) proved that there is no algebraic solution for a general polynomial of 5th degree or higher ([4]). Then only a few years later in 1832 Évariste Galois wrote down the basics of a theory that uses an interaction between groups and fields to help understand which polynomials can be solved "by radicals" ([4]). Understanding how groups become involved, called Galois Groups, is the goal of this chapter.

There are a few assumptions we need in place throughout the chapter. We assume that ***every field is either finite or has characteristic 0*** which guarantees that irreducible polynomials are separable. Also whenever we discuss a root field we assume it is the root field for a nonconstant polynomial.

10.1 Isomorphisms and Extension Fields

Homomorphisms between rings can be difficult to find, but polynomials will help us find a variety of *isomorphisms* between fields.

Example 10.1 Consider the polynomial $p(x) = 2 + x^2 \in \mathbb{Q}[x]$. Eisenstein's Criterion tells us $p(x)$ is irreducible over $\mathbb{Q}$, and the roots of this polynomial are $\sqrt{-2}$ and $-\sqrt{-2}$. Thus $p(x)$ is the minimum polynomial for $\sqrt{-2}$ and for $-\sqrt{-2}$, so by Theorem 9.13 we know that $\mathbb{Q}\left(\sqrt{-2}\right) = \left\{a + b\sqrt{-2} : a, b \in \mathbb{Q}\right\}$ and $\mathbb{Q}\left(-\sqrt{-2}\right) = \left\{a - b\sqrt{-2} : a, b \in \mathbb{Q}\right\}$. You should be able to verify that $\mathbb{Q}\left(-\sqrt{-2}\right) = \mathbb{Q}\left(\sqrt{-2}\right)$ since $\sqrt{-2} \in \mathbb{Q}(-\sqrt{-2})$ and $-\sqrt{-2} \in \mathbb{Q}(\sqrt{-2})$.

But why discuss both $\sqrt{-2}$ and $-\sqrt{-2}$ when they create the same field $\mathbb{Q}(\sqrt{-2})$? The two different roots show us how to create an isomorphism from the field $\mathbb{Q}(\sqrt{-2})$ to itself. Consider the function defined below.

$$f : \mathbb{Q}\left(\sqrt{-2}\right) \rightarrow \mathbb{Q}\left(\sqrt{-2}\right) \text{ where } f\left(a + b\sqrt{-2}\right) = a - b\sqrt{-2}$$

Suppose $y \in ker(f)$, then $y = a + b\sqrt{-2}$ for some $a, b \in \mathbb{Q}$, and $f(y) = 0$ or $a - b\sqrt{-2} = 0$. If $b \neq 0$ then we have a polynomial of degree 1 in $\mathbb{Q}[x]$ with $\sqrt{-2}$ as a root, which is impossible since $p(x)$ is the minimum polynomial for $\sqrt{-2}$ over $\mathbb{Q}$. Thus $b = 0$ and we must have $a = 0$ as well, so $y = 0$ and $ker(f) = \{0\}$. Hence f is one to one by Theorem 5.29. Also for any element $w \in \mathbb{Q}\left(-\sqrt{-2}\right)$ we know $w = a - b\sqrt{-2}$ for some $a, b \in \mathbb{Q}$. But using the same a and b we have $y = a + b\sqrt{-2} \in \mathbb{Q}\left(\sqrt{-2}\right)$ with $f(y) = w$. Hence f is onto and therefore f is a bijection.

An exercise at the end of the chapter asks you to prove that f is also a homomorphism, therefore f is an isomorphism.

Definition 10.2 Let K be a field and $f : K \rightarrow K$ be a function. If f is an isomorphism we say that f is an ***automorphism of K***.

The function we defined in Example 10.1 has many interesting properties. Notice that f is an automorphism of $\mathbb{Q}\left(\sqrt{-2}\right)$ that is not the identity function since $f\left(\sqrt{-2}\right) = -\sqrt{-2}$ and $f\left(-\sqrt{-2}\right) = \sqrt{-2}$. Also for any $a \in \mathbb{Q}$ we see $f(a) = a$. Thus every element of $\mathbb{Q}$ is *fixed* by f.

Definition 10.3 Let K be a field and E_1, E_2 be extension fields of K. Suppose $f : E_1 \rightarrow E_2$ is an isomorphism. If for every $a \in K$ we have $f(a) = a$, then we say that ***f fixes K***.

Notice in Example 10.1 that f maps each root of $p(x) = 2 + x^2$ *to another root* of $p(x)$. We will eventually show that automorphisms of the root field of a polynomial must have this property. The first step in the process is the next theorem.

Theorem 10.4

Suppose K_1, K_2 are fields, $f : K_1 \to K_2$ is an isomorphism, and $p(x) \in K_1[x]$ is irreducible over K_1. Then there exist extension fields $K_1(c_1)$ and $K_2(c_2)$ with the following properties:

(i) c_1 is a root of $p(x)$ and c_2 is a root of $\overline{f}(p(x))$ (as defined in Theorem 7.28).
(ii) There exists an isomorphism $g : K_1(c_1) \to K_2(c_2)$ with $g(c_1) = c_2$ and for any $a \in K_1$, $g(a) = f(a)$.

Proof Suppose K_1, K_2 are fields, $f : K_1 \to K_2$ is an isomorphism, and $p(x) \in K_1[x]$ is irreducible over K_1. We can assume $p(x)$ is monic since otherwise we replace $p(x)$ with its monic associate, which has the same roots by Theorem 8.5.

By Theorem 9.7 we know there exists an extension of K_1 that has a root for $p(x)$, call it c_1. Thus c_1 is algebraic over K_1 with minimum polynomial $p(x)$, so by Theorem 9.5 $K_1(c_1)$ exists. Similarly, we know by Theorem 8.14 that $\overline{f}(p(x))$ is irreducible over K_2, and by Theorem 6.15 that $\overline{f}(p(x))$ is also monic. Thus there exists c_2 a root of $\overline{f}(p(x))$ and extension field $K_2(c_2)$. Assume that $deg(p(x)) = n$ for some $n > 0$, and thus $deg(\overline{f}(p(x))) = n$ as mentioned after Theorem 7.30. Note that $\overline{f}(p(x))$ is the minimum polynomial for c_2 over K_2, as we will use this fact later.

The homomorphisms $\overline{f}$ and h_{c_2} (from Definition 7.3) will help us define our isomorphism g. Every nonzero element $y \in K_1(c_1)$ is uniquely defined as $u(c_1)$ for some $u(x) \in K_1[x]$ where $deg(u(x)) < n$. Thus we can define $g : K_1(c_1) \to K_2(c_2)$ by $g(y) = g(u(c_1)) = \left(h_{c_2} \circ \overline{f}\right)(u(x))$. If $y = 0_{K_1}$ then we use $u(x) = 0(x)$, the zero polynomial. But as $\overline{f}(0(x)) = 0(x)$ this guarantees that $g(0_{K_1}) = 0_{K_2}$.

The map g is uniquely defined since there is only one $u(x) \in K_1[x]$ where $deg(u(x)) < n$ (or $u(x) = 0(x)$) with $y = u(c_1)$. Also since $\overline{f}$ and h_{c_2} are both homomorphisms, it is clear that g is also a homomorphism. An exercise at the end of the chapter asks you to show that $g(c_1) = c_2$ and for any $a \in K_1$ we have $g(a) = f(a)$. Thus we only need to show that g is bijection to complete the proof.

Suppose $y \in ker(g)$. We already know that $0_{K_1} \in ker(g)$, so assume that $y \neq 0_{K_1}$. Then $y = b(c_1)$ for some $b(x) \in K_1[x]$ with $deg(b(x)) < n$ and $g(y) = 0_{K_2}$ as $g(b(c_1)) = \overline{f}(b(c_2))$. Now $\overline{f}(b(c_2)) = 0_{K_2}$ so c_2 is a root of $\overline{f}(b(x)) \in K_2[x]$. But $deg(\overline{f}(b(x)) < n$ which contradicts that $\overline{f}(p(x))$ is the minimum polynomial for c_2 over K_2 as noted above. Thus $ker(g) = \{0_{K_1}\}$ and g is one to one.

Finally, let $w \in K_2(c_2)$ with $w \neq 0_{K_2}$. There is a polynomial $b(x) \in K_2[x]$ with $w = b(c_2)$ and $deg(b(x)) < n$. But $\overline{f}$ is onto by Theorem 7.30 so there is $q(x) \in K_1[x]$ with $\overline{f}(q(x)) = b(x)$. Since $deg(q(x)) = deg(\overline{f}(q(x))$, then $deg(q(x)) < n$. Now $q(c_1) \in K_1(c_1)$ and $g(q(c_1)) = \overline{f}(q(c_2)) = b(c_2) = w$, so g is onto. Therefore g is an isomorphism. □

Example 10.5 Consider the field $K_1 = K_2 = \mathbb{Q}\left(\sqrt{-2}\right)$ and recall that $[K_1 : \mathbb{Q}] = 2$. In Example 10.1, we found an isomorphism $f : K_1 \to K_2$ defined by $f\left(a + b\sqrt{-2}\right) = a - b\sqrt{-2}$.

Consider the polynomial $p(x) = -5 + x^3 \in K_1[x]$. We know $p(x)$ is irreducible over $\mathbb{Q}$ by Eisenstein's Criterion and has root $c_1 = \sqrt[3]{5}$ so $[\mathbb{Q}(c_1) : \mathbb{Q}] = 3$. If $c_1 \in K_1$ then $\mathbb{Q}(c_1) \subseteq K_1$ so by Theorem 9.22 we would have $[K_1 : \mathbb{Q}] = [K_1 : \mathbb{Q}(c_1)][\mathbb{Q}(c_1) : \mathbb{Q}]$. But then $2 = [K_1 : \mathbb{Q}(c_1)](3)$ which is impossible. This also holds for any root of $p(x)$, so $p(x)$ is irreducible over K_1.

Now $\overline{f}(p(x)) = p(x)$ easily, so if we use $c_1 = \sqrt[3]{5}$ and $c_2 = -\frac{\sqrt[3]{5}}{2}\left(1 + \sqrt{-3}\right)$ (found in Example 9.24) then Theorem 10.4 tells us $K_1(\sqrt[3]{5}) \cong K_1(-\frac{\sqrt[3]{5}}{2}(1 + \sqrt{-3}))$. As $[K_1(c_1) : K_1] = 3$ we know every element $y \in K_1(c_1)$ is of the form $y = b(c_1) = b_0 + b_1c_1 + b_2c_1^2$ with $b_i \in K_1$. But also $b_0 = r_0 + r_1\sqrt{-2}$, $b_1 = d_0 + d_1\sqrt{-2}$ and $b_2 = q_0 + q_1\sqrt{-2}$ where $c_i, d_i, q_i \in \mathbb{Q}$. Thus we can calculate $g(y)$ as shown below.

$$\begin{aligned} g(y) &= g(b_0 + b_1(\sqrt[3]{5}) + b_2(\sqrt[3]{5})^2) \\ &= f(b_0) + f(b_1)(-\tfrac{\sqrt[3]{5}}{2}\left(1 + \sqrt{-3}\right)) + f(b_2)(-\tfrac{\sqrt[3]{5}}{2}\left(1 + \sqrt{-3}\right))^2 \\ &= (r_0 - r_1\sqrt{-2}) + (d_0 - d_1\sqrt{-2})(-\tfrac{\sqrt[3]{5}}{2}\left(1 + \sqrt{-3}\right)) \\ &\qquad +(q_0 - q_1\sqrt{-2})(-\tfrac{\sqrt[3]{5}}{2}\left(1 + \sqrt{-3}\right))^2 \end{aligned}$$

A simpler example brings up a very important special case of this theorem.

Example 10.6 Consider the field $\mathbb{Q} = K_1 = K_2$ with the identity function $\varepsilon : \mathbb{Q} \to \mathbb{Q}$. We again use roots of the irreducible polynomial $p(x) = -5 + x^3$, namely $\sqrt[3]{5}$ and $-\frac{\sqrt[3]{5}}{2}(1 + \sqrt{-3})$. Theorem 10.4 tells us that the fields $\mathbb{Q}\left(\sqrt[3]{5}\right)$ and $\mathbb{Q}\left(-\frac{\sqrt[3]{5}}{2}(1 + \sqrt{-3})\right)$ are isomorphic, and the isomorphism can be written as $g(a + b\sqrt[3]{5} + c\sqrt[3]{25}) = a + b(-\frac{\sqrt[3]{5}}{2}(1 + \sqrt{-3})) + c\left(-\frac{\sqrt[3]{5}}{2}(1 + \sqrt{-3})\right)^2$.

But unlike Example 10.1 the extension fields created here are *not* the same. In Example 9.24 we proved that $-\frac{\sqrt[3]{5}}{2}(1 + \sqrt{-3}) \notin \mathbb{Q}\left(\sqrt[3]{5}\right)$.

Having the identity function as the initial isomorphism in Theorem 10.4 will be so important we state the result as a separate theorem.

Theorem 10.7

Let K be a field and $p(x) \in K[x]$ an irreducible polynomial. If c_1 and c_2 are roots of $p(x)$ in some extension of K, then $K(c_1) \cong K(c_2)$ where the isomorphism $g : K(c_1) \to K(c_2)$ maps $g(c_1) = c_2$ and fixes K.

Example 10.8 Consider the finite field $K = \mathbb{Z}_3$ and $p(x) = 2 + 2x + x^2$ from Example 9.15. With a root c in an extension field we found the elements of $\mathbb{Z}_3(c)$.

$$\mathbb{Z}_3(c) = \{0, 1, 2, 0 + c, 1 + c, 2 + c, 0 + 2c, 1 + 2c, 2 + 2c\}$$

Since $1 + 2c$ is the other root of $p(x)$ by Theorem 10.7 there is an automorphism $g : \mathbb{Z}_3(c) \to \mathbb{Z}_3(c)$ where $g(c) = 1 + 2c$ and g fixes $\mathbb{Z}_3$. For any element $u + vc \in \mathbb{Z}_3(c)$ we can calculate $g(u + vc)$ as seen below.

$$g(u + vc) = g(u) + g(v)g(c) = u + vg(c) = u + v(1 + 2c) = (u + v) + (2v)c$$

In Theorem 10.7, we began with two roots of the same irreducible polynomial over the field K to create extensions of K and an isomorphism between them fixing K. Do we need to start with roots of the same polynomial to do this?

Theorem 10.9

Let K be any field and E_1, E_2 extension fields of K, with $f : E_1 \to E_2$ an isomorphism fixing K. If $p(x) \in K[x]$ and $c \in E_1$ is a root of $p(x)$, then $f(c) \in E_2$ is also a root of $p(x)$.

Proof Let K be a field and E_1, E_2 two extensions of K, with $f : E_1 \to E_2$ an isomorphism fixing K. Suppose $p(x) \in K[x]$ and $c \in E_1$ is a root of $p(x)$. Thus $p(c) = 0_{E_1} = 0_K$, and so by f a homomorphism $f(p(c)) = f(0_{E_1}) = 0_{E_2} = 0_K$ as well. But since f fixes K and $p(x) \in K[x]$, then $\overline{f}(p(x)) = p(x)$. Thus we have $0_K = f(p(c)) = p(f(c))$ so $f(c)$ is also a root of $p(x)$. □

The previous theorem tells us we could not find an isomorphism between $\mathbb{Q}(\sqrt{-2})$ and $\mathbb{Q}\left(\sqrt[3]{5}\right)$. If there was an isomorphism f then $f\left(\sqrt[3]{5}\right)$ must also be a root of $2 + x^2$. But the elements of $\mathbb{Q}\left(\sqrt[3]{5}\right)$ are of the form $a + b\sqrt[3]{5} + c\sqrt[3]{25}$, and you will show in an exercise that none of these elements can be a root of $2 + x^2$.

10.2 Automorphisms of Root Fields

Recall from Theorem 8.27 that for any nonconstant $a(x) \in K[x]$ we found that $a(x)$ can have at most $deg(a(x))$ distinct roots in an extension field of K. If we remove the restriction that the roots be *distinct* we can be more precise.

Theorem 10.10

Let K be a field and $a(x) \in K[x]$. If $deg(a(x)) = n > 0$, then $a(x)$ has exactly n roots in its root field.

Proof Let K be a field and $a(x) \in K[x]$ with $deg(a(x)) = n > 0$. Let E denote the root field of $a(x)$ over K. By Definition 9.23, we can write $a(x) = b_1(x)b_2(x) \cdots b_t(x)$ where $b_i(x) \in E[x]$ and $deg(b_i(x)) = 1$ for each i. But $deg(a(x)) = deg(b_1(x)) + deg(b_2(x)) + \cdots + deg(b_t(x))$ by Theorem 7.20 so $t = n$.

For each i we can write $b_i(x) = d_{i0} + d_{i1}x$ for some $d_{i0}, d_{i1} \in E$. As $deg(b_i(x)) = 1$ we have $d_{i1} \neq 0_E$ so $u_i = -d_{i0}(d_{i1})^{-1}$ is a root of $b_i(x)$. By Theorem 8.15 each u_i is a root of $a(x)$ so there are exactly n (not necessarily distinct) roots for $a(x)$. □

Example 10.11 Consider the polynomial $p(x) = -1 + x^3 \in \mathbb{Q}[x]$. The number 1 is clearly a root, but according to Theorem 10.10, there must exist two more roots of $p(x)$, i.e., two more "cube roots" of the number 1. How do we find them?

The first step is to factor out $-1+x$ since we know that 1 is a root, $p(x) = (-1+x)(1+x+x^2)$. Thus we need to find the roots of $1 + x + x^2$, and we use the quadratic formula since we are working over $\mathbb{Q}$. In this case we have $a = 1, b = 1, c = 1$ so the three roots are shown below.

$$1, \quad \tfrac{-1+\sqrt{-3}}{2}, \tfrac{-1-\sqrt{-3}}{2}$$

There seems to be a strong similarity between these and the roots of $p(x) = -5 + x^3$ found in Example 9.24. (Project 9.8 helps us see why.)

Root fields have special properties that we will use to create the needed automorphisms of those fields. The most important of these properties hinges on knowing a very precise way to write the elements of a root field. For a polynomial $a(x) \in K[x]$, its root field is a finite extension of K of the form $K(c_1, c_2, \ldots c_n)$ where the c_i are all of the roots of $a(x)$ in any extension of K. Also from the proof of Theorem 9.22 the basis for $K(c_1, c_2, \ldots c_n)$ over K is the set of all elements of the form $c_1^{i_1} c_2^{i_2} \ldots c_n^{i_n}$. Hence each element of $K(c_1, c_2, \ldots c_n)$ can be expressed as a finite sum such as the one below.

$$\sum k_i c_1^{i_1} c_2^{i_2} \ldots c_n^{i_n} \text{ with } k_i \in K \text{ and } i_j \geq 0$$

Theorem 10.12

Suppose that K is a field and E_1, E_2 are both finite extensions of K. If there exists an isomorphism $f : E_1 \to E_2$ which fixes K, and E_1 is the root field of a polynomial $p(x) \in K[x]$, then $E_1 = E_2$.

Proof Suppose that K is a field and E_1, E_2, are both extensions of K. Assume we have an isomorphism $f : E_1 \to E_2$ which fixes K, and that E_1 is the root field of a polynomial

$p(x) \in K[x]$. Let the distinct roots of $p(x)$ be denoted by $c_1, c_2, \ldots c_m$ so every element of E_1 has the form shown below. Using the fact that f is an isomorphism fixing K the following holds:

$$\begin{aligned} d &= \sum k_i c_1^{i_1} c_2^{i_2} \cdots c_m^{i_m} \\ f(d) &= f\left(\sum k_i c_1^{i_1} c_2^{i_2} \cdots c_m^{i_m}\right) \\ &= \sum k_i f(c_1)^{i_1} f(c_2)^{i_2} \cdots f(c_m)^{i_m}. \end{aligned}$$

From Theorem 10.9 each $f(c_j)$ is one of $c_1, c_2, \ldots c_m$. Thus we again have $f(d) \in E_1$, so $f(E_1) \subseteq E_1$. But since f is onto, $E_2 = f(E_1)$ so $E_2 \subseteq E_1$. An exercise at the end of the chapter asks you to verify that $E_1 \subseteq E_2$ and thus $E_1 = E_2$ as needed. □

The previous theorem tells us that the only way to make an isomorphism (fixing K) from the root field of $p(x) \in K[x]$ to another field is to make an automorphism of the root field. Obviously the identity function will always be such an automorphism, but is not very interesting. In Example 10.1 the root field of $2+x^2$ was found to be $\mathbb{Q}\left(\sqrt{-2}\right)$, and we found an automorphism of $\mathbb{Q}\left(\sqrt{-2}\right)$ that is not the identity. Can we always do this? The final step we need to guarantee it is next.

Theorem 10.13

Suppose K is a field and E is the root field for some nonconstant $p(x) \in K[x]$. Suppose L_1 and L_2 are finite extension fields of K with $K \subseteq L_1 \subseteq E$. If there exists $f : L_1 \to L_2$ an isomorphism fixing K, then $L_2 \subseteq E$ and there exist an automorphism g of E, with $g(a) = f(a)$ for all $a \in L_1$.

Proof Suppose K is a field and E is the root field for some nonconstant $p(x) \in K[x]$. Assume we have finite extension fields L_1 and L_2 of K with $K \subseteq L_1 \subseteq E$ and an isomorphism $f : L_1 \to L_2$ fixing K. Since E is a finite extension of K it is also a finite extension of L_1 (an exercise in Chapter 9), so $E = L_1(c)$ for some $c \in E$ by Theorem 9.34.

Let $q(x) \in L_1[x]$ be the minimum polynomial of c over L_1. Then $\overline{f}(q(x)) \in L_2[x]$ is irreducible over L_2 by Theorem 8.14 and has a root b in some extension of L_2. By Theorem 10.4 there is an isomorphism $g : L_1(c) \to L_2(b)$ which maps c to b, and for each $y \in L_1$, $g(y) = f(y)$. Since f fixes K, then g also fixes K. But $L_1(c)$ is the root field of $p(x)$, so by Theorem 10.12 $E = L_2(b)$. This tells us that $L_2 \subseteq E$, and we have g an automorphism of E fixing K with with $g(a) = f(a)$ for all $a \in L_1$. □

Putting together the results in Theorems 10.7 and 10.13 we can finally prove that a nonidentity automorphism always exists when K is not already the root field of the given irreducible polynomial. The proof of this theorem is an exercise at the end of the chapter.

Theorem 10.14

Suppose K is a field, E is the root field of a polynomial in $K[x]$, and $p(x) \in K[x]$ is irreducible over K with $deg(p(x)) > 1$. For any two distinct roots $c_1, c_2 \in E$ of $p(x)$, there exists an automorphism of E fixing K, mapping c_1 to c_2.

Example 10.15 In Example 10.11 we found the roots of $-1+x^3 \in \mathbb{Q}[x]$ to be 1, $\frac{-1+\sqrt{-3}}{2}$, and $\frac{-1-\sqrt{-3}}{2}$. Notice that the root field of $p(x)$ is $\mathbb{Q}\left(1, \frac{-1+\sqrt{-3}}{2}, \frac{-1-\sqrt{-3}}{2}\right) = \mathbb{Q}\left(\frac{-1+\sqrt{-3}}{2}\right)$ since $\frac{-1-\sqrt{-3}}{2} = -1 - \left(\frac{-1+\sqrt{-3}}{2}\right)$.

But Theorem 10.14 only guarantees we can create automorphisms with roots of an irreducible polynomial. We saw $-1 + x^3$ is reducible in Example 10.11, so any automorphism of $\mathbb{Q}\left(\frac{-1+\sqrt{-3}}{2}\right)$ fixing $\mathbb{Q}$ maps $\frac{-1+\sqrt{-3}}{2}$ to a root of $1 + x + x^2$ since $1 + x + x^2$ is the minimum polynomial for $\frac{-1+\sqrt{-3}}{2}$ over $\mathbb{Q}$.

One last fact about root fields, which we need later, completes this section.

Theorem 10.16

Suppose K is a field and E is the root field of $p(x) \in K[x]$. If the irreducible polynomial $a(x) \in K[x]$ has one root in E then every root of $a(x)$ is in E.

Proof Suppose K is a field and E is the root field of $p(x) \in K[x]$. Assume an irreducible polynomial $a(x) \in K[x]$ has one root in E, call it c, and let d be another root of $a(x)$. Recall that by Theorem 10.7 we know $K(c) \cong K(d)$ with an isomorphism f that fixes K and $f(c) = d$. Now $K \subseteq K(c) \subseteq E$ and $K \subseteq K(d)$ so by Theorem 10.13 we know $K(d) \subseteq E$. Thus $d \in E$ as well.

□

10.3 The Galois Group of a Polynomial

In this section we consider ***groups of automorphisms*** of a root field.

Example 10.17 In Example 10.8 we found that $E = \mathbb{Z}_3(c)$ is the root field for $p(x) = 2 + 2x + x^2$. We also have two automorphisms of E that fix $\mathbb{Z}_3$, the identity function ε, and the function α where $\alpha(u + vc) = (u + v) + (2v)c$. Could there be other automorphisms of E?

Suppose there is another automorphism of E that fixes $\mathbb{Z}_3$, call it β. Then for any element $u + vc \in \mathbb{Z}_3(c)$ we calculate $\beta(u + vc)$ as shown.

$$\beta(u + vc) = \beta(u) + \beta(v)\beta(c) = u + v\beta(c)$$

Theorem 10.9 tells us that $\beta(c)$ must be another root of the same minimum polynomial $2+2x+x^2$, so the only choices are $\beta(c) = c$ or $\beta(c) = 1+2c$. Hence either $\beta(u+vc) = u+vc$ or $\beta(u + vc) = (u + v) + (2v)c$ and thus $\beta \in \{\varepsilon, \alpha\}$.

Consider composition of functions on the set $\{\varepsilon, \alpha\}$. Since ε is the identity function $\varepsilon \circ \varepsilon = \varepsilon$ is clear, as well as $\varepsilon \circ \alpha = \alpha$ and $\alpha \circ \varepsilon = \alpha$. The next computation shows us that $\alpha \circ \alpha = \varepsilon$.

$$(\alpha \circ \alpha)(u + vc) = \alpha((u + v) + (2v)c) = (u + v + 2v) + 2(2v)c = u + vc$$

Thus composition is an operation on the set $\{\varepsilon, \alpha\}$. Composition of functions is associative and from our computations we know $\varepsilon^{-1} = \varepsilon$ and $\alpha^{-1} = \alpha$. Thus we have a group of automorphisms of $\mathbb{Z}_3(c)$ fixing $\mathbb{Z}_3$.

This example leads us to the next theorem.

Theorem 10.18

Let K be a field and $p(x) \in K[x]$. If E is the root field of $p(x)$ over K then the set of all automorphisms of E fixing K is a group under composition.

Proof Let K be a field and $p(x) \in K[x]$. Assume E is the root field of $p(x)$ over K. Define the set G as follows:

$$G = \{\alpha : \alpha \text{ is an automorphism of } E \text{ fixing } K\}.$$

The identity function ε is clearly in G, so $G \neq \emptyset$. Let $\alpha, \beta \in G$. As both α and β are bijections from E to E, then $\alpha \circ \beta$ is a bijection from E to E by Theorem 0.17 and $\alpha \circ \beta$ is a homomorphism (an exercise in Chapter 4). To see that $\alpha \circ \beta$ fixes K, let $a \in K$ and notice:

$$(\alpha \circ \beta)(a) = \alpha(\beta(a)) = \alpha(a) = a.$$

Thus K is fixed by $\alpha \circ \beta$. Hence $\alpha \circ \beta \in G$ and composition is an operation on G. Composition is associative (an exercise in Chapter 2), and so we only need to show each element of G has an inverse in G. For any $\alpha \in G$, as seen after Definition 0.14, there exists a bijective function $\alpha^{-1} : E \to E$ so that $\alpha \circ \alpha^{-1} = \varepsilon = \alpha^{-1} \circ \alpha$. In an exercise at the end of the chapter you are asked to verify that α^{-1} is also a homomorphism and fixes K. Hence $\alpha^{-1} \in G$, and thus G is a group under composition. $\square$

Definition 10.19 Let K be a field and $p(x) \in K[x]$ with root field E. The group of automorphisms of E fixing K is called the ***Galois group of E over K***, denoted by $Gal\left({}^{E}/_{K}\right)$. It can also be called the ***Galois group of p(x) over K***.

From Example 10.17 we can say that $Gal\left({}^{\mathbb{Z}_3(c)}/_{\mathbb{Z}_3}\right) = \{\varepsilon, \alpha\}$ where ε is the identity function and $\alpha(c) = 1 + 2c$. Notice that $[\mathbb{Z}_3(c) : \mathbb{Z}_3] = 2$, and there are exactly two elements in our Galois group. Is this a coincidence?

Theorem 10.20

Let K be a field and E the root field for some $p(x) \in K[x]$. The number of automorphisms of E fixing K is equal to $[E : K]$.

Proof Suppose K is a field, and E the root field for some $p(x) \in K[x]$. If $E = K$ then $Gal\left({}^{E}/_{K}\right) = \{\varepsilon\}$ since the automorphisms fix K. But also $[E : K] = 1$ so the conclusion of the theorem holds in this case. Now suppose that $[E : K] = n > 1$.

Since E is a finite extension of K (definition of a root field), by Theorem 9.34 $E = K(c)$ for some $c \notin K$. As $[E : K] = n$, then $\{1_K, c, c^2, \ldots, c^{n-1}\}$ is a basis for E over K by Theorem 9.20. For $y \in E$ and $\alpha \in Gal\left({}^{E}/_{K}\right)$ we see below that $\alpha(c)$ completely determines $\alpha(y)$.

$$\begin{aligned} y &= a_0 + a_1 c + \cdots + a_{n-1}c^{n-1} \\ \alpha(y) &= \alpha(a_0) + \alpha(a_1 c) + \cdots + \alpha(a_{n-1}c^{n-1}) \\ &= a_0 + a_1\alpha(c) + \cdots + a_{n-1}(\alpha(c))^{n-1} \end{aligned}$$

Let $s(x)$ be the minimum polynomial for c over K, then by Theorem 9.20 $deg(s(x)) = n$. Also, let L be the root field for $s(x)$ over K. By Theorem 10.16 we must have $L \subseteq E$. But $c \in L$, $K \subseteq L$, and $E = K(c)$, so $E \subseteq L$ as well or $E = L$. Thus E is also the root field for $s(x)$ over K.

Since $s(x)$ is separable it has exactly n distinct roots in E. By Theorem 10.7, for any root d_i of $s(x)$ there is an automorphism of E fixing K that maps c to d_i. This gives us n different automorphisms of E fixing K. However, if $\beta \in Gal\left({}^{E}/_{K}\right)$ then by Theorem 10.9 $\beta(c)$ must be a root of $s(x)$ as well, so β must have been one of the automorphisms we previously found. Thus, there are exactly n automorphisms of E fixing K. □

Example 10.21 In Example 9.27, we found the root field of $-5 + x^3$ over $K = \mathbb{Q}$ to be $E = \mathbb{Q}\left(\sqrt[3]{5}, \sqrt{-3}\right)$ and a basis of E as $\left\{1, \sqrt[3]{5}, \sqrt[3]{25}, \sqrt{-3}, (\sqrt[3]{5})(\sqrt{-3}), (\sqrt[3]{25})(\sqrt{-3})\right\}$. As in the previous proof, for $\alpha \in Gal\left({}^{E}/{}_{K}\right)$, the elements $\alpha\left(\sqrt[3]{5}\right)$ and $\alpha\left(\sqrt{-3}\right)$ completely determine $\alpha(y)$ for any $y \in E$. Since the minimum polynomial of $\sqrt[3]{5}$ over $\mathbb{Q}$ is $-5+x^3$ then we must have one of the following:

$$\alpha\left(\sqrt[3]{5}\right) = \sqrt[3]{5},\ \alpha\left(\sqrt[3]{5}\right) = \tfrac{-\sqrt[3]{5}}{2}(1+\sqrt{-3}),\text{ or } \alpha\left(\sqrt[3]{5}\right) = \tfrac{-\sqrt[3]{5}}{2}(1-\sqrt{-3}).$$

Similarly, the minimum polynomial for $\sqrt{-3}$ is $3+x^2$, so we know $\alpha\left(\sqrt{-3}\right) = \sqrt{-3}$ or $\alpha\left(\sqrt{-3}\right) = -\sqrt{-3}$. This helps us describe the automorphisms below. In an exercise at the end of the chapter you are asked to verify that $\{\varepsilon, \alpha, \beta, \delta, \varphi, \sigma\}$ is a group and create its Cayley table.

$$\begin{aligned} \varepsilon\left(\sqrt[3]{5}\right) &= \sqrt[3]{5} \\ \varepsilon\left(\sqrt{-3}\right) &= \sqrt{-3} \end{aligned} \qquad \begin{aligned} \alpha\left(\sqrt[3]{5}\right) &= \tfrac{-\sqrt[3]{5}}{2}(1+\sqrt{-3}) \\ \alpha\left(\sqrt{-3}\right) &= \sqrt{-3} \end{aligned}$$

$$\begin{aligned} \beta\left(\sqrt[3]{5}\right) &= \tfrac{-\sqrt[3]{5}}{2}(1-\sqrt{-3}) \\ \beta\left(\sqrt{-3}\right) &= \sqrt{-3} \end{aligned} \qquad \begin{aligned} \delta\left(\sqrt[3]{5}\right) &= \sqrt[3]{5} \\ \delta\left(\sqrt{-3}\right) &= -\sqrt{-3} \end{aligned}$$

$$\begin{aligned} \varphi\left(\sqrt[3]{5}\right) &= \tfrac{-\sqrt[3]{5}}{2}(1+\sqrt{-3}) \\ \varphi\left(\sqrt{-3}\right) &= -\sqrt{-3} \end{aligned} \qquad \begin{aligned} \sigma\left(\sqrt[3]{5}\right) &= \tfrac{-\sqrt[3]{5}}{2}(1-\sqrt{-3}) \\ \sigma\left(\sqrt{-3}\right) &= -\sqrt{-3} \end{aligned}$$

Each of these possibilities makes an automorphism of $E = \mathbb{Q}\left(\sqrt[3]{5}, \sqrt{-3}\right)$, fixing $\mathbb{Q}$. However, it is not obvious that these maps will send all of the roots of our polynomial $-5+x^3$ back to other roots. For φ the next steps show that $\varphi\left(\frac{-\sqrt[3]{5}}{2}\left(1+\sqrt{-3}\right)\right) = \sqrt[3]{5}$. Be sure you can calculate $\varphi\left(\frac{-\sqrt[3]{5}}{2}\left(1-\sqrt{-3}\right)\right)$ as well. In exercises at the end of the chapter, you will be asked to repeat this for the other automorphisms. Notice that we have six automorphisms of E fixing $\mathbb{Q}$ and that $[E:\mathbb{Q}] = 6$ as we expected.

$$\begin{aligned} \varphi\left(\tfrac{-\sqrt[3]{5}}{2}\left(1+\sqrt{-3}\right)\right) &= \varphi\left(\tfrac{-\sqrt[3]{5}}{2}\right)\varphi\left(1+\sqrt{-3}\right) \\ &= \varphi\left(\tfrac{-1}{2}\right)\varphi\left(\sqrt[3]{5}\right)\left[\varphi(1)+\varphi\left(\sqrt{-3}\right)\right] \\ &= \tfrac{-1}{2}\left[\tfrac{-\sqrt[3]{5}}{2}\left(1+\sqrt{-3}\right)\right]\left[1-\sqrt{-3}\right] \\ &= \tfrac{\sqrt[3]{5}}{4}\left(1+\sqrt{-3}\right)\left(1-\sqrt{-3}\right) \\ &= \sqrt[3]{5} \end{aligned}$$

Each automorphism of $E = \mathbb{Q}\left(\sqrt[3]{5}, \sqrt{-3}\right)$ corresponds to a specific *permutation* of the roots of $-5 + x^3$. We show this permutation for φ below, and will write "$\varphi =$" instead of constantly referring to it as the permutation corresponding to φ from now on.

$$\varphi = \begin{pmatrix} \sqrt[3]{5} & \frac{-\sqrt[3]{5}}{2}\left(1+\sqrt{-3}\right) & \frac{-\sqrt[3]{5}}{2}\left(1-\sqrt{-3}\right) \\ \frac{-\sqrt[3]{5}}{2}\left(1+\sqrt{-3}\right) & \sqrt[3]{5} & \frac{-\sqrt[3]{5}}{2}\left(1-\sqrt{-3}\right) \end{pmatrix}$$

Thus, the Galois group of a polynomial corresponds to a group of *permutations of the roots* of the polynomial. As there are only finitely many roots for a polynomial, we will always have a corresponding finite group of permutations of the roots. Note, however, that the degree of the polynomial is **not** always the number of automorphisms for the root field.

Now that we have defined the Galois group of a polynomial, a finite group of automorphisms, it is natural to consider subgroups of this group. Subgroups give us information about the root field that will be important in the final chapter.

Example 10.22 Let E denote the root field of $p(x) = -5 + x^3$, and $G = Gal\left({}^{E}/_{\mathbb{Q}}\right) = \{\varepsilon, \alpha, \beta, \delta, \varphi, \sigma\}$ as in the previous example. From the definition of the automorphisms ε, α, and β we see that they each map the element $\sqrt{-3}$ back to itself, so they also fix the entire field $\mathbb{Q}\left(\sqrt{-3}\right)$. The other three elements of $G = Gal\left({}^{E}/_{\mathbb{Q}}\right)$ do not fix $\mathbb{Q}\left(\sqrt{-3}\right)$. An exercise at the end of the chapter will ask you to verify that $H = \{\varepsilon, \alpha, \beta\}$ is a subgroup of $G = Gal\left({}^{E}/_{\mathbb{Q}}\right)$.

Notice that only the members of the subgroup H fix the subfield $\mathbb{Q}\left(\sqrt{-3}\right)$. Will every subgroup of the Galois group relate to a subfield this way?

Theorem 10.23

Let K be a field and $p(x) \in K[x]$ with root field E. Let $G = Gal\left({}^{E}/_{K}\right)$. If H is a subgroup of G then the set $E_H = \{y \in E : \alpha(y) = y \text{ for every } \alpha \in H\}$ is a subfield of E and $K \subseteq E_H \subseteq E$. The field E_H is called ***the fixed field for H***.

Proof Let K be a field and $p(x) \in K[x]$ with root field E. Let $G = Gal\left({}^{E}/_{K}\right)$ and suppose H is a subgroup of G. Define the set $E_H = \{y \in E : \alpha(y) = y \text{ for every } \alpha \in H\}$. We must show that E_H is a subfield of E. By definition of $Gal\left({}^{E}/_{K}\right)$ we know for each $\alpha \in Gal\left({}^{E}/_{K}\right)$ and every $y \in K$, $\alpha(y) = y$. Thus $K \subseteq E_H$ and $E_H \neq \emptyset$.

Suppose $y, z \in E_H$, and $\alpha \in H$. We know $\alpha(y) = y$, $\alpha(z) = z$, and that α is a homomorphism giving us the steps below. This shows E_H is a subring of E by Theorem 4.33.

$$\alpha(y+z)=\alpha(y)+\alpha(z)=y+z$$
$$\alpha(yz)=\alpha(y)\alpha(z)=yz$$
$$\alpha(-y)=-\alpha(y)=-y$$

Since for any $\alpha \in H$, $\alpha(1_E)=1_E$, by Theorem 6.15 then $1_E \in E_H$. Also multiplication in E is commutative, so E_H is a commutative ring with unity. We only need to show that each nonzero element of E_H has an inverse in E_H to complete the proof that E_H is a subfield of E. Suppose $y \in E_H$, $y \neq 0_E$, and $\alpha \in H$ then (an exercise in Chapter 5) we know $\alpha(y^{-1})=(\alpha(y))^{-1}=y^{-1}$, so $y^{-1} \in E_H$ as well. Thus, E_H is a subfield of E and $K \subseteq E_H \subseteq E$. □

In Example 10.22, we found that the automorphisms fixing $\mathbb{Q}(\sqrt{-3})$ create a subgroup H of $Gal\left({}^{E}/_{\mathbb{Q}}\right)$. If we begin with any field L with $K \subseteq L \subseteq E$, can we always find a subgroup of $G = Gal\left({}^{E}/_{K}\right)$ that fixes only those elements? A field L with $K \subseteq L \subseteq E$ is called an ***intermediate field*** between K and E.

Theorem 10.24

Let K be a field, $p(x) \in K[x]$ with root field E, and $G = Gal\left({}^{E}/_{K}\right)$. If L is a subfield of E with $K \subseteq L$, then $G_L = \{\alpha \in G : \text{ for every } y \in L, \alpha(y)=y\}$ is a subgroup of G. The subgroup G_L is called ***the fixer of L***.

Proof Let K be a field and $p(x) \in K[x]$ with root field E, and denote $G = Gal\left({}^{E}/_{K}\right)$. Suppose L is a subfield of E with $K \subseteq L$, and define $G_L = \{\alpha \in G : \text{ for every } y \in L, \alpha(y)=y\}$. We know that the identity automorphism ε is in G_L, so $G_L \neq \emptyset$. To see that G_L is closed under composition and inverses, let $\alpha, \beta \in G_L$ and $y \in L$.

The following steps guarantee $\alpha \circ \beta, \alpha^{-1} \in G_L$. Hence G_L is a subgroup of G:

$$(\alpha \circ \beta)(y)=\alpha(\beta(y))=\alpha(y)=y$$
$$y=(\alpha^{-1} \circ \alpha)(y)=\alpha^{-1}(\alpha(y))=\alpha^{-1}(y).$$

□

An interesting fact follows from the previous theorem. Suppose we begin with a subfield L where $K \subseteq L \subseteq E$ and E is the root field for some polynomial $p(x) \in K[x]$. Then E is also the root field of $p(x)$ over L since $p(x) \in L[x]$. Thus we can define the group $Gal\left({}^{E}/_{L}\right)$ just as we defined $G = Gal\left({}^{E}/_{K}\right)$. An exercise at the end of the chapter asks you to verify that in this case we also have $Gal\left({}^{E}/_{L}\right) \subseteq Gal\left({}^{E}/_{K}\right)$ and $G_L = Gal\left({}^{E}/_{L}\right)$.

10.4 The Galois Correspondence

Consider a field K, E the root field of a polynomial in $K[x]$, and $G = Gal\left({}^{E}/_{K}\right)$. If we have an intermediate field L, then by Theorem 10.24, L determines a subgroup G_L of G. Also by Theorem 10.23 there is a fixed field E_{G_L} between K and E. But did we create a new field or did we end up with L again?

Theorem 10.25

Let K be a field, $p(x) \in K[x]$ with root field E, and $G = Gal\left({}^{E}/_{K}\right)$. If L is a subfield of E with $K \subseteq L \subseteq E$, then L is the fixed field of G_L, i.e., $E_{G_L} = L$.

Proof Let K be a field, $p(x) \in K[x]$ with root field E, and $G = Gal\left({}^{E}/_{K}\right)$. Assume L is a subfield of E with $K \subseteq L \subseteq E$. We know that G_L is a subgroup of G by Theorem 10.24, so assume M is the fixed field of G_L. We must show that $L = M$.

Let $y \in L$. For every $\alpha \in G_L$ we know $\alpha(y) = y$ so by definition of M, $y \in M$ and $L \subseteq M \subseteq E$. We need to show that $M \subseteq L$ as well, so assume for a contradiction that we have some element $z \in M$ with $z \notin L$. Since z algebraic over K it is also algebraic over L, and thus there is a minimum polynomial for z over L, $b(x) \in L[x]$. Since $z \notin L$ then $deg(b(x)) > 1$, and so there is also a different root, w, for $b(x)$ (since $b(x)$ is separable) by Theorems 9.31 and 9.33.

By 10.16 we know that $w \in E$, so Theorem 10.7 tells us there is an isomorphism from $L(z)$ to $L(w)$, fixing L mapping z to w. By Theorem 10.13 there is an automorphism, α, of E fixing L that maps z to w. Since α fixes L we have $\alpha \in G_L$. But $\alpha(z) \neq z$ which means $z \notin M$, a contradiction. Thus $L = M$ as needed, and $L = E_{G_L}$. □

Example 10.26 Consider again the root field of $p(x) = -5 + x^3$, $E = \mathbb{Q}\left(\sqrt[3]{5}, \sqrt{-3}\right)$. The Galois group of $p(x)$ over $\mathbb{Q}$ is $G = \{\varepsilon, \alpha, \beta, \delta, \varphi, \sigma\}$ as found in Example 10.21. $H = \{\varepsilon, \delta\}$ is a subgroup of G, so how do we find the fixed field of H, i.e., E_H?

The basis for E over $\mathbb{Q}$ is $\left\{1, \sqrt[3]{5}, \sqrt[3]{25}, \sqrt{-3}, \sqrt[3]{5}\sqrt{-3}, \sqrt[3]{25}\sqrt{-3}\right\}$ so an arbitrary element $y \in \mathbb{Q}\left(\sqrt[3]{5}, \sqrt{-3}\right)$ can be expressed with $r, s, t, u, v, w \in \mathbb{Q}$ as:

$$y = r + s\left(\sqrt[3]{5}\right) + t\left(\sqrt[3]{25}\right) + u\left(\sqrt{-3}\right) + v\left(\sqrt[3]{5}\sqrt{-3}\right) + w\left(\sqrt[3]{25}\sqrt{-3}\right).$$

Using the fact that $\delta\left(\sqrt[3]{5}\right) = \sqrt[3]{5}$ we also know $\delta\left(\sqrt[3]{25}\right) = \sqrt[3]{25}$ so we calculate $\delta(y)$ as shown.

$$\begin{aligned}\delta(y) &= r + s\delta\left(\sqrt[3]{5}\right) + t\delta\left(\sqrt[3]{25}\right) + u\delta\left(\sqrt{-3}\right) + v\delta\left(\sqrt[3]{5}\right)\delta\left(\sqrt{-3}\right) + w\delta\left(\sqrt[3]{25}\right)\delta\left(\sqrt{-3}\right) \\ &= r + s\sqrt[3]{5} + t\sqrt[3]{25} - u\sqrt{-3} - v\sqrt[3]{5}\sqrt{-3} - w\sqrt[3]{25}\sqrt{-3}\end{aligned}$$

For y to be fixed by the automorphism δ it must be true that $y = \delta(y)$, or $y - \delta(y) = 0$.

Simplifying the expression $y - \delta(y) = 0$ gives us the equation below which must hold.

$$2u\left(\sqrt{-3}\right) + 2v\left(\sqrt[3]{5}\sqrt{-3}\right) + 2w\left(\sqrt[3]{25}\sqrt{-3}\right) = 0$$

Since the basis elements of E are linearly independent this implies $0 = 2u = 2v = 2w$, or $0 = u = v = w$. Thus the only elements that are fixed by δ are of the form $y = r + s\sqrt[3]{5} + t\sqrt[3]{25}$. We already know that $\{1, \sqrt[3]{5}, \sqrt[3]{25}\}$ is the basis for the field $\mathbb{Q}\left(\sqrt[3]{5}\right)$ over $\mathbb{Q}$. Thus $E_H = \mathbb{Q}\left(\sqrt[3]{5}\right)$. Exercises at the end of the chapter will ask you to find the fixed fields for other subgroups of G.

In the previous example only ε and δ fix $\sqrt[3]{5}$, which seems to imply that $G_{E_H} = H$ and leads to our next theorem.

Theorem 10.27

Let K be a field, $p(x) \in K[x]$ with root field E, and $G = Gal\left({}^{E}/_{K}\right)$. If H is a subgroup of G then the fixer of the field E_H is H, i.e., $G_{E_H} = H$.

Proof Let K be a field, $p(x) \in K[x]$ with root field E, and $G = Gal\left({}^{E}/_{K}\right)$. Assume H is a subgroup of G. By Theorem 10.23 we know that E_H is the subfield of E fixed by every element of H. Suppose that the fixer of E_H is the subgroup J. We will show $H = J$.

Let $\alpha \in H$. Since α fixes every element of E_H then $\alpha \in J$ and so $H \subseteq J$. We know H and J are finite sets since G is finite, and $J \subseteq H$ so by Theorem 0.15 we only need to show that $|H| = |J|$ to complete the proof.

Assume $|H| = m$ where $H = \{\alpha_1, \alpha_2, \ldots, \alpha_m\}$. Let $L = E_H$, then L is a field with $K \subseteq L \subseteq E$ and E is a root field over L as well. Thus $J = G_L = Gal\left({}^{E}/_{L}\right)$ and so by Theorem 10.20 $[E : L] = \left|Gal\left({}^{E}/_{L}\right)\right| = |J|$. Since $H \subseteq J$ we have $|J| \geq m$ so $[E : L] \geq m$.

Since E is a finite extension of K, it is also a finite extension of L so by Theorem 9.34 we can find $c \in E$ with $E = L(c)$. For each $\alpha_i \in H$ we have $\alpha_i(c) \in E$ and so we define the polynomial $d(x)$ in $E[x]$ shown below.

$$d(x) = (-\alpha_1(c) + x)(-\alpha_2(c) + x)\cdots(-\alpha_m(c) + x)$$

As H is a subgroup of G, $\varepsilon \in H$, thus one factor of $d(x)$ is $(-c+x)$ and thus c is a root of $d(x)$. Each α_i is an isomorphism from E to E so by Theorem 7.30 we have a corresponding isomorphism $\overline{\alpha}_i : E[x] \to E[x]$. For any $i \in \{1, 2, \ldots, m\}$ we calculate $\overline{\alpha}_i(d(x))$ as shown below.

$$\overline{\alpha}_i(d(x)) = (\alpha_i(-\alpha_1(c)) + x)(\alpha_i(-\alpha_2(c)) + x)\cdots(\alpha_i(-\alpha_m(c)) + x)$$

As H is a group, each $\alpha_i \circ \alpha_t$ is again in H, and $\alpha_i \circ \alpha_t \neq \alpha_i \circ \alpha_d$ when $\alpha_t \neq \alpha_d$. Thus $\overline{\alpha}_i(d(x)) = d(x)$ but with the factors possibly appearing in a different order and so the coefficients of $d(x)$ are fixed by every $\alpha_i \in H$ and $d(x) \in L[x]$. Since c is a root of $d(x)$, the minimum polynomial for c over L must divide $d(x)$ by Theorem 9.12. Hence degree of that minimum polynomial is no more than m and $[E : L] \leq m$. Therefore $[E : L] = m = |J|$ so $|H| = |J|$ and $H = J$ as we needed to show. □

Thus there is a one to one correspondence matching the subgroups of $Gal\left({}^{E}/_{K}\right)$ with the intermediate fields between K and E.

Example 10.28 Again, using the root field for $-5 + x^3$ over $\mathbb{Q}$ we found $G = Gal\left({}^{E}/_{\mathbb{Q}}\right) = \{\varepsilon, \alpha, \beta, \delta, \varphi, \sigma\}$. Subgroups of G must have a size dividing 6, by Theorem 3.9 (Lagrange's Theorem), so must have size 1, 2, 3, or 6. Clearly, we have $H_0 = \{\varepsilon\}$ and $H_5 = G$ as subgroups, as well as $H_1 = \{\varepsilon, \alpha, \beta\}$ from Example 10.22 and $H_2 = \{\varepsilon, \delta\}$ from Example 10.26. You can verify that the only other subgroups are $H_3 = \{\varepsilon, \sigma\}$, and $H_4 = \{\varepsilon, \varphi\}$.

Notice that this tells us there are exactly six different fields L with $\mathbb{Q} \subseteq L \subseteq \mathbb{Q}\left(\sqrt[3]{5}, \sqrt{-3}\right)$. A small group of six elements told us *all* of the intermediate fields between the infinite fields $\mathbb{Q}$ and $\mathbb{Q}\left(\sqrt[3]{5}, \sqrt{-3}\right)$.

In exercises at the end of the chapter, you will prove why $E_{H_0} = \mathbb{Q}\left(\sqrt[3]{5}, \sqrt{-3}\right)$ and $E_{H_5} = \mathbb{Q}$. We saw $E_{H_2} = \mathbb{Q}\left(\sqrt[3]{5}\right)$ and $E_{H_1} = \mathbb{Q}\left(\sqrt{-3}\right)$ in Examples 10.26 and 10.22. More exercises at the end of the chapter ask you to find the fixed fields for H_3 and H_4, completing the correspondence.

For E, a root field over K, and $G = Gal\left({}^{E}/_{K}\right)$, we saw that an intermediate field L between K and E gave us a subgroup $H = Gal\left({}^{E}/_{L}\right)$. If H is a normal subgroup or L is also a root field over K more can be said, giving us the final two theorems of this chapter.

Theorem 10.29

Let K be a field and E the root field for a polynomial over K. If L is a subfield of E with $K \subseteq L$ and L is a root field over K, then $Gal\left({}^{E}/_{L}\right) \triangleleft Gal\left({}^{E}/_{K}\right)$ and $Gal\left({}^{L}/_{K}\right) \cong {}^{Gal\left({}^{E}/_{K}\right)}/_{Gal\left({}^{E}/_{L}\right)}$.

Proof Let K be a field and E the root field for a polynomial over K. Assume L is a subfield of E with $K \subseteq L$ and L a root field over K. From an exercise in Chapter 9, we know E is also a root field over L, so each of the groups $Gal\left({}^{E}/_{K}\right)$, $Gal\left({}^{L}/_{K}\right)$, and $Gal\left({}^{E}/_{L}\right)$ are defined. The Fundamental Homomorphism Theorem (Theorem 3.34) will help us complete the proof once we define an onto homomorphism f from $Gal\left({}^{E}/_{K}\right)$ to $Gal\left({}^{L}/_{K}\right)$.

For each $\alpha \in Gal\left({}^{E}/_{K}\right)$ we need to find its image $f(\alpha)$, an automorphism of L that fixes

K. An exercise at the end of the chapter asks you to verify that for each $y \in L$ we have $\alpha(y) \in L$. Thus for any $\alpha \in Gal\left({}^{E}/_{K}\right)$, $\alpha(L) \subseteq L$. Thus we have fields L and $\alpha(L)$ with α, an isomorphism from L to $\alpha(L)$. Since L is a root field over K by Theorem 10.12, we know $L = \alpha(L)$. This tells us that any automorphism of E fixing K is also an automorphism of L fixing K, when its domain is restricted to L. We write α_L to denote the automorphism we get when the domain of α is restricted to L.

$$\text{Define } f : Gal\left({}^{E}/_{K}\right) \to Gal\left({}^{L}/_{K}\right) \text{ by } f(\alpha) = \alpha_L.$$

An exercise at the end of the chapter asks you to prove that f is onto and a homomorphism, so we will show that $ker(f) = Gal\left({}^{E}/_{L}\right)$. By definition $ker(f) = \{\alpha \in Gal\left({}^{E}/_{K}\right) : f(\alpha) = \varepsilon\}$ where ε denotes the identity function from L to L. Thus for every $\alpha \in ker(f)$, $\alpha_L = \varepsilon$ and α fixes L. Hence $\alpha \in Gal\left({}^{E}/_{L}\right)$ and $ker(f) \subseteq Gal\left({}^{E}/_{L}\right)$. But clearly if $\sigma \in Gal\left({}^{E}/_{L}\right)$ then σ fixes L and thus fixes K, so $\sigma \in Gal\left({}^{E}/_{K}\right)$ and $f(\sigma) = \sigma_L = \varepsilon$. Thus $\sigma \in ker(f)$, and we have $ker(f) = Gal\left({}^{E}/_{L}\right)$. Theorem 3.33 tells us that $Gal\left({}^{E}/_{L}\right) \lhd Gal\left({}^{E}/_{K}\right)$ and by Theorem 3.34 (FHT) we can conclude $Gal\left({}^{L}/_{K}\right) \cong {}^{Gal\left({}^{E}/_{K}\right)}/_{Gal\left({}^{E}/_{L}\right)}$. □

As part of the previous theorem we found: "If L is a root field over K then $Gal\left({}^{E}/_{L}\right) \lhd Gal\left({}^{E}/_{K}\right)$." The converse of this statement is also true, and completes the chapter.

Theorem 10.30

Let K be a field, E the root field for a polynomial over K, and L an intermediate field $K \subseteq L \subseteq E$. If $Gal\left({}^{E}/_{L}\right) \lhd Gal\left({}^{E}/_{K}\right)$ then L is a root field over K.

Proof Let K be a field, E the root field for a polynomial over K, and L an intermediate field $K \subseteq L \subseteq E$. Suppose we have $H = Gal\left({}^{E}/_{L}\right) \lhd Gal\left({}^{E}/_{K}\right) = G$, then we need to show that L is a root field over K. If $L = K$ or $L = E$, the statement is true, so assume instead that $L \neq K$ and $L \neq E$. The first step of this proof is to show that for any $\beta \in G$ we have $\beta(L) = L$. This is left as an exercise at the end of the chapter.

We know L is a finite extension of K, so $L = K(u)$ for some $u \in E$ by Theorem 9.34. Let $p(x) \in K[x]$ be the minimum polynomial for u over K. Suppose L is not the root field of $p(x)$, then there is a root w of $p(x)$ that is not in L. By Theorem 10.16, $w \in E$, so $K(w)$ is also an intermediate field between K and E. Now using Theorems 10.7 and 10.13 there is an automorphism β of E fixing K which maps u to w. Thus $\beta \in G$ and $\beta(L) \neq L$, giving us a contradiction. Therefore L is the root field of $p(x)$ over K as we needed to show. □

We can see these theorems in action in the examples used throughout this chapter, with $E = \mathbb{Q}\left(\sqrt[3]{5}, \sqrt{-3}\right)$. Notice that $\mathbb{Q}\left(\sqrt{-3}\right)$ is the root field of $p(x) = 3 + x^2$ over $\mathbb{Q}$, and the subgroup $H_1 = \{\varepsilon, \alpha, \beta\}$ is a normal subgroup of $Gal\left({}^{E}/_{\mathbb{Q}}\right)$ by Theorem 3.20. However, $\mathbb{Q}\left(\sqrt[3]{5}\right)$ is not a root field over $\mathbb{Q}$ since it does not contain the other two roots of the minimum polynomial for $\sqrt[3]{5}$, and the subgroup $H_2 = \{\varepsilon, \delta\}$ is not a normal subgroup of $Gal\left({}^{E}/_{\mathbb{Q}}\right)$ since $\alpha\delta\alpha^{-1} = \sigma \notin H_2$.

In the final chapter of this text, we will see how the relationship between permutation groups and automorphisms of the root field of a polynomial helped Galois to discover which polynomials are "solvable by radicals."

Exercises for Chapter 10

Section 10.1 Isomorphisms and Extension Fields

1. Prove that the function f defined in Example 10.1 is a homomorphism.
2. Prove that the function $f : \mathbb{Q}(i) \to \mathbb{Q}(i)$ defined by $f(a+bi) = a-bi$ is a homomorphism.
3. Consider the field $\mathbb{Z}_3(c)$ where c is a root of the irreducible polynomial $p(x) = 2+2x+x^2$. The elements are of the form $a + bc$ with $a, b \in \mathbb{Z}_3$ (as discovered in Example 9.18). Prove that the function $f : \mathbb{Z}_3(c) \to \mathbb{Z}_3(c)$ defined by $f(a + bc) = (a +_3 b) + (2 \cdot_3 b)c$ is a homomorphism. Remember that $c^2 = 1 + c$.
4. Consider the field $\mathbb{Z}_5(c)$ where c is a root of the irreducible polynomial $p(x) = 2 + x^2$ in $\mathbb{Z}_5[x]$. Prove that the function $f : \mathbb{Z}_5(c) \to \mathbb{Z}_5(c)$ defined by $f(r + sc) = r + (4 \cdot_5 s)c$ is a homomorphism. Remember that $2 + c^2 = 0$.
5. Give a counterexample to show that the function $f : \mathbb{Q}(i) \to \mathbb{Q}(i)$ defined by $f(a+bi) = a + (2b)i$ is *not* a homomorphism.
6. Give a counterexample to show that the function $f : \mathbb{Q}(\sqrt{2}) \to \mathbb{Q}(\sqrt{3})$ defined by $f(a + b\sqrt{2}) = a + b\sqrt{3}$ is *not* a homomorphism.
7. Give a counterexample to show that the function $f : \mathbb{Q}(\sqrt[3]{2}) \to \mathbb{Q}(\sqrt[3]{5})$ defined by $f(a + b\sqrt[3]{2} + c(\sqrt[3]{2})^2) = a + b\sqrt[3]{5} + c(\sqrt[3]{5})^2$ is *not* a homomorphism.
8. Give a counterexample to show that the function $f : \mathbb{Q}(\sqrt{7}) \to \mathbb{Q}(2 + i)$ defined by $f(a + b\sqrt{7}) = a + b(2 + i)$ is *not* a homomorphism.
9. Give a counterexample to show that the function $f : \mathbb{Q}(i) \to \mathbb{Q}(\sqrt{3})$ defined by $f(a + bi) = a + b\sqrt{3}$ is *not* a homomorphism.
10. Complete the proof of Theorem 10.4 by showing that $g(c_1) = c_2$ and $g(a) = f(a)$ for any $a \in K_1$.
11. To see why it was necessary to have an irreducible polynomial $p(x) \in K[x]$ for Theorem 10.7, consider the polynomial $p(x) = (1 + 2x + x^2)(1 + x^2)$ in $\mathbb{Q}[x]$. Show there are roots c_1, c_2 of $p(x)$ but there is no isomorphism between $\mathbb{Q}(c_1)$ and $\mathbb{Q}(c_2)$ which fixes $\mathbb{Q}$ and maps c_1 to c_2.
12. Show that no element of $\mathbb{Q}\left(\sqrt[3]{5}\right)$ can be a root of $p(x) = -2 + x^2$ as discussed after Theorem 10.9.

13. Explain how Theorem 10.9 tells us it is impossible to have an isomorphism $g : \mathbb{Q}(3+2i) \to \mathbb{Q}(\sqrt{13})$ which fixes $\mathbb{Q}$.
14. Explain how Theorem 10.9 tells us it is impossible to have an isomorphism $g : \mathbb{Q}(1 - \sqrt{-5}) \to \mathbb{Q}(\sqrt{7})$ which fixes $\mathbb{Q}$.
15. Explain how Theorem 10.9 tells us it is impossible to have an isomorphism $g : \mathbb{Q}(\sqrt{-5}) \to \mathbb{Q}(i)$ which fixes $\mathbb{Q}$.

Section 10.2 Automorphisms of Root Fields

16. Complete the proof of Theorem 10.12 by proving that $E_1 \subseteq E_2$. (Don't forget $E_1 = K(c_1, c_2, \ldots, c_m)$.)
17. Explain how Theorem 10.12 guarantees there is no isomorphism between $\mathbb{Q}(\omega_3)$ and $\mathbb{Q}(\sqrt[3]{2})$ which fixes $\mathbb{Q}$.
18. Explain how Theorem 10.12 guarantees there is no isomorphism between $\mathbb{Q}(\sqrt{5})$ and $\mathbb{Q}(\sqrt{3})$ which fixes $\mathbb{Q}$.
19. Explain how Theorem 10.12 guarantees there is no isomorphism between $\mathbb{Q}(i)$ and $\mathbb{Q}(\sqrt{7})$ which fixes $\mathbb{Q}$.
20. Explain how Theorem 10.12 guarantees there is no isomorphism between $\mathbb{Q}(\sqrt{-5})$ and $\mathbb{Q}(\sqrt[3]{5})$ which fixes $\mathbb{Q}$.
21. Explain how Theorem 10.12 guarantees there is no isomorphism between $\mathbb{Q}(\sqrt[4]{5})$ and $\mathbb{Q}(\sqrt{5})$ which fixes $\mathbb{Q}$. Think about the degrees of the extensions.
22. Explain how Theorem 10.12 guarantees there is no isomorphism between $\mathbb{Z}_3(c)$ and $\mathbb{Z}_3(d)$ which fixes $\mathbb{Z}_3$, where c is a root of $p(x) = 1+x^2$ and d is a root of $q(x) = 2+x+2x^3$. Think about the degrees of the extensions, and be sure to show that $\mathbb{Z}_3(c)$ is a root field.
23. Prove Theorem 10.14. Use Theorems 10.7 and 10.13 to help, and remember that $p(x)$ is separable.
24. Use the polynomials $p(x) = -3 + x^2$ and $a(x) = (-3 + x^2)(1 + x^2)$ to show why $a(x)$ is required to be irreducible in Theorem 10.16.
25. Prove: If E is a root field over K and $p(x) \in K[x]$ is irreducible over K with $deg(p(x)) > [E : K]$ then $p(x)$ has no root in E.
26. Suppose p is a prime integer. Explain why the root field of $-p + x^n$ over $\mathbb{Q}$ must also contain every n^{th} root of unity.

Section 10.3 The Galois Group of a Polynomial

27. Complete the proof of Theorem 10.18, showing that if $\alpha \in G$ then α^{-1} is a homomorphism and fixes K.
28. Use $E = \mathbb{Q}(\sqrt[3]{5})$ to show that E must be the *root field* of a polynomial $p(x) \in K[x]$ in Theorem 10.20. How many automorphisms of E that fix $\mathbb{Q}$ are there?
29. Find all of the automorphisms for the root field of $p(x) = 1 + x^2$ over $\mathbb{Z}_3$ which fix $\mathbb{Z}_3$. Use Theorem 10.20 to help show you have all of them.

30. Find all of the automorphisms for the root field of $p(x) = 1 + x + x^2$ over $\mathbb{Q}$ which fix $\mathbb{Q}$. Use Theorem 10.20 to help show you have all of them. Does this polynomial look familiar?
31. Consider the polynomial $p(x) = (-2 + x^2)(-7 + x^2)$ in $\mathbb{Q}[x]$. Find the root field E of $p(x)$ and find all of the automorphisms of E fixing $\mathbb{Q}$.
32. Find the Cayley table under composition for the Galois group found in Exercise 31 and identify all of its subgroups.
33. Consider the polynomial $p(x) = (1+x^2)(-5+x^2)$ in $\mathbb{Q}[x]$. Find the root field E of $p(x)$ and find all of the automorphisms of E fixing $\mathbb{Q}$.
34. Find the Cayley table under composition for the Galois group found in Exercise 33 and identify all of its subgroups.
35. Consider the polynomial $p(x) = (1 + x + x^2)(-2 + x^2)$ in $\mathbb{Q}[x]$. Find the root field E of $p(x)$ and find all of the automorphisms of E fixing $\mathbb{Q}$.
36. Find the Cayley table under composition for the Galois group found in Exercise 35 and identify all of its subgroups.
37. Consider the polynomial $p(x) = (1 + x + x^2 + x^3)(3 + x^2)$ in $\mathbb{Q}[x]$. Find the root field E of $p(x)$ and find all of the automorphisms of E fixing $\mathbb{Q}$.
38. Find the Cayley table under composition for the Galois group found in Exercise 37 and identify all of its subgroups.
39. Consider the polynomial $p(x) = 1 + x^4$ in $\mathbb{Q}[x]$. Find the root field E of $p(x)$ and find all of the automorphisms of E fixing $\mathbb{Q}$. Hint: One of the roots for $p(x)$ is $\frac{1}{\sqrt{2}}(1 + i)$.
40. Find the Cayley table under composition for the Galois group found in Exercise 39 and identify all of its subgroups.
41. Consider the polynomial $p(x) = -1 + x^5$ in $\mathbb{Q}[x]$. Find the root field E of $p(x)$ and find all of the automorphisms of E fixing $\mathbb{Q}$. Look back at Project 9.8 for some help.
42. Find the Cayley table under composition for the Galois group found in Exercise 41 and identify all of its subgroups.
43. Create the Cayley table for $G = \{\varepsilon, \alpha, \beta, \delta, \varphi, \sigma\}$ defined in Example 10.21 and verify that G is a group under composition.
44. Using β from Example 10.21, calculate $\beta\left(\frac{-\sqrt[3]{5}}{2}\left(1+\sqrt{-3}\right)\right)$, $\beta\left(\frac{-\sqrt[3]{5}}{2}\left(1-\sqrt{-3}\right)\right)$, and $\beta\left(\sqrt[3]{5}\right)$ then write β as a permutation of the roots of $-5 + x^3$ as was done for φ.
45. Using α from Example 10.21, calculate $\alpha\left(\frac{-\sqrt[3]{5}}{2}\left(1+\sqrt{-3}\right)\right)$, $\alpha\left(\frac{-\sqrt[3]{5}}{2}\left(1-\sqrt{-3}\right)\right)$, and $\alpha\left(\sqrt[3]{5}\right)$ then write α as a permutation of the roots of $-5 + x^3$ as was done for φ.
46. Using δ from Example 10.21, calculate $\delta\left(\frac{-\sqrt[3]{5}}{2}\left(1+\sqrt{-3}\right)\right)$, $\delta\left(\frac{-\sqrt[3]{5}}{2}\left(1-\sqrt{-3}\right)\right)$, and $\delta\left(\sqrt[3]{5}\right)$ then write δ as a permutation of the roots of $-5 + x^3$ as was done for φ.
47. Using σ from Example 10.21, calculate $\sigma\left(\frac{-\sqrt[3]{5}}{2}\left(1+\sqrt{-3}\right)\right)$, $\sigma\left(\frac{-\sqrt[3]{5}}{2}\left(1-\sqrt{-3}\right)\right)$, and $\sigma\left(\sqrt[3]{5}\right)$ then write σ as a permutation of the roots of $-5 + x^3$ as was done for φ.
48. Show the steps to compose each pair of automorphisms to verify that H in Exam-

ple 10.22 is a subgroup.

49. Suppose K is a field and E is the root field of some $p(x) \in K[x]$. Prove: If L is a field with $K \subseteq L \subseteq E$ then $Gal\left({}^{E}/_{L}\right) \subseteq Gal\left({}^{E}/_{K}\right)$ and $G_L = Gal\left({}^{E}/_{L}\right)$.

Section 10.4 The Galois Correspondence

50. Suppose K is a field, E a root field over K, and $G = Gal\left({}^{E}/_{K}\right)$. Prove: If $H = \{\varepsilon\}$ (where ε is the identity automorphism) then $E_H = E$.
51. Suppose K is a field, E a root field over K, and $G = Gal\left({}^{E}/_{K}\right)$. Prove: $E_G = K$.
52. Find the fixed field for the subgroup H_3 of $G = Gal\left({}^{E}/_{\mathbb{Q}}\right) = \{\varepsilon, \alpha, \beta, \delta, \varphi, \sigma\}$ defined in Example 10.28.
53. Find the fixed field for the subgroup H_4 of $G = Gal\left({}^{E}/_{\mathbb{Q}}\right) = \{\varepsilon, \alpha, \beta, \delta, \varphi, \sigma\}$ defined in Example 10.28.
54. Find the fixed fields for all of the subgroups found in Exercise 32.
55. Find the fixed fields for all of the subgroups found in Exercise 34.
56. Find the fixed fields for all of the subgroups found in Exercise 36.
57. Find the fixed fields for all of the subgroups found in Exercise 38.
58. Find the fixed fields for all of the subgroups found in Exercise 40.
59. Find the fixed fields for all of the subgroups found in Exercise 42.
60. Suppose K is a field, E the root field for a polynomial over K, and L is a field with $K \subseteq L \subseteq E$. Prove: If L is a root field over K then for every $y \in L$ and $\alpha \in Gal\left({}^{E}/_{K}\right)$ we must have $\alpha(y) \in L$.
61. Complete the proof of Theorem 10.29 by showing that the function f defined in the proof is an onto homomorphism.
62. Complete the proof of Theorem 10.30 by showing that for any $\beta \in G$ we have $\beta(L) = L$. Use use the fact that $H \lhd G$ and the group $G_{\beta(L)}$ to help.

Projects for Chapter 10

Project 10.1

In Project 9.4 you found the root field for $a(x) = (-(i+2)+x^2)((i-2)+x^2)$ over $K = \mathbb{Q}(i)$, namely $E = \mathbb{Q}(u, w)$ where $u = \sqrt{i+2}$ and $w = \sqrt{-i+2}$.

1. How many distinct automorphisms of $\mathbb{Q}(u, w)$ which fix $\mathbb{Q}(i)$ must exist? Explain how you knew this number.
2. Describe each automorphism in $G = Gal({}^{E}/_{K})$ by its action on u and w, then show the Cayley table for G.
3. Find all of the subgroups of G, verifying that each is in fact a subgroup.

Project 10.2

Consider the polynomial $a(x) = -1 + x^7$. We know its root field is $\mathbb{Q}(\omega_7)$ from Project 9.8. Let $G = Gal\left(^{\mathbb{Q}(\omega_7)}/_{\mathbb{Q}}\right)$.

1. Explain why $[\mathbb{Q}(\omega_7) : \mathbb{Q}] = 6$ and find a basis for $\mathbb{Q}(\omega_7)$ over $\mathbb{Q}$.
2. Explain why each $\alpha \in G$ is completely described by its action on ω_7.
3. Since ω_7 must map back to a root of its minimum polynomial over $\mathbb{Q}$, we can easily describe the elements of G. Finish the descriptions and the Cayley table for G started below.

$$\begin{aligned} \varepsilon(\omega_7) &= \omega_7 \\ f(\omega_7) &= (\omega_7)^2 \\ g(\omega_7) &= (\omega_7)^3 \\ h(\omega_7) &= \\ j(\omega_7) &= \\ k(\omega_7) &= \end{aligned}$$

$\circ$	ε	f	g	h	j	k
ε	ε	f	g	h	j	k
f	f					
g	g					
h	h					
j	j					
k	k					

Don't forget that the operation is composition, so $(f \circ g)(\omega_7) = f(g(\omega_7)) = f((\omega_7)^3) = (f(\omega_7))^3 = ((\omega_7)^2)^3 = (\omega_7)^6$. Also $(\omega_7)^7 = 1$.

4. Show that G is a cyclic group and thus abelian.

In fact, any root field of the form $E = \mathbb{Q}(\omega_n)$ will have $Gal(^E/_{\mathbb{Q}})$ abelian since if we have $f(\omega_n) = (\omega_n)^t$ and $g(\omega_n) = (\omega_n)^v$ then $(g \circ f)(\omega_n) = (\omega_n)^{tv} = (f \circ g)(\omega_n)$.

5. Find all of the subgroups of G.

Project 10.3

Consider the root field over $\mathbb{Q}$ for the polynomial $a(x) = -2 + x^3$ which we know is $E = \mathbb{Q}(\sqrt[3]{2}, \omega_3)$ from Project 9.8. Recall that any $\alpha \in Gal(^E/_{\mathbb{Q}})$ is completely determined by $\alpha(\sqrt[3]{2})$ and $\alpha(\omega_3)$.

1. If $\alpha \in Gal(^E/_{\mathbb{Q}})$ what are the possible choices for $\alpha(\sqrt[3]{2})$? Don't forget what the minimum polynomial is for $\sqrt[3]{2}$.
2. If $\alpha \in Gal(^E/_{\mathbb{Q}})$ what are the possible choices for $\alpha(\omega_3)$? Don't forget what the minimum polynomial is for ω_3.
3. Write each of the automorphisms in $G = Gal(^E/\mathbb{Q})$ by its action on $\sqrt[3]{2}$ and ω_3, following the pattern started next.

$$\varepsilon: \begin{array}{ccc} \sqrt[3]{2} & \to & \sqrt[3]{2} \\ \omega_3 & \to & \omega_3 \end{array} \qquad \alpha: \begin{array}{ccc} \sqrt[3]{2} & \to & \sqrt[3]{2} \\ \omega_3 & \to & (\omega_3)^2 \end{array} \qquad \beta: \begin{array}{ccc} \sqrt[3]{2} & \to & \sqrt[3]{2}(\omega_3) \\ \omega_3 & \to & \omega_3 \end{array}$$

$$\delta: \begin{array}{ccc} \sqrt[3]{2} & \to & \sqrt[3]{2}(\omega_3) \\ \omega_3 & \to & \end{array} \qquad \gamma: \begin{array}{ccc} \sqrt[3]{2} & \to & \\ \omega_3 & \to & \end{array} \qquad \kappa: \begin{array}{ccc} \sqrt[3]{2} & \to & \\ \omega_3 & \to & \end{array}$$

4. Complete the Cayley table for G. Is G abelian?
   Notice $(\alpha \circ \beta)(\sqrt[3]{2}) = \alpha(\beta(\sqrt[3]{2})) = \alpha(\sqrt[3]{2}\omega_3) = \alpha(\sqrt[3]{2})\alpha(\omega_3) = \sqrt[3]{2}(\omega_3)^2$.

$\circ$	ε	α	β	δ	γ	κ
ε	ε	α	β	δ	γ	κ
α	α					
β	β					
δ	δ					
γ	γ					
κ	κ					

5. Find all of the subgroups of G.

Project 10.4

Use the Galois group $G = Gal(^{\mathbb{Q}(\omega_7)}/_{\mathbb{Q}})$ discovered in Project 10.2 to solve the problems of this project.

1. List the four subgroups of G.
2. Prove that for any field E which is a root field over K, the identity automorphism of E, ε, has $E_{\{\varepsilon\}} = E$. (This is also Exercise 50.)
3. Prove that for any field E which is a root field over K, with $G = Gal(^E/_K)$, we have $E_G = K$. (This is also Exercise 51.) Use Theorems 10.7 and 10.13 to help in a contradiction proof.

   Thus two of our four fixed fields for $G = Gal(^{\mathbb{Q}(\omega_7)}/_{\mathbb{Q}})$ are clear.

4. Write an arbitrary element $y \in E$ in terms of the basis for E, then compute $f(y)$, $h(y)$, $k(y)$ simplifying the answers to again be written in terms of the basis elements. Don't forget that ω_7 is a root of $1 + x + x^2 + x^3 + x^4 + x^5 + x^6$ so that $(\omega_7)^6 = -1 - \omega_7 - (\omega_7)^2 - (\omega_7)^3 - (\omega_7)^4 - (\omega_7)^5$.
5. For an arbitrary $y \in E$, set $y = f(y)$ and simplify to determine what an element of E fixed by f must look like. (See Example 10.26.)
6. Repeat the previous question using g and k to see what elements they each fix.
7. Find the fixed fields of your subgroups defined in the first question of this project.

Project 10.5

Consider the group $G = Gal(^E/_K)$, $K = \mathbb{Q}(i)$, and $E = \mathbb{Q}(u, w)$ where $u = \sqrt{i+2}$ and $w = \sqrt{-i+2}$. The elements of G as well as the subgroups were found in Project 10.1.

1. List the subgroups of G again. Which two fixed fields can we already identify? (See Project 10.4 or Exercises 50 and 51.)
2. Show how an arbitrary element $y \in E$ can be expressed in terms of u and w (what is the basis for E over $\mathbb{Q}(i)$?), then for $\alpha, \beta, \delta \in G$, find $\alpha(y), \beta(y), \delta(y)$. Remember to simplify each image to be written in terms of the basis for E.
3. Determine what elements of E have $\alpha(y) = y$, by equating correct coefficients of the basis elements. Then find the appropriate fixed field this determines.
4. Repeat the previous problem for each of β and δ.
5. Use the fact that $i^2 = -1$ to compute a simplified version of uw. Then use this to rewrite the fixed field for $\{\varepsilon, \delta\}$.

Project 10.6

Consider the group $G = Gal(^E/_K)$, $E = \mathbb{Q}(\sqrt[3]{2}, \omega_3)$, and $K = \mathbb{Q}$ The elements of G as well as the subgroups were found in Project 10.3.

1. List the subgroups of G again. Which two fixed fields can we already identify? (See Project 10.4 or Exercises 50 and 51.)
2. Show how an arbitrary element $y \in E$ can be expressed in terms of the basis for E over $\mathbb{Q}$, then for each automorphism α, δ, κ, find $\alpha(y), \delta(y), \kappa(y)$ simplifying to express each image in terms of the basis elements of E again. Remember that ω_3 is a root of $1 + x + x^2$ so we know $(\omega_3)^2 = -1 - \omega_3$.
3. Use the previous answers to determine what elements of E are fixed by each of α, δ, κ. What are the appropriate fixed fields based on these results?
4. Repeat the previous steps for β and γ to determine the final fixed field for G.

Project 10.7

Suppose E is the root field of a polynomial over K and $G = Gal\left(^E/_K\right)$.
Define $L = \{u \in E$: for all $\sigma, \tau \in G$ we have $(\sigma \circ \tau)(u) = (\tau \circ \sigma)(u)\}$.

1. Prove that L is an intermediate field with $K \subseteq L \subseteq E$.
2. Let $H = Gal(^E/_L)$. Show $H \triangleleft G$.
3. Show that as a group, $^G/_H$ is abelian. (Use cosets.)
4. Use all of the above information with Theorems 10.29 and 10.30 to explain why $Gal(^L/_K)$ is abelian.

Chapter 11

Solvability

Degrees of field extensions and Galois groups can be used to show that some famous mathematical problems have no solution, as well as helping us determine what polynomials are "solvable by radicals." We begin with three construction problems from Euclidean Geometry. Each field in this chapter is an extension of $\mathbb{Q}$ unless otherwise stated.

11.1 Three Construction Problems

In Euclidean Geometry there are many "constructions," i.e., theorems describing how to construct points, from which a specific figure is created, using only an unmarked ruler (straightedge) and compass (to draw circles). Three such problems evaded the efforts of mathematicians for over 1000 years [4], but the theorems of field extensions finally showed they cannot be constructed. The construction problems are stated here first.

1. ***Doubling the Cube***: If points already existed (were previously constructed) in space as vertices of a cube, construct the points (vertices) for a cube with exactly double the volume of the original cube.
2. ***Squaring the Circle***: If a circle exists (its center and radius already constructed), construct the points (vertices) needed to form a square with exactly the same area as the circle.
3. ***Trisecting an Angle***: For any angle that has been constructed (three points have already been constructed with one as the vertex of the angle and the other two creating rays from that vertex to define the angle), construct a line though the same vertex that creates an angle of exactly one third the measure of the original angle.

Geometric constructions of this type follow very precise axioms and theorems (see [2]). Points are found as intersections of lines, a line with a circle, or two circles, each constructed from previous points. We can assume we have any point with rational coordinates constructed, as exercises at the end of the chapter will help justify.

The coordinates of the points that are constructed, when adjoined to $\mathbb{Q}$, create finite extensions of $\mathbb{Q}$ with very special properties. In each situation we begin with $\mathbb{Q}$ or K, a finite extension of $\mathbb{Q}$ created when the coordinates of previously constructed points are adjoined.

1. ***Two Lines Intersecting***: Suppose we have constructed two distinct, nonparallel lines $ax + by = c$, and $ex + fy = g$, where the coefficients a, b, c, d, e, f, g already exist in our field K ($\mathbb{Q}$ or a previous finite extension of $\mathbb{Q}$), and that these two lines intersect at some point (u, v). But this point (u, v) already has coordinates in the field K (an exercise at the end of the chapter), so we do not need to extend K to include these points.
2. ***A Line Intersecting a Circle***: Suppose we have constructed a circle and a line which intersect at a point (u, v). We can write the equations as $x^2 + ax + y^2 + by + c = 0$ and $dx + ey = f$, where a, b, c, d, e, f are in our field K. Notice that at least one of d or e must be nonzero, so assume $d \neq 0$. You will show, in an exercise at the end of the chapter, that v is a root of the polynomial $p(x) \in K[x]$ defined by:

   $$p(x) = [(d^{-1}f)^2 + afd^{-1} + c] + [-2ef(d^{-1})^2 - aed^{-1} + b]x + [(d^{-1}e)^2 + 1]x^2.$$

   If $p(x)$ is irreducible over K then $[K(v) : K] = 2$, but if $p(x)$ is reducible over K then $[K(v) : K] = 1$. Once v is adjoined to K, $u = d^{-1}(f - ev)$ and so $u \in K(v)$. Thus adjoining the coordinates of the intersection point to K results in a field extension $K(v)$ with degree 1 or 2 over K.
3. ***Two Intersecting Circles***: Suppose we have constructed distinct circles which intersect at a point (u, v). The circles can be expressed as $x^2 + ax + y^2 + by + c = 0$ and $x^2 + dx + y^2 + ey + f = 0$, where $a, b, c, d, e, f \in K$. We can then subtract the two equations to get $(a - d)x + (b - e)y + (c - f) = 0$ which is the equation of a line that also intersects the two circles at (u, v).

   If $a = d$ and $b = e$ then also $c = f$ so the circles were not distinct. Hence we can assume at least one of $a \neq d$ or $b \neq e$. Notice that if $a = d$ but $b \neq e$ then we have the horizontal line $y = -(c - f)(b - e)^{-1}$, which can be constructed from the points $(0, -(c-f)(b-e)^{-1})$ and $(1, -(c-f)(b-e)^{-1})$ whose coordinates are in K. Similarly, if $a \neq d$ but $b = e$ the line is vertical and constructed using the points $(-(c-f)(a-d)^{-1}, 0)$ and $(-(c - f)(a - d)^{-1}, 1)$ whose coordinates are in K.

Finally, if $a \neq d$ and $b \neq e$ then the line can be constructed from the points $(0, -(c-f)(b-e)^{-1})$ and $\left(-(c-f)(a-d)^{-1}, 0\right)$, whose coordinates are already in K. Thus, we could instead use this line and one of the circles to construct (u, v). Hence, the degree of the extension containing the coordinates u and v is either 1, or 2 over K as seen in the previous case.

A complete construction in Euclidean Geometry involves constructing a finite number of points, thus creating a field extension E of $\mathbb{Q}$ with $[E : \mathbb{Q}] = 2^n$ for some nonnegative integer n, by Theorem 9.22. We will find that the three constructions discussed at the beginning of the section require extensions of other degrees, and thus they cannot be solved by straightedge and compass.

1. ***Doubling the Cube***: Consider the cube of side length 1, one of whose corners is at the origin (0,0,0) in space. This cube has vertices whose coordinates are in $\mathbb{Q}$, and its volume is 1 cubic unit. If it can be doubled we can construct a cube with volume exactly 2 cubic units, so each of its sides measures $\sqrt[3]{2}$ units. Now we can construct the cube with one corner at (0,0,0) and another at $\left(\sqrt[3]{2}, 0, 0\right)$, and the field extension K containing the coordinates of this point must contain $\mathbb{Q}\left(\sqrt[3]{2}\right)$ by Theorem 9.5. However, the minimum polynomial for $\sqrt[3]{2}$, over $\mathbb{Q}$ is $p(x) = -2 + x^3$ (irreducible over $\mathbb{Q}$ by Eisensteins Criterion 8.35), and thus $\left[\mathbb{Q}\left(\sqrt[3]{2}\right) : \mathbb{Q}\right] = 3$. By Theorem 9.22 $[K : \mathbb{Q}] = \left[K : \mathbb{Q}\left(\sqrt[3]{2}\right)\right]\left[\mathbb{Q}\left(\sqrt[3]{2}\right) : \mathbb{Q}\right]$ is a multiple of 3. But $[K : \mathbb{Q}]$ is a power of 2 so $\mathbb{Q}\left(\sqrt[3]{2}\right)$ is not a subfield of K, a contradiction. Thus we cannot construct this cube by straightedge and compass.
2. ***Squaring the Circle***: The unit circle in the xy-plane can be constructed using the points (0,0) and (1,0) whose coordinates are already in $\mathbb{Q}$. The area of this circle is π square units, so if there exists a square with the same area, the side length for the new square is $\sqrt{\pi}$ units. We now construct the square with one corner at $(0, \sqrt{\pi})$. However, π is transcendental over $\mathbb{Q}$ as explained in Example 9.3, so an extension of $\mathbb{Q}$ containing $\sqrt{\pi}$ would also contain π and have infinite dimension over $\mathbb{Q}$. Thus we cannot construct the square by straightedge and compass.
3. ***Trisecting an Angle***: Consider a 60° angle with the ray from $A = (0, 0)$ to $B = (1, 0)$ as the base of the angle. In an exercise you will show this angle can be constructed by ruler and compass. If we can trisect this 60° angle, we can construct a line through (0,0) and a point D on that line so that the rays $\overrightarrow{AB}$ and $\overrightarrow{AD}$ form a 20° angle. The new line intersects the unit circle at a point, (u, w) where $u = \cos(20°)$. You will show in a exercise that this implies $2u$ is a root of the irreducible polynomial $p(x) = 1 + 3x - x^3 \in \mathbb{Q}[x]$. Hence $[\mathbb{Q}(2u) : \mathbb{Q}] = 3$. Since $2u \in K(u)$ and $\mathbb{Q} \subseteq K(u)$, then $\mathbb{Q}(2u)$ is a subfield of $K(u)$. Thus $[K(u) : \mathbb{Q}] = [K(u) : \mathbb{Q}(2u)][\mathbb{Q}(2u) : \mathbb{Q}]$ by Theorem 9.22. This tells us $[K(u) : \mathbb{Q}]$ is a multiple of 3, but $[K(u) : \mathbb{Q}]$ is a power of 2, a contradiction. So the 60° angle cannot be trisected by straightedge and compass.

You will find other applications to geometric constructions in the exercises.

11.2 Solvable Groups

Definition 11.1 A group G is called a ***solvable group*** if there are subgroups $\{e_G\} = H_0, H_1, \ldots, H_n = G$, so that for each $0 \leq i \leq n-1$, $H_i \lhd H_{i+1}$ and ${}^{H_{i+1}}/_{H_i}$ is an abelian group.

Example 11.2 Consider $p(x) = -5 + x^3 \in \mathbb{Q}[x]$ and its root field $E = \mathbb{Q}\left(\sqrt[3]{5}, \sqrt{-3}\right)$ from Example 9.27. We found the Galois group of $p(x)$, $G = Gal\left({}^{E}/_{\mathbb{Q}}\right) = \{\varepsilon, \alpha, \beta, \delta, \varphi, \sigma\}$ in Example 10.21 and subgroup $H = \{\varepsilon, \alpha, \beta\}$ in Example 10.22. Notice that the quotient group ${}^{G}/_{H} = \{\varepsilon H, \delta H\}$ since $\delta H = \{\delta, \sigma, \varphi\}$.

At the end of Chapter 10, we saw that $H \lhd G$. Also, ${}^{G}/_{H}$ is a group of order 2, so must be cyclic by Theorem 3.10. Using the subgroup $\{\varepsilon\}$ we see that $\{\varepsilon\} \lhd H$ and ${}^{H}/_{\{\varepsilon\}} = \{\varepsilon\{\varepsilon\}, \alpha\{\varepsilon\}, \beta\{\varepsilon\}\}$ is a group of order 3, and again cyclic. Thus we have $\{\varepsilon\} \lhd H \lhd G$ with ${}^{G}/_{H}$ and ${}^{H}/_{\{\varepsilon\}}$ abelian groups by Theorem 2.13. Therefore our group $Gal\left({}^{E}/_{\mathbb{Q}}\right) = \{\varepsilon, \alpha, \beta, \delta, \varphi, \sigma\}$ is a solvable group.

Not every group is solvable as seen in the next theorem.

Theorem 11.3

The permutation group S_5 is not solvable.

Proof Suppose that S_5 is solvable. We know S_5 is not abelian as $(12)(123) \neq (123)(12)$, so there must exist $n > 1$ and subgroups $\{\varepsilon\} = H_0, H_1, H_2, \cdots, H_n = S_5$ so that for each j, $H_j \lhd H_{j+1}$ and ${}^{H_{j+1}}/_{H_j}$ is abelian. We know $H_n = S_5$ contains all three cycles but $\{\varepsilon\}$ does not. Hence there is some $0 < j \leq n$ which is the smallest j for which H_j contains all three cycles. Thus H_{j-1} does not contain some three cycle. The next steps will be virtually identical using any three cycle (abc) that is not in H_{j-1}, so we will assume $(123) \notin H_{j-1}$. Thus we know $(123)H_{j-1} \neq \varepsilon H_{j-1}$.

Notice $(123) = (143)(253)(341)(352)$ and $(143), (253), (341), (352) \in H_j$ so $(123)H_{j-1} = (143)(253)(341)(352)H_{j-1}$. Using coset rules and the fact that ${}^{H_j}/_{H_{j-1}}$ is an abelian group we find the following contradiction. Hence S_5 is not solvable.

$$
\begin{aligned}
(123)H_{j-1} &= ((143)H_{j-1}) * ((253)H_{j-1}) * ((341)H_{j-1}) * ((352)H_{j-1}) \\
&= ((143)H_{j-1} * (341)H_{j-1}) * ((253)H_{j-1} * ((352)H_{j-1}) \\
&= \varepsilon H_{j-1}
\end{aligned}
$$

□

Since automorphisms of the root field of a polynomial can be identified with permutations of the roots, it will be important to know if S_n is solvable. It turns out that 5 is a critical value of n since S_5 is not solvable, but S_4 is solvable.

Theorem 11.4

The permutation group S_4 is a solvable group.

Proof The group S_4 has 24 elements by Theorem 2.40, and the subgroup A_4 of even permutations has 12 elements by the same theorem. We know that $A_4 \triangleleft S_4$ by Theorem 3.20. Also, since ${}^{S_4}/_{A_4}$ has exactly two elements, it is a cyclic group by Theorem 3.10 and thus abelian.

Now consider $H = \{\varepsilon, (12)(34), (13)(24), (14)(23)\}$ which is a subset of A_4. You will show, in exercises at the end of the chapter, that H is a subgroup of A_4, H is abelian, and $H \triangleleft A_4$. Then ${}^{A_4}/_{H}$ is a group of exactly three elements, and thus cyclic by Theorem 3.10 and abelian by Theorem 2.13. Also, $\{\varepsilon\} \triangleleft H$ with ${}^{H}/_{\{\varepsilon\}}$ abelian. Thus we have subgroups $\{\varepsilon\} \triangleleft H \triangleleft A_4 \triangleleft S_4$, and each quotient ${}^{H}/_{\{\varepsilon\}}, {}^{A_4}/_{H}, {}^{S_4}/_{A_4}$ is abelian, showing that S_4 is solvable. □

There is no guarantee that a Galois group will always be the entire group S_n, so the next two theorems are frequently used. We will not prove (ii) of Theorem 11.5 here as it requires material beyond the scope of this text. See [1] for details.

Theorem 11.5

Let G be a group and J a subgroup of G.

(i) If G is a solvable group then J is a solvable group.
(ii) If $J \triangleleft G$ and both J and ${}^{G}/_{J}$ are solvable groups, then G is a solvable group.

Proof In an exercise you will show that if $G = \{e_G\}$ then (i) and (ii) hold, thus for the rest of the proof assume $G \neq \{e_G\}$.

(i) Suppose we have a subgroup J of G and G is solvable. There exist subgroups $\{e_G\} = H_0, H_1, \ldots, H_n = G$ where for each $0 \leq i \leq n-1$, $H_i \triangleleft H_{i+1}$, and ${}^{H_{i+1}}/_{H_i}$ is an abelian group. Then for each $0 \leq i \leq n$ we know that $H_i \cap J$ is a subgroup of G by Theorem 1.33. But since $H_i \cap J \subseteq J$ then $H_i \cap J$ is a subgroup of J as well. Thus, we have subgroups $\{e_G\} = (H_0 \cap J), (H_1 \cap J), \ldots, (H_n \cap J) = J$. In exercises, you will show that for each $0 \leq i \leq n-1$, $(H_i \cap J) \triangleleft (H_{i+1} \cap J)$, and ${}^{H_{i+1} \cap J}/_{H_i \cap J}$ is abelian. Hence, J is a solvable group. □

Theorem 11.6

Suppose G and B are groups. If there is an onto homomorphism $f : G \to B$ and G is a solvable group, then B is a solvable group.

Proof Suppose G and B are groups, $f : G \to B$ is an onto homomorphism, and G is a solvable group. If $G = \{e_G\}$ and f is onto then $B = \{e_B\}$ and so B is easily solvable. Thus assume $G \neq \{e_G\}$ for the rest of the proof.

As G is solvable there exist subgroups $\{e_G\} = H_0, H_1, \ldots, H_n = G$ where for each $0 \leq i \leq n-1$, $H_i \lhd H_{i+1}$, and ${}^{H_{i+1}}/_{H_i}$ is an abelian group. Define $B_i = f(H_i)$ for each i. By Theorem 3.33, B_i is a subgroup of B, for each i. As $f(e_G) = e_B$ by Theorem 1.37, then $B_0 = \{e_B\}$ as we need, and by f onto we know $B_n = f(G) = B$. In exercises at the end of the chapter you will show that for each $0 \leq i \leq n-1$, $B_i \lhd B_{i+1}$, and ${}^{B_{i+1}}/_{B_i}$ is abelian. Therefore B is a solvable group. □

From Theorem 11.5 we know that every subgroup of S_4 is solvable. Now the subgroup $H = \{\alpha \in S_4 : \alpha(4) = 4\}$ of S_4 is solvable, and in an exercise at the end of the chapter you will show that this subgroup H is isomorphic to S_3. Thus by Theorem 11.5 S_3 is also solvable (and similarly S_2). However, when $n > 4$ we can say the opposite.

Theorem 11.7

For each $n \geq 5$, S_n is not a solvable group.

Proof Suppose we have a positive integer n with $n \geq 5$. We know from Theorem 11.3 that S_5 is not solvable. Now consider $n > 5$. In S_n, define $H = \{\alpha \in S_n : \alpha(j) = j \text{ for all } j > 5\}$. In an exercise you will verify that H is a subgroup of S_n. Consider the function $f : H \to S_5$ defined by $f(\alpha) = \beta$ where for $1 \leq i \leq 5, \beta(i) = \alpha(i)$. The function f is well defined, since any $\alpha \in H$ is completely determined by its action on 1, 2, 3, 4, 5, so there is a unique permutation in S_5 which maps 1, 2, 3, 4, 5 in exactly the same way. To verify that f is onto, notice that for any $\beta \in S_5$ we can find $\alpha \in H$ with:

$$\alpha(j) = \begin{cases} \beta(j) & 1 \leq j \leq 5 \\ j & j > 5. \end{cases}$$

Thus $f(\alpha) = \beta$, so f is onto. To see that f is a homomorphism, suppose $\delta, \varphi \in H$. Then for $1 \leq j \leq 5$, $f(\delta \circ \varphi)(j) = (\delta \circ \varphi)(j)$. Also, for $1 \leq j \leq 5$, $[f(\delta) \circ f(\varphi)](j) = f(\delta)(\varphi(j))$, but since $1 \leq \varphi(j) \leq 5$ then $[f(\delta) \circ f(\varphi)](j) = (\delta \circ \varphi)(j)$ so $f(\delta \circ \varphi) = f(\delta) \circ f(\varphi)$. Thus f is an onto homomorphism. If S_n was a solvable group, then by Theorem 11.5 H would also be solvable. But S_5 is the homomorphic image of H so again by Theorem 11.5 we would have S_5 solvable. This is a contradiction, therefore for $n > 5$, S_n is not solvable. □

We must be careful however when we use Theorems 11.5 and 11.7. Theorem 11.5 does not say a subgroup of S_5 cannot be solvable, only that the whole group S_5 is not solvable.

Example 11.8 Consider the following set of permutations in S_5, $H = \{\varepsilon, (12345), (13524), (14253), (15432)\}$. We can see that H is a subgroup of S_5 since $H = \langle (12345) \rangle$. But since H is cyclic it is also abelian. Thus using the subgroups $\{\varepsilon\}$ and H we see that $\{\varepsilon\} \triangleleft H$ and $H/\{\varepsilon\}$ is a abelian, showing that H is a solvable group.

11.3 Solvable by Radicals

Before proving anything about solving polynomials "by radicals" we need to make this idea precise.

Definition 11.9 Let K be a field. A ***radical extension*** of K is a finite extension of the form $K(c_1, \ldots, c_n)$ where for each $1 \leq i \leq n$, there is a positive integer $m_i \geq 2$ so that $(c_1)^{m_1} \in K$ and for $1 < i \leq n$, $(c_i)^{m_i} \in K(c_1, \ldots, c_{i-1})$.

Example 11.10 Notice that $\mathbb{Q}(\sqrt{7}, \sqrt[3]{5})$ is a radical extension of $\mathbb{Q}$ since $(\sqrt{7})^2 = 7 \in \mathbb{Q}$ and $(\sqrt[3]{5})^3 = 5 \in \mathbb{Q}(\sqrt{7})$.

Definition 11.11 Let K be a field and $p(x) \in K[x]$. We say that $p(x)$ is ***solvable by radicals*** if the root field of $p(x)$ is contained in a radical extension of K.

Example 11.12 We have used the polynomial $p(x) = -5 + x^3$ from $\mathbb{Q}[x]$ in examples throughout the previous chapters as it illustrated many of the ideas we learned. Once again it is a perfect choice. We found in Example 9.27 that the root field is $E = \mathbb{Q}\left(\sqrt[3]{5}, \sqrt{-3}\right)$. This satisfies the definition of a radical extension since $5 \in \mathbb{Q}$ and $-3 \in \mathbb{Q}\left(\sqrt[3]{5}\right)$. Thus $p(x) = -5 + x^3$ is solvable by radicals.

The most familiar method of solving a polynomial involves the quadratic formula, a formula that shows us how to find the roots of a quadratic polynomial, $c + bx + ax^2 = 0$, *using radicals.*

$$x = \frac{-b+\sqrt{b^2-4ac}}{2a} \text{ or } x = \frac{-b-\sqrt{b^2-4ac}}{2a}$$

The roots of $p(x) = c + bx + ax^2$ are in the extension $\mathbb{Q}\left(\sqrt{b^2-4ac}\right)$ where $b^2 - 4ac \in \mathbb{Q}$.

We need to have radical extensions of $\mathbb{Q}$ that are also root fields for one of the main results of this section. Thus we prove that we will always be able to find these.

Theorem 11.13

Let L be a radical extension of $\mathbb{Q}$. Then there exists a radical extension E of $\mathbb{Q}$, with $\mathbb{Q} \subseteq L \subseteq E$, where E is also a root field over $\mathbb{Q}$.

Proof Suppose L is a radical extension of $\mathbb{Q}$, so $L = \mathbb{Q}(c_1, c_2, \ldots, c_n)$ where for each i there is a positive integer $m_i \geq 2$ with $(c_1)^{m_1} \in \mathbb{Q}$ and for $i > 1$, $(c_i)^{m_i} \in \mathbb{Q}(c_1, c_2, \ldots, c_{i-1})$. For each i we know c_i is algebraic over $\mathbb{Q}$ since L is a finite extension of $\mathbb{Q}$. Thus there is a minimum polynomial $s_i(x)$ for c_i over $\mathbb{Q}$. Consider the polynomial $t(x) = s_1(x)s_2(x)\cdots s_n(x)$ in $\mathbb{Q}[x]$. Let E be the root field for $t(x)$ over $\mathbb{Q}$. Clearly, $L \subseteq E$. We only need to guarantee that E is a radical extension of $\mathbb{Q}$.

First, consider $s_1(x)$. The group $G = Gal({}^E/_{\mathbb{Q}})$ is finite since E is the root field of $t(x)$. We will assume $G = \{\varepsilon, \sigma_1, \sigma_2, \ldots, \sigma_d\}$. Thus every root of $s_1(x)$ can be found as one of $\varepsilon(c_1), \sigma_1(c_1), \sigma_2(c_1), \ldots, \sigma_d(c_1)$ by Theorem 10.14. Consider the extensions shown below:

$$\mathbb{Q} \subseteq \mathbb{Q}(c_1), \subseteq \mathbb{Q}(c_1, \sigma_1(c_1)) \subseteq \cdots \subseteq \mathbb{Q}(c_1, \sigma_1(c_1), \ldots, \sigma_d(c_1))$$

By definition $\sigma_i(\mathbb{Q}) = \mathbb{Q}$ for each $1 \leq i \leq d$. But $(c_1)^{m_1} \in \mathbb{Q}$, so for any σ_i we know $(\sigma_i(c_1))^{m_1} = \sigma_i((c_1)^{m_1}) \in \mathbb{Q}$. Hence $(\sigma_i(c_1))^{m_1} \in \mathbb{Q}(c_1, \sigma_1(c_1), \ldots, \sigma_{i-1}(c_1))$. Thus each extension in the sequence above is a radical extension. Let $V_1 = \mathbb{Q}(c_1, \sigma_1(c_1), \ldots, \sigma_d(c_1))$, which we know is a radical extension of $\mathbb{Q}$. Also, V_1 is the root field of $s_1(x)$ over $\mathbb{Q}$.

We can now repeat this process for $s_2(x)$, by finding its roots among $\varepsilon(c_2), \sigma_1(c_2), \sigma_2(c_2), \ldots, \sigma_d(c_2$ and writing the following sequence of extensions:

$$V_1 \subseteq V_1(c_2), \subseteq V_1(c_2, \sigma_1(c_2)) \subseteq \cdots \subseteq V_1(c_2, \sigma_1(c_2), \ldots, \sigma_d(c_2)).$$

As V_1 is a root field over $\mathbb{Q}$ we know by Theorem 10.12 that $\sigma_i(V_1) = V_1$. Thus as before $(\sigma_i(c_2))^{m_2} \in V_1(c_2, \sigma_1(c_2), \ldots, \sigma_{i-1}(c_2))$, and each field is a radical extension of the previous one. So $V_2 = V_1(c_2, \sigma_1(c_2), \ldots, \sigma_d(c_2))$ is a radical extension of $\mathbb{Q}$ and the root field of $s_1(x)s_2(x)$. Continuing this way for each $s_i(x)$ we eventually find that $V_{n-1}(c_n, \sigma_1(c_n), \ldots, \sigma_d(c_n)) = E$ is a radical extension of $\mathbb{Q}$, as needed. □

We have used roots of unity to help find roots of polynomials of the form $-a + x^n$ over $\mathbb{Q}$ throughout Chapters 9 and 10 (especially in projects). The roots of unity turn out to be a useful tool, so we discuss some properties of the *primitive* roots of unity next.

Definition 11.14 For $n > 1$, the root $\omega_n = \cos\left(\frac{2\pi}{n}\right) + i\sin\left(\frac{2\pi}{n}\right)$ for the polynomial $-1 + x^n \in \mathbb{Q}[x]$ is called a ***primitive*** n^{th} ***root of unity***.

As seen in Project 9.8 $(\omega_n)^j \neq (\omega_n)^k$ for $1 \leq j, k \leq n$ and $j \neq k$. Thus we have n distinct n^{th} roots of unity for each n, namely $1, \omega_n, \omega_n^2, \ldots, \omega_n^{n-1}$. The following theorem is left as an exercise at the end of the chapter.

Theorem 11.15

For each positive integer n, the polynomial $p(x) = -1 + x^n$ in $\mathbb{Q}[x]$ is solvable by radicals.

Theorem 11.16

If n is a positive integer with $n \geq 2$ then $Gal\left({}^{\mathbb{Q}(\omega_n)}/_{\mathbb{Q}}\right)$ is abelian.

Proof Let n be a positive integer $n \geq 2$. The minimum polynomial for ω_n over $\mathbb{Q}$ is a factor of $a(x) = -1 + x^n$ so $[\mathbb{Q}(\omega_n) : \mathbb{Q}] \leq n - 1$. We will assume that $[\mathbb{Q}(\omega_n) : \mathbb{Q}] = t$. From Theorem 9.20 $\{1, \omega_n, \omega_n^2, \ldots, \omega_n^{t-1}\}$ is a basis for $\mathbb{Q}(\omega_n)$ over $\mathbb{Q}$. For any $y \in \mathbb{Q}(\omega_n)$ we have $y = c_0 + c_1\omega_n + c_2\omega_n^2 + \cdots + c_{t-1}\omega_n^{t-1}$ with $c_i \in \mathbb{Q}$.

Since $\mathbb{Q}(\omega_n)$ is a root field over $\mathbb{Q}$, $Gal\left({}^{\mathbb{Q}(\omega_n)}/_{\mathbb{Q}}\right)$ is a group by Theorem 10.18, so we only need to show it is an abelian group. Suppose $\alpha, \beta \in Gal\left({}^{\mathbb{Q}(\omega_n)}/_{\mathbb{Q}}\right)$, then as these automorphisms must map roots of the minimum polynomial for ω_n back to other roots, $\alpha(\omega_n) = \omega_n^k$ and $\beta(\omega_n) = \omega_n^r$ for some $0 \leq k, r \leq t - 1$. An exercise at the end of the chapter asks you to verify that $(\alpha \circ \beta)(y) = (\beta \circ \alpha)(y)$ for any $y \in \mathbb{Q}(\omega_n)$. Thus $\alpha \circ \beta = \beta \circ \alpha$ and so $Gal\left({}^{\mathbb{Q}(\omega_n)}/_{\mathbb{Q}}\right)$ is abelian. □

For any finite extension K of $\mathbb{Q}$, a similar argument would show that $K(\omega_n)$ is a root field over K and $Gal\left({}^{K(\omega_n)}/_{K}\right)$ is abelian. An extension of this theorem is next, and the proof is an exercise at the end of the chapter.

Theorem 11.17

Let $m_1, m_2, \ldots, m_r$ be distinct positive integers, and $m_j \geq 2$ for each j. Then the field $L = \mathbb{Q}(\omega_{m_1}, \omega_{m_2}, \ldots, \omega_{m_r})$ is a root field over $\mathbb{Q}$, and $Gal(^L/_\mathbb{Q})$ is a solvable group.

Suppose we have a radical extension of $\mathbb{Q}$ of the form $K = \mathbb{Q}(c_1, \ldots, c_n)$ where $(c_1)^{m_1} \in \mathbb{Q}$ and for each $0 < j \leq n$, $(c_j)^{m_j} \in \mathbb{Q}(c_1, c_2, \ldots, c_{j-1})$. For each j, ω_{m_j} may or may not be in $\mathbb{Q}(c_1, \ldots, c_n)$. But adjoining the elements $\omega_{m_1}, \omega_{m_2}, \ldots, \omega_{m_n}$ simply creates another radical extension $U = \mathbb{Q}(\omega_{m_1}, \omega_{m_2}, \ldots, \omega_{m_n}, c_1, \ldots, c_n)$. Thus when $p(x)$ is solvable by radicals its root field is contained in an extension of the form $U = \mathbb{Q}(\omega_{m_1}, \omega_{m_2}, \ldots, \omega_{m_n}, c_1, \ldots, c_n)$ as well.

We now prove the first half of the statement that will complete our study of solvable groups and polynomials: *p(x) is solvable by radicals if and only if the Galois group of its root field is a solvable group.*

Theorem 11.18

If the polynomial $p(x) \in \mathbb{Q}[x]$ is solvable by radicals, then the Galois group G for $p(x)$ is a solvable group.

Proof Suppose $p(x) \in \mathbb{Q}[x]$ is solvable by radicals. Thus its root field E is contained in a radical extension of the form $K = \mathbb{Q}(c_1, \ldots, c_n)$ where $(c_1)^{m_1} \in \mathbb{Q}$ and for each $0 < j \leq n$, $(c_j)^{m_j} \in \mathbb{Q}(c_1, c_2, \ldots, c_{j-1})$. From our discussion before this theorem we can also adjoin the needed roots of unity, $\omega_{m_1}, \omega_{m_2}, \ldots, \omega_{m_n}$, so we can assume that the radical extension is of the form $K = \mathbb{Q}(\omega_{m_1}, \omega_{m_2}, \ldots, \omega_{m_n}, c_1, \ldots, c_n)$ and is also a root field by Theorem 11.13.

Let $L = \mathbb{Q}(\omega_{m_1}, \omega_{m_2}, \ldots, \omega_{m_n})$, then by Theorem 11.17 $Gal(^L/_\mathbb{Q})$ is a solvable group. Consider the group $G = Gal(^K/_\mathbb{Q})$ and subgroup $H = Gal(^K/_L)$. Since L is a root field over $\mathbb{Q}$ we know by Theorem 10.29 that $H \lhd G$ and $^G/_H \cong Gal(^L/_\mathbb{Q})$. Thus $H \lhd G$, and $^G/_H$ is a solvable group. By Theorem 11.5 to complete the proof we only need to verify that H is a solvable group.

For each $1 < i \leq n$ define $W_{i-1} = L(c_1, \ldots, c_{i-1})$ then $W_i = L(c_1, \ldots, c_i) = W_{i-1}(c_i)$. Since $(c_i)^{m_i} \in W_{i-1}$ then c_i is a root of $-(c_i)^{m_i} + x^{m_i}$. However, since $\omega_{m_i} \in W_{i-1}$ all of the roots of $-(c_1)^{m_i} + x^{m_i}$ are in W_i. Thus W_i is a root field over W_{i-1}. Now look at the sequence of subgroups of $Gal(^K/_L)$ shown below.

$$\{\varepsilon\} \subseteq Gal(^K/_{W_{n-1}}) \subseteq Gal(^K/_{W_{n-2}}) \subseteq \cdots \subseteq Gal(^K/_{W_1}) \subseteq Gal(^K/_L)$$

By Theorem 10.29 $Gal(^K/_{W_i}) \lhd Gal(^K/_{W_{i-1}})$ and $^{Gal(^K/_{W_{i-1}})}/_{Gal(^K/_{W_i})} \cong Gal(^{W_i}/_{W_{i-1}})$. In an exercise at the end of the chapter you will show for each i, $Gal(^{W_i}/_{W_{i-1}})$ is abelian.

Hence $Gal(^K/_L)$ and $Gal(^L/_{\mathbb{Q}})$ are solvable so $Gal(^K/_{\mathbb{Q}})$ is solvable. Thus by Theorem 11.5 $Gal(^E/_{\mathbb{Q}})$ is also solvable. □

We now know that a polynomial will not be solvable by radicals if its Galois group isomorphic to S_5. The following example finds such a polynomial.

Example 11.19 Consider the polynomial $p(x) = 6 - 12x + x^5$, with root field E over $\mathbb{Q}$ which is irreducible over $\mathbb{Q}$ by Eisenstein's Criterion 8.35. By Theorem 9.31 there are five distinct roots of $p(x)$ so the Galois group of $p(x)$ is a subgroup of S_5. For a root u of $p(x)$ we know $[\mathbb{Q}(u) : \mathbb{Q}] = 5$ since $p(x)$ is the minimum polynomial for u over $\mathbb{Q}$. Thus since $[E : \mathbb{Q}] = [E : \mathbb{Q}(u)][\mathbb{Q}(u) : \mathbb{Q}]$ then $[E : \mathbb{Q}]$ is divisible by 5. If we denote $G = Gal(^E/_{\mathbb{Q}})$, then the cardinality of the finite group G is divisible by 5. Theorem 3.43 (Cauchy's Theorem) guarantees us that our group G must contain an element of order 5, which in S_5 can only be a 5-cycle. If we could also show that there was a transposition in G, we would be finished by Example 2.41.

In one exercise at the end of the chapter you are asked to show that only three roots of $p(x)$ are real numbers, and so two of the roots must be complex numbers that are not real. In another exercise you will show that those roots must be of the form $a + bi$ and $a - bi$ in the root field E. Now $\varphi : E \to E$ defined by $\varphi(a + bi) = a - bi$ is a $\mathbb{Q}$ fixing automorphism of E, and thus is in G. As a permutation of the roots of $p(x)$, φ fixes the three real number roots and $a+bi, a-bi$ are interchanged. Thus φ is a transposition. Hence we have a 5-cycle and a transposition in G and so $G = S_5$ by Example 2.41. Therefore G is not solvable, so by Theorem 11.18 $p(x)$ is not solvable by radicals.

Recall that a group G is solvable if there is a sequence of subgroups $\{e_G\} = H_0, H_1, \ldots, H_n = G$ where for each j, $H_j \triangleleft H_{j+1}$ and $^{H_{j+1}}/_{H_j}$ is abelian. When G is a finite group this can be strengthened. The steps to prove the next theorem are left as a series of exercises at the end of the chapter.

Theorem 11.20

If G is a finite solvable group then there is a sequence of subgroups $\{e_G\} = H_0, H_1, \ldots, H_n = G$ where for each $0 \leq j < n$, $H_j \triangleleft H_{j+1}$ and $^{H_{j+1}}/_{H_j}$ is cyclic of prime order.

Our final theorem is the converse of Theorem 11.18, and tells us that a solvable Galois group guarantees a polynomial is solvable by radicals.

Theorem 11.21

If $p(x) \in \mathbb{Q}[x]$ has root field E and $Gal(^E/_\mathbb{Q})$ is solvable, then $p(x)$ is solvable by radicals.

Proof Suppose $p(x) \in \mathbb{Q}[x]$ with root field E where the Galois group $Gal(^E/_\mathbb{Q})$ is a solvable group. We must show that E is contained in a radical extension of $\mathbb{Q}$. Assume we have $L = \mathbb{Q}(\omega_{k_1}, \omega_{k_2}, \ldots, \omega_{k_r})$, which contains any of the primitive roots of unity we need. You will show in an exercise at the end of the chapter that $E' = E(\omega_{k_1}, \omega_{k_2}, \ldots, \omega_{k_r})$ also has $Gal(^{E'}/_\mathbb{Q})$, a solvable group. Thus E' is also a root field over L, and $G = Gal(^{E'}/_L)$ is a solvable group.

Since G is solvable there is a series of subgroups $\{\varepsilon\} = H_0, H_1, \ldots, H_n = G$, so that for each $0 \leq j \leq n-1$, $H_j \lhd H_{j+1}$, and $^{H_{j+1}}/_{H_j}$ is cyclic of prime order. (Note that $H_j \neq H_{j+1}$.)

Define for each i, $K_i = E'_{H_i}$ the fixed field of H_i.

We will show that for each i, K_i is a radical extension of K_{i+1}, and thus $\mathbb{Q} \subseteq L = K_n \subseteq K_{n+1} \subseteq \cdots \subseteq K_0 = E'$ shows us that E' is a radical extension of $\mathbb{Q}$.

As $H_i \neq H_{i+1}$ we know $K_i \neq K_{i+1}$. You will show in an exercise that $[K_i : K_{i+1}]$ must be prime. Since $H_i \lhd H_{i+1}$, by Theorem 10.30 we also know K_i is a root field over K_{i+1}. Also $Gal(^{K_i}/_{K_{i+1}})$ has some prime p, and so order p, and must be cyclic. Let $Gal(^{K_i}/_{K_{i+1}}) = \langle\sigma\rangle$, and let ω_p be the pth root of unity which is already in K_{i+1}. There must exist some $a \in K_i - K_{i+1}$, so let $u = a + (\omega_p)\sigma^{-1}(a) + (\omega_p)^2\sigma^{-2}(a) + \cdots + (\omega_p)^{p-1}\sigma^{-p+1}(a)$. In an exercise at the end of the chapter you will show that $\sigma(u) = \omega_p u$ and thus u is not fixed by σ. Hence $u \notin K_{i+1}$.

Notice that $\sigma(u^p) = (\sigma(u))^p = (\omega u)^p = \omega^p u^p = u^p$. You will show in an exercise at the end of the chapter that for every positive integer t, $\sigma^t(u^p) = u^p$ and so every element of $Gal(^{K_i}/_{K_{i_1}})$ fixes u^p. Thus $u^p \in K_{i+1}$, and we have $K_{i+1}(u)$ a subfield of K_i with $K_{i+1} \subseteq K_{i+1}(u) \subseteq K_i$. But by Theorem 9.22 we know

$$[K_i : K_{i+1}] = [K_i : K_{i+1}(u)][K_{i+1}(u) : K_{i+1}].$$

As $[K_i : K_{i+1}]$ is prime then $[K_{i+1}(u) : K_{i+1}] = 1$ or $[K_i : K_{i+1}(u)] = 1$. Thus $[K_i : K_{i+1}(u)] = 1$ and $K_i = K_{i+1}(u)$ where $u^p \in K_{i+1}$. Hence K_i is a radical extension of K_{i+1} as we needed. Therefore E' is a radical extension of $\mathbb{Q}$ containing E and $p(x)$ is solvable by radicals! □

Exercises for Chapter 11

Section 11.1 Three Construction Problems

1. In Euclidean Geometry, an axiom guarantees there are two different points. We can consider them to be (0,0) and (1,0). Use the facts that any two points form a line, and with any two points a circle can be constructed using the points as center and point on the circle, to show how to construct all points of the form $(m, 0)$ for m an integer.
2. A theorem of Euclidean Geometry asserts that given a line and a point on the line, there is a line perpendicular to the given line and through the point. Use this to explain how all points of the form (m, n) can be constructed where $m, n \in \mathbb{Z}$.
3. Another Euclidean theorem allows us to construct midpoints of previously constructed segments. Explain how this will help us construct any point of the form $\left(\frac{m}{2}, \frac{n}{2}\right)$ for integers m, n.
4. Consider the acute angle $\angle ABC$ with $A = (1, 1)$, $B = (0, 0)$, and $C = (1, 0)$ since those can be constructed. Explain how to find a point D on the ray $\overrightarrow{BA}$ with a distance from B to D exactly 1 unit, as well as a point E on $\overrightarrow{BA}$ with length from B to E exactly n units. Then use the line $\overleftrightarrow{EC}$, and a line through D parallel to $\overleftrightarrow{EC}$ with similar triangles to construct $\left(\frac{1}{n}, 0\right)$.
5. Using a construction similar to the previous exercise show how to construct a point with coordinates $\left(\frac{m}{n}, 0\right)$ for positive integers $m < n$.
6. Using the results of the previous exercises, explain how to construct any point with coordinates of the form $\left(0, \frac{m}{n}\right)$ for positive integers $m < n$.
7. Explain how to construct any point of the form $\left(\frac{m}{n}, 0\right)$ or $\left(0, \frac{m}{n}\right)$ for any nonzero integers m, n.
8. Use the results of the previous exercises to construct any point in $\mathbb{Q} \times \mathbb{Q}$.
9. Suppose we have two distinct, nonparallel lines $ax + by = c$, and $ex + fy = g$, where the coefficients a, b, c, d, e, f, g already exist in our field K ($\mathbb{Q}$ or a previous extension of $\mathbb{Q}$), and that these two lines intersect at some point (u, v). Prove that $u, v \in K$.
10. Suppose we have constructed a circle and a line which intersect at a point (u, v). We can write the equations as $x^2 + ax + y^2 + by + c = 0$ and $dx + ey = f$, where a, b, c, d, e, f are in our field K. Notice that at least one of d or e must be nonzero, so assume $d \neq 0$. Prove: v is a root of the polynomial $p(x) \in K[x]$ defined by:
$p(x) = [(d^{-1}f)^2 + afd^{-1} + c] + [-2ef(d^{-1})^2 - aed^{-1} + b]x + [(d^{-1}e)^2 + 1]x^2$.
11. Explain how a $60°$ angle can be constructed using a vertical line through $\left(\frac{1}{2}, 0\right)$ and the unit circle. You may assume the vertex of the angle is at (0,0).
12. If $u = \cos(20°)$, use trigonometric identities to show that $2u$ is a root of the irreducible polynomial $p(x) = 1 + 3x - x^3$ in $\mathbb{Q}[x]$.

Section 11.2 Solvable Groups

13. Consider the set $H = \{\varepsilon, (12)(34), (13)(24), (14)(23)\}$ contained in the group S_4. Prove that H is a subgroup of A_4 and is abelian.
14. For $H = \{\varepsilon, (12)(34), (13)(24), (14)(23)\}$ prove that $H \triangleleft A_4$.
15. Prove: If G is an abelian group then G is a solvable group.
16. Prove: If $G = \{e_G\}$ then (i), and (ii) of Theorem 11.5 hold.
17. Let G be a solvable group with $\{e_G\} = H_0, H_1, \ldots, H_n = G$ where for each $0 \leq i \leq n-1$, $H_i \triangleleft H_{i+1}$ and ${}^{H_{i+1}}/_{H_i}$ is an abelian group. Prove: If J is a subgroup of G then for each $0 \leq i \leq n-1$, $H_i \cap J \triangleleft H_{i+1} \cap J$.
18. Let G be a solvable group with $\{e_G\} = H_0, H_1, \ldots, H_n = G$ where for each $0 \leq i \leq n-1$, $H_i \triangleleft H_{i+1}$, and ${}^{H_{i+1}}/_{H_i}$ is an abelian group. Prove: If J is a subgroup of G then for each $0 \leq i \leq n-1$, ${}^{H_{i+1}\cap J}/_{H_i \cap J}$ is abelian.
19. Suppose we have a solvable group G, with $\{e_G\} = H_0, H_1, \ldots, H_n = G$ where for each $0 \leq i \leq n-1$, $H_i \triangleleft H_{i+1}$, and ${}^{H_{i+1}}/_{H_i}$ is an abelian group. Assume we have a group B and $f : G \to B$ an onto homomorphism. Prove: For each $0 \leq i \leq n-1$, $f(H_i) \triangleleft f(H_{i+1})$.
20. Suppose we have a solvable group G, with $\{e_G\} = H_0, H_1, \ldots, H_n = G$ where for each $0 \leq i \leq n-1$, $H_i \triangleleft H_{i+1}$, and ${}^{H_{i+1}}/_{H_i}$ is an abelian group. Assume we have a group B and $f : G \to B$ an onto homomorphism. Prove: For each $0 \leq i \leq n-1$, ${}^{f(H_{i+1})}/_{f(H_i)}$ is abelian.
21. Prove: The subgroup $H = \{\alpha \in S_4 : \alpha(4) = 4\}$ of S_4 is isomorphic to S_3.
22. In S_n, where $n > 5$ define $H = \{\alpha \in S_n : \alpha(j) = j \text{ for all } j > 5\}$. Prove that H is a subgroup of S_n.

Section 11.3 Solvable by Radicals

23. Prove Theorem 11.15.
24. Let n be a positive integer, $n \geq 2$. Prove: If $\alpha, \beta \in Gal\left({}^{\mathbb{Q}(\omega_n)}/_{\mathbb{Q}}\right)$ and $y = c_0 + c_1\omega_n + c_2\omega_n^2 + \cdots + c_{t-1}\omega_n^{t-1}$ with $c_i \in \mathbb{Q}$ then $(\alpha \circ \beta)(y) = (\beta \circ \alpha)(y)$ to complete Theorem 11.16.
25. Prove Theorem 11.17.
26. Using the notation of the proof of Theorem 11.18, prove that $Gal({}^{W_i}/_{W_{i-1}})$ is abelian.
27. Suppose an irreducible polynomial $p(x) \in \mathbb{Q}[x]$ has exactly two complex (nonreal) roots $a + bi$ and $c + di$ where $b \neq 0$ and $d \neq 0$. Prove: $c + di = a - bi$.
28. Prove: Suppose G is a finite group and $H \triangleleft G$ with $\left|{}^{G}/_{H}\right| = n$. Prove: If p is a prime dividing n then there is a subgroup J with $H \triangleleft J \triangleleft G$ so that $\left|{}^{G}/_{J}\right| = p$.
29. Use the result of Exercise 28 to prove Theorem 11.20.
30. Explain using calculus why $p(x) = 2 - 4x + x^5$ has exactly three real roots and two complex (nonreal) roots.
31. In the proof of Theorem 11.21 show why we know $[K_{i+1} : K_i]$ is prime.
32. Let E be a rot field over $\mathbb{Q}$ so that $Gal({}^{E}/_{\mathbb{Q}})$ is solvable. Prove: If $E' = E(\omega_{m_1}, \omega_{m_2}, \ldots, \omega_{m_r})$ then $Gal({}^{E'}/\mathbb{Q})$ is also a solvable group.
33. Using the notation of Theorem 11.21, show $\sigma(u) = \omega_p u$, so that u is not fixed by σ.

34. Using the notation of Theorem 11.21, show by induction that for every positive integer t, $\sigma^t(u^p) = u^p$.
35. Show that the Galois group G found in Exercise 32 of Chapter 10 is solvable to verify that the polynomial $p(x) = (-2 + x^2)(-7 + x^2)$ is solvable by radicals.
36. Show that the Galois group G found in Exercise 34 of Chapter 10 is solvable to verify that the polynomial $p(x) = (1 + x^2)(-5 + x^2)$ is solvable by radicals.
37. Show that the Galois group G found in Exercise 36 of Chapter 10 is solvable to verify that the polynomial $p(x) = (1 + x + x^2)(-2 + x^2)$ is solvable by radicals.
38. Show that the Galois group G found in Exercise 38 of Chapter 10 is solvable to verify that the polynomial $p(x) = (1 + x + x^2 + x^3)(3 + x^2)$ is solvable by radicals.
39. Show that the Galois group G found in Exercise 40 of Chapter 10 is solvable to verify that the polynomial $p(x) = 1 + x^4$ is solvable by radicals.
40. Show that the Galois group G found in Exercise 42 of Chapter 10 is solvable to verify that the polynomial $p(x) = -1 + x^5$ is solvable by radicals.

Projects for Chapter 11

Project 11.1

We can construct, by ruler and compass, any point of the form (m, n) where $m, n \in \mathbb{Q}$. We will use this fact to help us show that the number $\sqrt{d}$ can be constructed as well for any positive rational number d.

1. Let $d \in \mathbb{Q}^+$. Explain why we can construct a circle with center (0,0) through the point $(\frac{d+1}{2}, 0)$.
2. Explain why we can construct a vertical line through $(\frac{d-1}{2}, 0)$.
3. Explain why the vertical line and circle of the previous questions must intersect at a point $(\frac{d-1}{2}, a)$ for some real number a.
4. Use the fact that the point $(\frac{d-1}{2}, a)$ is on the circle with equation $x^2 + y^2 = (\frac{d+1}{2})^2$ to show that $a = \sqrt{d}$.
5. What can we conclude from the previous questions?
6. If we began with a positive number d which was already known to be *constructible*, would similar steps tell us that $\sqrt{d}$ is constructible as well?

Project 11.2

Definition: A regular n-gon is a polygon with exactly n sides, all of whose angles have equal measure and all of whose side lengths are equal.

Definition: A regular n-gon is constructible by ruler and compass if and only if the angle of measure $\frac{2\pi}{n}$ is constructible by ruler and compass, i.e., a vertex of the angle and the lines needed to create the angle are all constructible.

(Note: The angle inside each corner of the n-gon measures $\pi - \frac{2\pi}{n}$, but the angle of measure π is already constructed by the x-axis, so both angles will in fact be constructible as long as we can construct the angle of measure $\frac{2\pi}{n}$.)

In order to construct the angle, we only need to show we can construct the point on the unit circle that defines the angle (with the positive x-axis), i.e., the point $\left(\cos\left(\frac{2\pi}{n}\right), \sin\left(\frac{2\pi}{n}\right)\right)$.

1. Explain why the regular 3-gon, 4-gon, and 6-gon are constructible. What are the points we need to construct and why do we know they are constructible? Project 11.1 may help.

   The regular pentagon (5-gon) is also constructible, but we will have to work harder to discover why. The following steps will guide you through it. They may seem like strange steps to do, but keep at it until the end.

2. Recall that the complex number $\omega_5 = \cos\left(\frac{2\pi}{5}\right) + i\sin\left(\frac{2\pi}{5}\right)$ is a root of the polynomial $b(x) = -1 + x^5$. Use this fact to show $(\omega_5)^{-1} = (\omega_5)^4$ and $(\omega_5)^{-2} = (\omega_5)^3$.
3. Show that $(\omega_5)^{-1} = \cos\left(\frac{2\pi}{5}\right) - i\sin\left(\frac{2\pi}{5}\right)$ then explain why $\omega_5 + (\omega_5)^{-1} = 2\cos\left(\frac{2\pi}{5}\right)$. [Hint: Multiply them to see if you have the correct inverse.]

   Recall from Project 9.8 that we know $(\omega_5)^2 = \cos\left(\frac{4\pi}{5}\right) + i\sin\left(\frac{4\pi}{5}\right)$.

4. Show that $(\omega_5)^{-2} = \cos\left(\frac{4\pi}{5}\right) - i\sin\left(\frac{4\pi}{5}\right)$ (again, multiply to check), then use the trigonometric identity $\cos(2\theta) = 2(\cos(\theta))^2 - 1$ to show that $(\omega_5)^2 + (\omega_5)^{-2} = 4\left(\cos\left(\frac{2\pi}{5}\right)\right)^2 - 2$.
5. Recall from Project 8.6 that ω_5 is a root of $p(x) = 1 + x + x^2 + x^3 + x^4$. Use this to explain why $1 + \omega_5 + (\omega_5)^2 + (\omega_5)^{-2} + (\omega_5)^{-1} = 0$.
6. Simplify the equation found in the previous question using all you have discovered so far to show that $\cos\left(\frac{2\pi}{5}\right)$ is a root of the polynomial $a(x) = -1 + 2x + 4x^2$.
7. Why do the previous questions guarantee us that $\cos\left(\frac{2\pi}{5}\right)$ is constructible? [Can you discover the actual value of $\cos\left(\frac{2\pi}{5}\right)$ now?]

   Now $\sin\left(\frac{2\pi}{5}\right) = \sqrt{1-c^2}$ using $c = \cos\left(\frac{2\pi}{5}\right)$. Since c is constructible, we also know that $1 - c^2$ is constructible so the last question of Project 11.1 tells us that $\sin\left(\frac{2\pi}{5}\right)$ is also constructible. Hence the regular pentagons is constructible.

   Unfortunately, the regular 7-gon (heptagon) is NOT constructible since the minimum polynomial for $\sin\left(\frac{2\pi}{7}\right)$ over $\mathbb{Q}$ is $-7 + 56x^2 - 112x^4 + 64x^6$. Thus $\sin\left(\frac{2\pi}{7}\right)$ cannot be constructible since the degree of its extension over $\mathbb{Q}$ would be 6, not a power of 2.

Project 11.3

In this project we will find the Galois group of the polynomial $p(x) = -2 + x^4$, and then show that the group is solvable with a sequence of subgroups.

1. Show that the root field of $p(x)$ is $E = \mathbb{Q}(\sqrt[4]{2}, i)$ and find $[\mathbb{Q}(\sqrt[4]{2}, i) : \mathbb{Q}]$.
2. We can define each automorphism of E which fixes $\mathbb{Q}$ by its action on each of $\sqrt[4]{2}$ and i as shown below. Complete the Cayley table for this group G $= \{\varepsilon, \alpha, \beta, \delta, \gamma, \kappa, \varphi, \theta\}$.

$$\varepsilon: \begin{array}{ccc} \sqrt[4]{2} & \to & \sqrt[4]{2} \\ i & \to & i \end{array} \qquad \alpha: \begin{array}{ccc} \sqrt[4]{2} & \to & \sqrt[4]{2} \\ i & \to & -i \end{array} \qquad \beta: \begin{array}{ccc} \sqrt[4]{2} & \to & -\sqrt[4]{2} \\ i & \to & i \end{array}$$

$$\delta: \begin{array}{ccc} \sqrt[4]{2} & \to & -\sqrt[4]{2} \\ i & \to & -i \end{array} \qquad \gamma: \begin{array}{ccc} \sqrt[4]{2} & \to & \sqrt[4]{2}i \\ i & \to & i \end{array} \qquad \kappa: \begin{array}{ccc} \sqrt[4]{2} & \to & \sqrt[4]{2}i \\ i & \to & -i \end{array}$$

$$\varphi: \begin{array}{ccc} \sqrt[4]{2} & \to & -\sqrt[4]{2}i \\ i & \to & i \end{array} \qquad \theta: \begin{array}{ccc} \sqrt[4]{2} & \to & -\sqrt[4]{2}i \\ i & \to & -i \end{array}$$

$\circ$	ε	α	β	δ	γ	κ	φ	θ
ε	ε	α	β	δ	γ	κ	φ	θ
α	α							
β	β							
δ	δ							
γ	γ							
κ	κ							
φ	φ							
θ	θ							

3. Prove that the sets $H_1 = \{\varepsilon, \beta\}$ and $H_2 = \{\varepsilon, \gamma, \beta, \varphi\}$ are subgroups of G.
4. Use cosets to prove that $\{\varepsilon\} \lhd H_1$, $H_1 \lhd H_2$, and $H_2 \lhd G$.
5. Compute the elements of the quotient groups ${}^{G}/_{H_2}$, ${}^{H_2}/_{H_1}$, and ${}^{H_1}/_{\{\varepsilon\}}$ and show each quotient group is abelian.
6. What can we conclude about the group G?
7. Why do we now have two reasons that $p(x) = -2 + x^4$ is solvable by radicals?

Hints for Selected Exercises

Chapter 0

0.3 Use the equation $2x - 6 = 2$ and solve for x to see that only one integer is in this set.

0.5 Don't forget when solving $4x^2 + 1 = 10$ that negative rational numbers are also possible solutions.

0.11 Don't forget that $\emptyset$ is a subset of C so is one of the elements of $\wp(C)$.

0.15 Look for a counterexample here, and create a set A that has some elements from B and some from C.

0.19 This is a true statement so needs a proof. First, show that if $x \in A \cup (B \cap C)$ then $x \in (A \cup B) \cap (A \cup C)$ using the rules defining intersection and union. Don't forget that saying $x \in A \cup (B \cap C)$ means that either $x \in A$ OR $x \in B \cap C$ so you will need to consider cases. Then prove the reverse; if $x \in (A \cup B) \cap (A \cup C)$ then $x \in A \cup (B \cap C)$.

0.24 Remember that the base case here is when $n = 4$. It can be helpful to notice that if $k \geq 4$ then $k^2 > k + 1$ (Exercise 23). This will help if you multiply the statement $P(k)$ by $k + 1$.

0.28 This one seems complicated, but it really comes down to rewriting $P(k+1)$ as $P(k)+(k+1)((k+1)!)$. Then use the induction hypothesis to rewrite $P(k)$ and simplify carefully.

0.32 You should be able to quickly see that $\sim$ is an equivalence relation, but when finding the equivalence classes don't forget that squaring a negative real number is the same as squaring the positive one.

0.36 Although the elements must all be written as ordered pairs, (m, n), the relation only looks at the first coordinate. But don't forget to keep the ordered pairs in the proof that $\sim$ is an equivalence relation. Also notice that $(1, 2) \sim (1, 7)$ and $(1, 2) \sim (1, 146)$. You will not be able to list each element of an equivalence class, so use set builder notation to describe the classes carefully.

0.44 Note that to calculate $f\left(\frac{a}{b}\right)$ it is assumed that $\frac{a}{b}$ is in lowest terms. Thus $f\left(\frac{4}{6}\right)$ is defined to be $f\left(\frac{2}{3}\right) = -1$. Don't use two ways to write the same fraction (such as $\frac{4}{6}$ and $\frac{2}{3}$) to find a counterexample for either property. You should be able to find that f is surjective but not injective.

0.48 To show that $g \circ f$ is surjective, assume you have an element $z \in C$, and use the fact that g is surjective to find $y \in B$ with $g(y) = z$. Now you only need to see why there is $x \in A$ with $g(f(x)) = z$.

0.54 The key here is that the two matrices BC and CB are different. One calculation is shown here to help you practice.

$$BC = \begin{bmatrix} \frac{1}{2} & -1 & 0 \\ 2 & -\frac{2}{3} & 1 \\ 2 & -3 & -1 \end{bmatrix} \begin{bmatrix} 1 & \frac{1}{3} & -1 \\ -1 & 0 & 4 \\ \frac{3}{4} & 0 & 1 \end{bmatrix} = \begin{bmatrix} \frac{3}{2} & \frac{1}{6} & -\frac{9}{2} \\ \frac{41}{12} & \frac{2}{3} & -\frac{11}{3} \\ \frac{17}{4} & \frac{2}{3} & -15 \end{bmatrix}$$

Note that if we had rounded off $-\frac{2}{3}$ to -0.67 then $4 \cdot -0.67 = -2.68$. But $4 \cdot -\frac{2}{3} = -\frac{8}{3}$ which if rounded to two places after the decimal point is -2.67. Thus it is better to leave the fractions in the calculations.

Chapter 1

1.1 Is the square root of an integer always an integer? Try a few examples to see.

1.9 The key here is to determine if the answer to $x * y$ will ever be 0.

1.13 Look for counterexamples for associativity and commutativity. Try a contradiction to see why no identity exists in $\mathbb{Z}$. If such an identity e existed it would satisfy $e * 1 = 1$ and $1 * e = 1$. Can any integer e satisfy this?

1.19 There are various answers. Consider what it might mean to have two inverses for an element and how you would recognize this situation in the Cayley table.

1.23 Be careful since the elements are functions. We need to be sure that for functions $f, g, h \in \Im(\mathbb{R})$ the new functions $f + (g + h)$ and $(f + g) + h$ are equal which can be shown by verifying $[f+(g+h)](x) = [(f+g)+h](x)$ for an arbitrary real number x. To find the identity it must be a function $e : \mathbb{R} \to \mathbb{R}$ so you must define $e(x)$ for every real number x. Once you define this function be sure to verify that for any $f \in \Im(\mathbb{R})$ it is true that $e * f = f = f * e$. Once you have the correct function e, you need to determine how to define the inverse of an arbitrary function f, i.e., the correct function to make $f * f^{-1} = e = f^{-1} * f$. Do not be fooled into thinking $f^{-1}(x) = \frac{1}{f(x)}$.

1.25 This requires cases within a PMI proof. Notice that k is a fixed integer, but we don't know if it is positive, negative, or 0. This is important for the base case since to show $P(1)$ is true you must calculate $a^1 a^k$. Thus you need to consider all three possibilities for k. Since k is being used for this fixed integer, use a different letter in the induction step assuming $P(m)$ instead of writing it as $P(k)$. Thus in the inductive step assume that $a^m a^k = a^{m+k}$. Once again you will need to consider the possibilities of $m + k = 0$, $m + k < 0$, and $m + k > 0$ to show that $a^{m+1} a^k = a^{m+k+1}$.

1.30 Remember that to show $ord(aba^{-1}) = n$ you must show that $(aba^{-1})^n = e_G$ AND that any smaller positive integer k has $(aba^{-1})^k \neq e_G$. Use the previous exercise and the definition of $ord(a) = n$ to help.

1.37 Be sure to explain why the given conditions guarantee that $a \neq e_G$ help rule out possible values for $ord(a)$. Theorem 1.26 helps find the only choices.

1.43 Remember that each element is an ordered pair (a, b). Create the Cayley table for H to see if it is closed under the given operation. Be sure to identify the inverse of each element to be sure it is in H as well.

1.46 The key to this one is inverses. If $n \in \mathbb{Z}$ and $n \neq 0$, what rational number is the inverse of $\frac{1}{n}$? Is that inverse always of the form $\frac{1}{m}$ for a nonzero integer m?

1.53 Recall that in the proof of Theorem 1.30 we saw that a subgroup of a group G must always contain e_G. This will help you show that $H_1 \cap H_2 \neq \emptyset$. The rest of the proof is just carefully using the definition of intersection.

1.54 You need to verify that for $a, b \in \mathbb{Z}_8$ we have $f(a +_8 b) = f(a) +_4 f(b)$. Note that $f(a +_8 b) = [2(a +_8 b)] \bmod 4$, where first $a +_8 b$ is calculated, then the answer multiplied by 2 (usual integer multiplication), and finally this answer is divided by 4 to find the remainder. For example, $f(3 +_8 7) = [2(3 +_8 7)] \bmod 4 = [2(2)] \bmod 4 = 4 \bmod 4 = 0$.

1.58 For $x, y \in \mathbb{Z}$ we have $f(x + y) = (x + y)^3$. Is it always the same as $x^3 + y^3$? Try some examples to see.

1.61 The statement you need to prove is for any integer n. Use PMI to prove that the statement is true for any $n \in \mathbb{N}$, then use parts (i) and (ii) of Theorem 1.37 to see why it must also hold if $n = 0$ or $n < 0$.

1.64 Suppose $a \in G$ has finite order n. Use Theorem 1.37 to see that $(f(a))^n = e_K$, but you must also know that n is the smallest positive integer with this property. As f is one to one, if $(f(a))^m = e_K$, then why must we also have $a^m = e_G$?

1.65 The elements of $f(G)$ are actually elements of K. Thus you need to see that the set is closed under the operation of K, and each element has an inverse in K. Once you have elements $y, z \in f(G)$ you can use the definition of $f(G)$ to know there are $a, b \in G$ with $f(a) = y$ and $f(b) = z$. Don't forget to explain why $f(G)$ must be nonempty.

Chapter 2

2.3 Don't forget that the elements in $U(14)$ must be relatively prime to 14, so they cannot have either 2 or 7 as a factor. There should be six elements in the set. When finding the Cayley table remember that you are using $\cdot_{14}$ as your operation. Thus for example you will find $5 \cdot_{14} 11 = (55) \bmod 14 = 13$.

2.10 Don't forget that the elements in $U(30)$ must be relatively prime to 30, so they cannot have 2, 3, or 5 as factors. Even though $\mathbb{Z}_{30}$ is fairly large, there should only be eight element here. Be sure to use $\cdot_{30}$ as your operation.

2.15 Look at the six elements one at a time. Don't forget that according to the definition of order for each a you must find a positive integer n with $a^n = 1$ (using $\cdot_{14}$ of course) AND verify that no smaller value of n will also have this property.

2.21 Try some examples here for different primes to see if you think it is true.

2.22 Simplify Euler's formula carefully and remember that $n = pq$ to simplify it even more.

2.29 For each element $a \in U(12)$, find the set of powers $a, a^2, a^3, \ldots$ (using $\cdot_{12}$, of course) stopping when you get 1 as an answer. The set of all of these answers gives the cyclic subgroup $\langle a \rangle$. To decide if $U(12)$ is cyclic, look to see if one of the subgroups you found contains every element of $U(12)$.

2.33 Use Theorems 2.16 and 0.15.

2.37 You will need to use that f is both one to one and onto in each half of this "if and only if" proof. The homomorphism property is key, especially Theorem 1.37.

2.39 Since $G \times K$ is cyclic there must be $a \in G$ and $b \in K$ with $G \times K = \langle (a, b) \rangle$. Show that a and b are also generators of G and K respectively. Don't forget that if $x \in G$ then we know $(x, b) \in G \times K$.

2.45 Don't forget that you must show that $f(x)$ will never be equal to 0 or 1. To see that f is one to one don't forget that usual multiplication and addition on $\mathbb{R}$ are both associative and commutative. To show f is onto assume you have $y \in A$ and find the correct element $x \in A$ with $f(x) = y$. Be sure to prove that the element you found, x, is really in A.

2.56 To be sure you are on the right track, notice that $\alpha^3 = (15)(2763)$ by first computing $\alpha \circ \alpha = (2367)(15)(2367)(15) = (26)(37)$ then $\alpha^3 = \alpha \circ \alpha^2 = (2367)(15)(26)(37) = (15)(2763)$.

2.66 Two examples are below; it is possible to find others as well.
$\omega = (56)(14)(13)(12)(56)(68)(67)$ and $\omega = (14)(13)(12)(68)(67)(13)(13)(46)(46)$.

2.71 Use symbols to represent the 3-cycle such as $(a_1 a_2 a_3)$ or (abc) which will allow you to compute α, α^2, and α^3 to solve this problem.

2.72 Show that if m is even, say $m = 2n$, then the cycle α of length m has α^2, a product of two cycles of length n.

Chapter 3

3.6 You first need to find the elements of $H = \langle (123) \rangle$; there will be only three of them. Then use Theorem 3.6 to make the job easier. For example, once you find that $(1234) \in (14)H$ then you need not calculate $(1234)H$ since Theorem 3.6 tells you that $(14)H = (1234)H$. Hence you should only need to calculate 8 cosets instead of 24.

3.9 Don't forget that the elements of $\mathbb{Z}_4 \times \mathbb{Z}_4$ are ordered pairs, so each coset should be of the form $(a, b) + H$.

3.15 Use (iii) of this theorem to find $G = \langle a \rangle$ for some $a \in G$. Now suppose there is a subgroup H which is not one of $\{e_G\}$ or G and see why this leads to a contradiction. Think about what Lagrange's Theorem tells us and look at Theorem 0.15.

3.22 This is a direct application of Lagrange's Theorem but don't forget it does not just mean primes dividing 30. Thus 6 is also a possible size. You should find a total of eight possible sizes for subgroups.

3.27 Notice that $(cb)^{-1}(ab) = b^{-1}c^{-1}ab$ so unless G is abelian you cannot cancel the b's. This might give you the idea to consider some familiar groups that are not abelian to see if you think the statement is true.

3.28 Use the definition of normal, i.e., closed under conjugates.

3.38 Don't forget that the set $a^{-1}Ha$ as given is only a set, not a coset of G. Thus do not try to use coset properties when showing it is a subgroup of G. Also look at nonabelian groups to see if $a^{-1}Ha = H$ for any group G and subgroup H.

3.40 Do not assume you have $H \triangleleft G$, thus you *cannot* use $(xy)H = (xH)(yH)$. Also, the properties of Theorem 3.6 do not apply since we do not have both left or both right cosets. Instead, to prove that K is closed under the operation of G suppose $x, y \in G$ and show that $(xy)H \subseteq (yx)H$ and $(xy)H \supseteq (yx)H$. To use $xH = Hx$ you can say that any element $u \in xH$ can be written as $u = xa$ for some $a \in H$ but also $u \in Hx$ so $u = bx$ for some $b \in H$.

3.41 When you have $x, y \in HK$ there are $a, c \in H$ and $b, d \in K$ with $x = ab$ and $y = cd$. To see that $xy \in HK$, look for where you can insert $b^{-1}b$ to help find a new way to write xy. Don't forget that $H \triangleleft G$ so we know $bcb^{-1} \in H$.

3.46 Since the operation on $U(20)$ is $\cdot_{20}$ the cosets of $H = \langle 11 \rangle$ are of the form aH. There are only four distinct cosets here and in the Cayley table you create should have row and column headings that are cosets. For example note that $7H * 7H = (7 \cdot_{20} 7)H = 9H$.

3.52 When finding the order of the coset xH as a group element in ${}^{U(9)}/_{\langle 8 \rangle}$, remember you must find the smallest positive integer with $(xH)^n = 1H$. Thus start calculating the powers of each coset using the operation $*$ to see the first time H is the answer. This is not always the same as finding $x^n = 1$ in $U(9)$.

3.57 Since G is abelian we have normal subgroups as needed so that the groups ${}^{G}/_{K}$ and ${}^{H}/_{K}$ are defined. To see that ${}^{H}/_{K}$ is closed under $*$, assume there are $xK, yK \in {}^{G}/_{K}$. Thus $x, y \in H$. Why can you easily see that $(xy)K \in {}^{G}/_{K}$ as well?

3.59 Suppose $f(x) = f(y)$ and use rules from Theorem 1.37 to help you show $f(xy^{-1}) = e_K$. Why does this tell us $xy^{-1} = e_G$?

3.64 You need to find an onto homomorphism $f : \mathbb{Z}_3 \times \mathbb{Z}_3 \to \mathbb{Z}_3$ which also has the property that $ker(f) = K$. You need to say what each element $f(a, b)$ is equal to, and don't forget to prove that your function is a homomorphism. Remember you want to be sure that elements of the same coset xK are all mapped to the same answer by f.

3.69 Use that $U(9)$ is cyclic to know that the only possible subgroups are cyclic. Thus use each element $a \in U(9)$ to create $H = \langle a \rangle$. For each of the subgroups (why are they all normal?) consider the possible size of the quotient group ${}^{U(9)}/_{H}$ using Theorem 3.10. Then use Theorem 2.18 to know what each of these quotient groups is isomorphic to.

3.70 Consider the function $f : K \to {}^{HK}/_{H}$ defined by $f(x) = Hx$. Be sure to show f is a homomorphism. Remember that when you choose an arbitrary element $Hy \in {}^{HK}/_{H}$ you cannot assume that $y \in K$. By Exercise 42, $yH = Hy$, then use the result of Exercise 58 to help show f is onto. Also remember that H is the identity of the group ${}^{HK}/_{H}$ to help you show that $ker(f) = H \cap K$.

Chapter 4

4.3 You need to verify the properties of a ring carefully here. Don't be fooled into expecting the zero for this ring be the number 0 since $3 \oplus 0 = 3(0) - (3+0) + 2 = -3 + 2 = -1$. Since $3 \oplus 0 \neq 3$ then 0 cannot be the zero for this ring. You need to discover which integer b will have $a \oplus b = a$. Usual addition and multiplication in the integers are commutative.

4.7 Although the two operations are the same you cannot automatically assume that the distributive laws hold. Be careful checking them.

4.8 The multiplication is just like using the distributive law and then gathering the terms containing $\sqrt{2}$ together. Don't forget to use $(\sqrt{2})^2 = 2$. For example:

$$\begin{aligned}(3+2\sqrt{2}) \otimes (5+(-1)\sqrt{2}) &= 3(5) + 3(-1)\sqrt{2} + (2\sqrt{2})(5) + (2\sqrt{2})(-1\sqrt{2}) \\ &= 15 - 3\sqrt{2} + 10\sqrt{2} - 2(\sqrt{2})^2 \\ &= (15-4) + (10-3)\sqrt{2} = 11 + 7\sqrt{2}.\end{aligned}$$

4.19 The statement here says "at least" so you only need to identify that many elements you know will be zero divisors. Remember that $(0_A, b)$ exists in $A \times B$ for every $b \in B$. When $b \neq 0_B$ we know $(0_A, b)$ is a nonzero element of $A \times B$.

4.22 Assume that $(\mathbb{Z}_n, +_n, \cdot_n)$ is an integral domain, but that n is not prime. How can you use the fact that $n = mk$ for some $m, k, \in \mathbb{Z}$ to find a zero divisor?

4.25 The conclusion contains "or" so you need to be careful. In the proof you can say that either a is a zero divisor or a is not a zero divisor giving you two cases to consider. Somehow knowing a is not a zero divisor must guarantee that $b = c$. Remember that in an arbitrary ring having $xy = 0_A$ does *not* guarantee that one of x or y must be 0_A.

4.30 Consider the element $(a+b)^2$ for $a, b \in A$. Use the result of Exercise 28 to help.

4.38 Be careful here, saying that ab is a zero divisor does NOT say $ab = 0_A$. It means there is some element $c \in A$ so that either $(ab)c = 0_A$ or $c(ab) = 0_A$.

4.47 To see if the set S is closed under addition, you need to determine if having $x^2, y^2 \in S$ will tell you that $x^2 + y^2$ is also in S, i.e., is there an element $z^2 \in S$ so that $z^2 = x^2 + y^2$?

4.53 You only need a subgroup under multiplication so the set S does *not* have to be closed under $+_A$. If you have invertible elements $a, b \in S$, how do you find the inverse of ab?

4.60 Since the rings are finite, use the Cayley tables to determine if $f((a,b) + (c,d)) = f(a,b) +_6 f(c,d)$ and $f((a,b)(c,d)) = f(a,b) \cdot_6 f(c,d)$. Don't forget that the operations on $\mathbb{Z}_3 \times \mathbb{Z}_2$ act coordinate-wise. Thus $f((2,1)+(0,1)) = f(2,0)$ and $f((2,1)(0,1)) = f(0,1)$.

4.68 Think about $h(x)h(y) = (ax)(ay)$. How can it be equal to $h(xy)$?

4.69 First, use the bijective property to see why $f(1_A) \neq 0_K$, then use that K is a field to know $f(1_A)^{-1}$ is in K. Now use the homomorphism property on $f(1_A \cdot_A 1_A)$ to help show $f(1_A) = 1_K$.

Chapter 5

5.1 Since S only contains four elements, create the Cayley table for S to see if it is closed under $\oplus$ and $\otimes$.

5.3 Notice that every element of $\mathbb{Z} \times \mathbb{Z}$ whose second coordinate is even is in S. Thus you should only have two cosets.

5.17 To see if the set S is closed under multiplication, suppose you have $\frac{a}{3}, \frac{b}{3} \in S$ and determine if the product can be written in the form $\frac{c}{3}$. Try various examples.

5.19 How does a being a unit help you show $1_A \in S$? Then use the result of Exercise 5.18.

5.30 Show that $\langle 4 \rangle = \mathbb{Z}_{13}$ so there is only one coset. The Cayley tables will be easy after that.

5.40 Remember that the zero of a ring is *not* a zero divisor. Thus you must know $a + S \neq 0_A + S$ as well as $(a + S) * (b + S) = 0_A + S$ for a nonzero $b + S$ to say that $a + S$ is a zero divisor.

5.43 Consider a principal ideal of $\mathbb{Z}$ whose generator is not prime.

5.45 Don't forget that we can rewrite $f(a) - f(b) = f(a - b)$. Use the definition of $ker(f)$.

5.47 Use the function defined by $f(x) = 0_K$ for one of them (which one?) and determine the correct function for the other one. Be sure to prove you have an onto homomorphism with the correct kernel in each case.

5.49 Use the result from Exercise 5.19.

5.53 Use the result of Exercise 5.48 since $ker(f)$ must be an ideal of $\mathbb{Q}$.

5.55 In order to define the correct onto homomorphism, remember that every element of S must map to (0,0). Then once you decide which element (a, b) of $\mathbb{Z}_4 \times \mathbb{Z}_4$ to map to (0,1) then every element of the coset $(a, b) + S$ will also map to (0,1), and so on.

5.59 Be sure you find every subring of $\mathbb{Z}_2 \times \mathbb{Z}_4$ (why are they ideals?) then form each quotient ring. For each subring, S, you need to know what familiar ring it is isomorphic to; not all of them will be of the form $\mathbb{Z}_k$. There are two distinct groups of order 4 so be careful.

Chapter 6

6.1-7 Find the unity of the ring A, and compute $n \bullet 1_A$ to find the least positive n with $n \bullet 1_A = 0_A$.

6.10 Consider the ring in Example 6.8. Can you find a nonzero element of finite order?

6.13 Consider the ring $(\mathbb{Z}, +, \cdot)$ and pick a principal ideal whose generator is not 1, -1, or 0. What is the characteristic of the quotient ring?

6.19 You need to show that the unity, $(1_A, 1_B)$ satisfies $m \bullet (1_A, 1_B) = (0_A, 0_B)$ but for any smaller positive k you have $k \bullet (1_A, 1_B) \neq (0_A, 0_B)$. Remember that $n = mt$ for some positive integer t.

6.27 Don't forget to use Theorem 6.12. Explain how Theorem 6.6 guarantees the characteristic cannot be 0.

6.31 Why can you rule out the possibilities of $char(A) = 2$ or $char(A) = 3$?

6.36 Use that f is one to one and the homomorphism property to see that if a is a zero divisor in A then $f(a)$ will also be a zero divisor in K.

6.44 What are the elements of S equal to? Can you find two elements $(a,b),(u,v) \in \mathbb{Z}_4 \times \mathbb{Z}_4$ with $(a,b),(u,v) \notin S$ but $(a,b)(u,v) = (0,0)$?

6.47 Every ideal is principal in $\mathbb{Z}_{20}$ so see if you can find $b \in \mathbb{Z}_{20}$ with $\langle 12 \rangle \subset \langle b \rangle$ but $\langle b \rangle \neq \mathbb{Z}_{20}$.

6.51 This is a maximal ideal. Explain why having an ideal T with $S \subset T$ will force us to conclude $T = \mathbb{Z}_2 \times \mathbb{Z}_6$. For example, suppose $(0,3) \in T$. Notice that $5 \bullet (1,2) = (1,4)$ so we have $(1,4) \in S$, and thus both $(0,3),(1,4) \in T$. But as T is closed under addition we then have $(0,3)+(1,4) = (1,1) \in T$.

6.55 Assuming for a contradiction there is an ideal T with $S \subset T \subset A$, first explain why $1_A \notin T$ and there is $b \in T$ with $b \notin S$. Use the coset $b+S$ to contradict that ${}^{A}/_{S}$ is a field.

6.59 Don't forget that a field is also an integral domain.

6.61 Be careful since in the rules for an ordered integral domain you must add the same element to both sides of an inequality. First find $a+c < b+c$ and continue.

6.65 This is a true statement so you need to prove it. Assume for a contradiction that $a \geq c$ and $b \geq d$. Don't forget to consider the possibility of $a = c$ as well as $b = d$ during the proof.

6.67 Consider the element $(-a)(-a)$ to see if it is positive. How does this help?

Chapter 7

7.2 The coefficients for these polynomials must be ordered pairs from the ring $\mathbb{Z}_3 \times \mathbb{Z}_3$ so an example of a polynomial is $a(x) = (2,1)+(1,1)x+(0,2)x^2$.

7.9 You should have $a(x)+b(x) = 2+2x+0x^2+0x^3 = 2+2x$. Be careful using $+_3$ and $\cdot_3$ with the coefficients when calculating $a(x)b(x)$.

7.15 Your polynomials must have degree 2, so do not to include a polynomial with $a_2 = 0_A$. You can write out the elements of A as $\{0_A, 1_A, a\}$ to help you actually create the polynomials.

7.18 Don't forget how to use PMI and remember that $(n+1) \bullet a(x) = n \bullet a(x) + a(x)$.

7.22 To show that $a(x) = 2+3x$ is *not* a zero divisor you must show why no nonzero polynomial $b(x) \in \mathbb{Z}_6[x]$ can have $a(x)b(x) = 0(x)$. Assume such a $b(x)$ existed, and show why it is impossible. Be careful not to assume that $deg(b(x)) = 1$ as well, since it could potentially have any degree. Think about the smallest k which has $b_k \neq 0$ to help.

7.27 Is it possible to create polynomials over a commutative ring with unity with $a_0 \neq 0_A$, $b_0 \neq 0_A$ but $a_0 b_0 = 0_A$? Think about easy rings you have used.

7.31 To determine $char(A[x])$ recall that the unity of $A[x]$ is the polynomial $1(x) = 1_A + 0x$, and calculate $n \bullet 1(x)$. Exercise 19 can be helpful.

7.32 Remember you want $a(x) = b(x)q(x)+r(x)$ and $deg(r(x)) < deg(b(x))$, but $deg(a(x)) < deg(b(x))$.

7.46 This cannot be solved if $deg(q(x)) \leq 4$ as asked. There are two possible choices for q_4; be sure to check why neither can be used to solve it.

7.49 Be sure to use arbitrary polynomials, not just examples. Use the notation $a(x) = \sum_{i=0}^{n} a_i x^i$ and $b(x) = \sum_{i=0}^{m} b_i x^i$ as was used in the proof of associativity and commutativity of our operations.

7.50 Don't forget that in order for two polynomials to be equal they must have all of the same coefficients. So when you assume that $\overline{f}(a(x)) = \overline{f}(b(x))$ in the proof that $\overline{f}$ is one to one, it tells you that the coefficients of the two polynomials $\overline{f}(a(x))$ and $\overline{f}(b(x))$ are the same. What are the coefficients?

7.59 You should find several roots; be sure you are calculating the answers in the correct ring.

7.68 The definition of h_c is critical here; use arbitrary $a(x) = \sum_{i=0}^{n} a_i x^i$ and $b(x) = \sum_{i=0}^{m} b_i x^i$ in $A[x]$ to see how to calculate $h_c(a(x)) +_A h_c(b(x))$ and $h_c(a(x) + b(c))$.

7.70 The definition of S can cause some confusion. It is a set of polynomials, but every polynomial in the set must have both 0 and 2 as a root. For example, $a(x) = 1 - 5x$ cannot be in S since $a(0) = 1$.

Chapter 8

8.3 Try a contradiction proof here, so assume that the polynomials $a(x)$ and $b(x)$ have different degrees. Thus one of them has smaller degree than the other. How could that happen?

8.5 It is critical that K is a field (or at least an integral domain) here. Consider the leading coefficients of both $b(x)$ and $c(x)$ to see why $deg(b(x)) \geq deg(a(x))$ leads to a contradiction.

8.12 Use Theorem 7.35 to help, and that there are no zero divisors in an integral domain.

8.19 Find a nonconstant polynomial in $\mathbb{Z}_5[x]$ which has no roots in $\mathbb{Z}_5$. How can you use it to create polynomials that are not associates but have the same roots in $\mathbb{Z}_5$?

8.22 You need two proofs here since the statement has "if and only if" in it. Assume you have associate polynomials $a(x), b(x) \in K[x]$. For one direction suppose that $a(x)$ is reducible and use that they are associates to see that $b(x)$ will also be reducible.

8.25 Use Theorems 8.11 and 6.23. Remember that $b(x) \in \langle p(x) \rangle$ means that $p(x)$ is a factor of $b(x)$.

8.29 Assume that $\overline{f}(a(x))$ is irreducible over E, but that $a(x)$ is reducible over K. See how this results in a contradiction. Which half does this help you prove?

8.38 Just use Theorem 8.15 to help, but don't forget you must know the factors you find are both nonconstant to say $a(x)$ is reducible.

8.46 Use Theorem 8.17. To see there are at most n roots you just need to rule out the possibility of having more than n distinct roots, so suppose there were more than n of them.

8.50 After you use the Rational Roots Theorem on an associate of $b(x)$ in the previous exercise, simply plug each possibility into $b(x)$ to see if the answer is 0. You should only find two roots.

8.64 You need an associate of $a(x)$ that is in V to use h. Now consider using Eisenstein's Criterion to help complete it.

8.67 Show that $\overline{f}(a(x))$ is irreducible over $\mathbb{Z}_5$. Don't forget that it could factor into polynomials of degree 2 so you need to explain why that is impossible too.

8.70 After finding the possible roots, factor out the polynomials of degree 1 you get from the roots. Now you need to determine how to factor (or prove it is irreducible) the rest of the polynomial.

Chapter 9

9.2 Suppose $c \in K$. Can you find a factor of $a(x)$ in $K[x]$ that will contradict the irreducibility of $a(x)$?

9.5 Can you use Eisenstein's Criterion here?

9.9 Suppose instead that $deg(b(x)) = 0$. How is $b(c)$ computed? Is c a root of $b(x)$?

9.11 You already know $K \subseteq S$ and $c \in S$. Think about why an arbitrary element $u \in K(c)$ must be of the form $w(c)$ for some $w(x) \in K[x]$. How does that help you complete the proof?

9.18 Your minimum polynomial should have degree 4, but you must show it cannot factor into polynomials of degree 2 in $\mathbb{Q}[x]$ as well as having no roots in $\mathbb{Q}$.

9.25 Find the minimum polynomial $p(x)$ for the element, being sure to show it is irreducible, then show how to factor $a(x) = p(x)q(x)$ for some $a(x) \in \mathbb{Q}[x]$.

9.39 Remember that $p(c) = 0$, so you can write $6+6c+c^2 = 0$. Now when you are multiplying elements you can use the fact that $c^2 = 1 + c$ since in $\mathbb{Z}_7$ we have $-6 = 1$.

9.49 Once you list the elements of $\mathbb{Z}_3(c)$ you simply need to plug them into $p(x)$ to find the roots. Don't forget that in this case we have $c^2 = 1 + c$ since in $\mathbb{Z}_3$ we have $-2 = 1$.

9.55 Why is $\{1_K\}$ a basis for $K(c)$ over K when $c \in K$?

9.58 This is a direct application of the result in Exercise 57.

9.64 Can you factor the minimum polynomial for c over K? What do the factors look like? The very first theorem of this chapter will help you know the roots are back in $K(c)$.

9.67 Find a way to write a polynomial of $\mathbb{Q}[x]$ which has $\sqrt{2}$ as the answer when $\sqrt{2} + \sqrt{3}$ is plugged in. (This polynomial will have degree 3.) How does this tell us that $\sqrt{2} \in \mathbb{Q}(\sqrt{2} + \sqrt{3})$?

Chapter 10

10.1 Don't forget that you need to prove both $f(a+b) = f(a) + f(b)$ and $f(ab) = f(a)f(b)$ for any $a, b \in \mathbb{Q}(\sqrt{-2})$.

10.4 Remember that in $\mathbb{Z}_5$ we have $4 \cdot_5 4 = 1$, $4 +_5 4 = 3$, and in $\mathbb{Z}_5(c)$ we have $c^2 = 3$.

10.7 Assuming f was a homomorphism find two elements $u, v \in \mathbb{Q}(\sqrt[3]{2})$ which make $f(uv) \neq f(u)f(v)$ as a contradiction.

10.11 Find roots of the separate factors to see what happens.

10.15 Show that no element of the form $a + bi$ with $a, b \in \mathbb{Q}$ can be a root of the minimum polynomial for $\sqrt{-5}$.

10.19 You must show you have a root field, and that the two fields are not equal.

10.23 Assume there are two different roots c_1 and c_2 of $p(x)$ in E and consider the fields $K(c_1)$ and $K(c_2)$. See how Theorem 10.7 gives you an isomorphism and then use it with Theorem 10.13 to complete the proof.

10.27 Remember that if we have $\alpha^{-1}(y) = u$ then also $\alpha(u) = y$. Thus you can use that α is a homomorphism to help you show that α^{-1} is a homomorphism.

10.30 This polynomial should remind you of the roots of unity used in Example 10.15.

10.37 Find the roots of each factor and don't forget that $\omega_4 = i$. The final root field should be fairly simple.

10.41 Use the fact that ω_5 is one of the roots and and look at Project 9.8 to help find all of the roots.

10.45 Don't forget that to compute $\alpha\left(\frac{-\sqrt[3]{5}}{2}\left(1+\sqrt{-3}\right)\right)$ you must use the fact that α is a homomorphism, which means you are computing $\alpha(\frac{-1}{2})\alpha(\sqrt[3]{5})[\alpha(1) + \alpha(\sqrt{-3})]$.

10.49 First, suppose there is an element $\alpha \in Gal\left({}^{E}/_{L}\right)$ and show it must fix every element of K. You need to prove two subsets to show $G_L = Gal\left({}^{E}/_{L}\right)$ but only the definition of each set in the proof.

10.52-59 Don't forget how to write an arbitrary element y of the root field, using the basis you can find over $\mathbb{Q}$. Then calculate $\alpha(y)$ for each $\alpha \in G$, and if $\alpha(y) = y$ it will tell you what form y must have. All elements of the fixed field of a subgroup must be fixed by every automorphisms in that subgroup.

10.60 Consider the minimum polynomial for y over K for help.

Chapter 11

11.2 Think about how to construct the y-axis (which is perpendicular to the x-axis), and then the points of the form $(0, n)$ for each integer n. How can $(0, n)$ and $(m, 0)$ be used to construct (m, n)?

11.4 Don't forget that similar triangles have a special ratio which holds between the side lengths. What will $\frac{BF}{BC}$ be equal to when F is the point where the line parallel to $\overleftrightarrow{EC}$ intersects $\overleftrightarrow{BC}$?

11.8 Use the previous exercises to know that $\left(\frac{m}{n}, 0\right)$ and $\left(0, \frac{r}{s}\right)$ can be constructed. How do you use these to help you complete the problem?

11.9 Don't forget to handle cases where some of the coefficients are 0 as well as the case where all are nonzero. Since the coefficients are in a field you must know they are nonzero before using inverses.

11.12 There is an identity involving $\cos(3\theta)$ that will be useful here since for $\theta = 20°$ we see $3\theta = 60°$.

11.14 Use cosets and Theorem 3.17 to help.

11.17 This time show that each subgroup $H_i \cap J$ is closed under conjugates from $H_{i+1} \cap J$ (not from G).

11.18 Use Theorem 3.6 (ii) to help show that $(aH_i \cap J) * (bH_i \cap J) = (bH_i \cap J) * (aH_i \cap J)$.

11.20 A process similar to the proof of 11.18 is the key, but don't forget that the subgroups are of the form $f(H_i)$.

11.24 Remember that $\alpha(\omega_n) = (\omega_n)^k$ for some $1 \leq k \leq n$.

11.25 Consider the subfields $L_j = \mathbb{Q}(\omega_{m_1}, \omega_{m_2}, \ldots, \omega_{m_j})$ to help. Why are they all root fields over $\mathbb{Q}$?

11.26 You only need to show that for any $\alpha, \beta \in Gal({}^{W_i}/_{W_{i-1}})$ the functions $\alpha \circ \beta$ and $\beta \circ \alpha$ map c_i to the same element of W_i.

11.27 Remember that since the two complex roots are assumed to be nonreal, we know $b \neq 0$ and $d \neq 0$. Factoring $p(x)$ using these roots must leave a factor with real coefficients.

11.29 Apply Cauchy's Theorem to the quotient group to find a cyclic subgroup.

11.35-40 Find subgroups that create a solvable series in each question. Theorem 3.20 can be helpful.

Bibliography

[1] David Dummit and Richard Foote. *Abstract Algebra.* Wiley & Sons, Inc., 2nd edition, 1999.

[2] Euclid. *The Thirteen Books of the Elements, Vol I.* Dover Publications, Inc., 1956.

[3] Israel N. Hierstein. *Noncommutative Rings.* The Mathematical Association of America, 1968.

[4] Victor J. Katz. *A History of Mathematics: An Introduction.* Addison Wesley, 2nd edition, 1998.

[5] Bernard Kolman and David R. Hill. *Introductory Linear Algebra with Applications.* Prentice Hall, Inc., 7th edition, 2001.

[6] Cheryl Chute Miller and Blair F. Madore. Carry groups: Abstract algebra projects. *PRIMUS*, XIV(3), September 2004.

[7] Ivan Niven and Herbert S. Zuckerman. *The Theory of Numbers.* Wiley & Sons, 4th edition, 1980.

[8] Michael L. O'Leary. *The Structure of Proof: With Logic and Set Theory.* Prentice Hall, Inc., 2002.

[9] Joseph J. Rotman. *Advanced Modern Algebra.* Pearson Education, Inc., 2002.

Index